Lecture Notes in Geoinformation and Cartography

Publications of the International Cartographic Association (ICA)

Series editors

William Cartwright, Melbourne, Australia
Georg Gartner, Wien, Austria
Michael P. Peterson, Omaha, USA
Liqiu Meng, Munich, Germany

For further volumes:
http://www.springer.com/series/10036

Dirk Burghardt · Cécile Duchêne
William Mackaness
Editors

Abstracting Geographic Information in a Data Rich World

Methodologies and Applications of Map Generalisation

Springer

Editors
Dirk Burghardt
Department of Geosciences
Institute of Cartography
Dresden
Germany

William Mackaness
Geographical Information Science
School of Geosciences
Edinburgh
UK

Cécile Duchêne
IGN COGIT Laboratory
Saint-Mande
France

ISSN 2195-1705 ISSN 2195-1713 (electronic)
ISBN 978-3-319-00202-6 ISBN 978-3-319-00203-3 (eBook)
DOI 10.1007/978-3-319-00203-3
Springer Cham Heidelberg New York Dordrecht London

Library of Congress Control Number: 2013954034

Printed on acid-free paper

Springer is part of Springer Science+Business Media (www.springer.com)

Preface

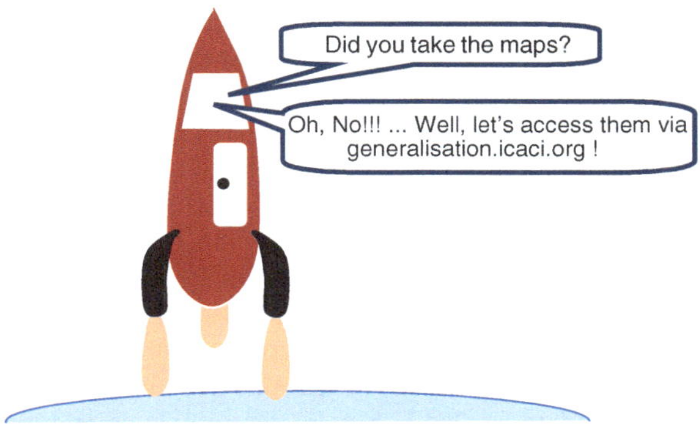

As I carefully read this book, I am wondering if we have reached the happy moment of generalisation 3.0... well not exactly, but perhaps we are not so far from it. Indeed, if you start reading this book from the end (the very impressive Chap. 11) you would see the tremendous progress made by our scientific community since 1991. Why 1991? Because 1991 is THE landmark year of our generalisation bible: 'Map Generalisation' edited by Buttenfield and McMaster. Suggesting that the history of map generalisation began in 1991 is of course very unfair but this book was simply fantastic since it contained all the very necessary seedlings from which today's results have grown. Written by many young geographers and computer scientists, this 1991 book was definitely full of ideas (please read it).

Then the ICA played an important role. From the beginning, in 1992, the ICA working group on generalisation led by Robert Weibel was very dynamic. Nearly every year a workshop was organised where researchers, engineers and even vendors came from all over the world, to share ideas, debate and even compete on the subject. After Robert Weibel's term (1992–2003), William Mackaness and myself carried on organizing meetings through the umbrella of the ICA (2003–2007). This synergy carried on thanks to Sébastien Mustière and William Mackaness (2007–2011) and is still going on with Dirk Burghardt, Cécile Duchêne and William Mackaness from 2011.

In 1999, during the ICA conference in Ottawa, a project was proposed to write a collective ICA book on generalisation that recorded the progress made between 1991 and 1999. The project was postponed but William Mackaness and I decided to take on the challenge. We added Tiina Sarjakoski to the team to boost our efforts and arbitrate over our usual French and English confrontations! "Generalisation of geographic information: cartographic modelling and applications" was born in 2007, after three very pleasant years of collaborative working (please read it too). It contained 17 chapters written by people from a very wide variety of nationalities: Canada, Denmark, England, Finland, France, Germany, Netherland, Switzerland and USA. The 2007 generalisation book was completely different from the 1991 one! For more than ten years (from 1991 to 2003), many algorithms and even platforms—from vendors or research laboratories—were developed, more teams were working on the subject, and we started also to include ideas and the first results related to real-time generalisation, open generalisation systems, on demand mapping and 3D generalisation. The concept of Multi-agent systems was used by the generalisation research community and the first results from National Mapping Agency production lines were presented. Many of the results presented in 2007 have been widely used and improved upon. Thus if you read Chap. 11 of the 2013 book, you will definitely see the progress made since 1991 and even since 2007. The 65 pages of this Chap. 11 are delicious because many of our propositions are today used to produce maps in different countries. Progress is ongoing reflected in the quote from this chapter: "Please be aware that the facts reported in this chapter are up to date in 2013, but might evolve quite quickly since the developments in generalisation are currently particularly active in several NMAs". But enough pleasure! Let us reflect on some other salient points of this new generalisation book.

The first interesting point to note is the discussions in several chapters of how generalisation connects back to cartography. Of course, generalisation is a cartographic process—if not THE cartographic process (see the publications of E. Imhof, J. Bertin, R. Cuenin or E. Spiess for example). But over these last 20 years, the complexity of processing digital data and developing sophisticated algorithms and processes shadowed the cartographic inheritance of generalisation. Thus it is interesting to read Chap. 2 or Chap. 10 where even R. Brunet and the Chorematic maps are quoted. Here, we touch on the point that we represent the geographical space for humans, not for computers. A very different task from this is to use digital data to compute important information (such as the shortest path from A to B). These are two very different tasks. Our goal is to propose the best representation of space according to specific needs and this requires optimal generalisation and symbolisation. Generalisation is necessary for human cognition.

The second point I want to make is the imminent arrival of generalisation 3.0 (the one that includes not only people in contact with one another but also the semantic web and the Internet of things). Chap. 5 for example illustrates new needs and challenges coming from the multitude of data sources, and the heterogeneity of data. Chap. 7 proposes ideas to use and chain processes wherever they are coming from. This requires new ontologies (such as those described in

Chap. 3). Thus these chapters are proposing new ideas that might be the seedlings for a 2020 book on generalisation!

Last but not least, is the quality of the various states of the art contained in this book. It is enhanced by the original structure of most chapters where the first part is devoted to state of the art and the second is centred on the presentation of some current works which illustrate very nicely the state of the art. Chapters on operators Chap. 6, evaluation Chap. 9 and terrain generalisation Chap. 8 show great maturity of our discipline and will definitely help researchers and engineers wishing to learn more about our domain.

What else is there to say? Read and enjoy!

Paris, France, July 2013 Anne Ruas

Acknowledgments

This book reflects the energies and endeavours of the many members of the ICA Commission on Generalisation and Multiple Representation. This international group is made up of researchers, practitioners, academic institutions and Mapping Agencies. In addition to all those who have variously contributed to this book, the editors would particularly like to thank Anne Ruas for her review of the book and the writing of the Preface. We would like to thank the ICA Executive Committee for encouraging us to write this book. We are also grateful for funding from the ICA and the British Cartographic Society for covering the costs of an editorial meeting in Dresden. The editors are very grateful to Jantien Stoter as Chair of the EuroSDR Commission on Data Specifications for co-organising the NMA Symposium in Barcelona in 2013. All participants from this symposium helped provide some of the material presented in Chap. 11. We would like to acknowledge contributions from the many participants of past ICA Generalisation Workshops—particularly those in Moscow 2007, Montpellier 2008, Zurich 2010, Paris 2011, Istanbul 2012 and Dresden 2013; this book reflects many of the ideas and discussions presented at these events.

The book reflects a collective endeavour between authors, co-authors, editors, internal reviewers and publishers. We are particularly grateful to the external reviewers of the book—Liqiu Meng, Sébastien Mustière and Robert Olszewski. Thanks too to the technical employees of TU Dresden Uta Heidig and Peggy Thiemt for their administrative services and support in editing the book. We are also grateful for the professional involvement and cooperation of Agata Oelschläger, Gopinath Chandrasekar and Shine David from Springer. Finally we would like to thank our colleagues, friends and family for their forbearance in this somewhat lengthy journey!

Contents

1 Map Generalisation: Fundamental to the Modelling and Understanding of Geographic Space . . . 1
William Mackaness, Dirk Burghardt and Cécile Duchêne
1.1 Map Generalisation: Why So Complex? . . . 1
1.2 The Map as a System of Relationships . . . 3
1.3 The Importance of Data Enrichment . . . 6
1.4 Alternate Paradigms to Map Generalisation . . . 8
1.5 Structure and Content of this Book . . . 10
References . . . 14

2 Map Specifications and User Requirements . . . 17
Sandrine Balley, Blanca Baella, Sidonie Christophe, Maria Pla, Nicolas Regnauld and Jantien Stoter
2.1 Introduction . . . 18
2.2 Key Concepts: Needs, Requirements and Specifications . . . 19
2.3 Inferring Map Specifications from User Requirements . . . 25
2.4 Collecting User Requirements . . . 29
2.5 Case Study I: Specifying Generic and Specific Map Specifications—A EuroSDR Case Study . . . 31
2.6 Case Study II: A Map Specifications Model for On-Demand Mapping at Ordnance Survey . . . 38
2.7 Case Study III: COLorLEGend—Design of Personalised and Original Maps . . . 43
2.8 Conclusions . . . 48
References . . . 49

3 Modelling Geographic Relationships in Automated Environments . . . 53
Guillaume Touya, Bénédicte Bucher, Gilles Falquet, Kusay Jaara and Stefan Steiniger
3.1 Introduction . . . 54
3.2 Spatial Relations Classification . . . 55
3.3 An Ontology of Spatial Relations . . . 58
3.4 Spatial Relations Ontology to Support Automatic Processes . . . 62

3.5 Case Study I: Spatial Relations for Urban 3D Models 64
3.6 Case Study II: Relations for the Extraction of Groups of Objects. 69
3.7 Case Study III: Data Migration of User Data 74
3.8 Conclusions 79
References 80

4 Data Structures for Continuous Generalisation: tGAP and SSC 83
Peter van Oosterom, Martijn Meijers, Jantien Stoter and Radan Šuba
4.1 Introduction 83
4.2 Principles of the Generalised Area Partitioning Structure 84
4.3 Space Scale Cube for Vario-Scale 89
4.4 Case Study I: Dutch Large Scale Basic Topographic Data in Constraint tGAP 97
4.5 Case Study II: German Land Cover Data in Constraint'tGAP 104
4.6 Case Study III: Corine and ATKIS Data in the Space Scale Cube 107
4.7 Conclusions 112
References 114

5 Integrating and Generalising Volunteered Geographic Information 119
Monika Sester, Jamal Jokar Arsanjani, Ralf Klammer, Dirk Burghardt and Jan-Henrik Haunert
5.1 Introduction 120
5.2 The Potential and Characteristics of User-Generated Content 121
5.3 Aspects of Data Integration. 124
5.4 The Visualisation and Generalisation of VGI 128
5.5 Case Study I: VGI Platforms and Data Generalisation 131
5.6 Case Study II: Generalisation within the OpenStreetMap Project Compared to the Generalisation of Authoritative Data. 139
5.7 Case Study III: Matching GPS Trajectories with Incomplete User-Generated Road Data 145
5.8 Conclusions 150
References 151

6 Generalisation Operators 157
Lawrence V. Stanislawski, Barbara P. Buttenfield, Pia Bereuter, Sandro Savino and Cynthia A. Brewer
6.1 Introduction 158

6.2 Generalisation Operators: Chronology of Typologies 159
6.3 Operators in Commercial Software 163
6.4 Recent Advances in Operator Development 166
6.5 Case Study I: Generalisation of Road Networks 169
6.6 Case Study II: River Network Pruning by Enrichment and Density Analysis 175
6.7 Case Study III: Algorithms for On-the-Fly Generalisation of Point Data Using Quadtrees 181
6.8 Conclusions 189
References 190

7 **Process Modelling, Web Services and Geoprocessing** 197
Nicolas Regnauld, Guillaume Touya, Nicholas Gould and Theodor Foerster
7.1 Introduction and State of the Art 198
7.2 Deciding on the Components of the Generalisation Process . . . 200
7.3 Formalising the Procedural Knowledge 203
7.4 Chaining Processes 204
7.5 Future Opportunities 206
7.6 Case Study I: Collaborative Generalisation 207
7.7 Case Study II: An Ontological Approach to On-Demand Mapping and Generalisation 212
7.8 Case Study III: Live Geoinformation with Standardised Geoprocessing Services 217
7.9 Conclusions 221
References 222

8 **Terrain Generalisation** 227
Eric Guilbert, Julien Gaffuri and Bernhard Jenny
8.1 Introduction 228
8.2 Issues in Terrain Generalisation. 229
8.3 Object-Oriented Classification of Landforms. 232
8.4 Generalisation Methods 236
8.5 Case Study I: Hypsometric Colouring 242
8.6 Case Study II: Isobathic Line Generalisation. 245
8.7 Case Study III: Preserving Relations with Other Objects During Generalisation 251
8.8 Conclusions 255
References 256

9 **Evaluation in Generalisation** 259
Jantien Stoter, Xiang Zhang, Hanna Stigmar and Lars Harrie
9.1 Introduction 260
9.2 The Purposes of Evaluation 260

9.3 Visual and Quantitative Evaluation on Map Generalisation. . . . 263
9.4 Frameworks of Automated Evaluation 266
9.5 Components of Automated Evaluation 267
9.6 Map Readability Formulas 273
9.7 Case Study I: Automated Evaluation of Generalised Building Patterns 276
9.8 Case Study II: Map Readability Formulas 283
9.9 Case Study III: The EuroSDR Project 287
9.10 Conclusions and Further Research 291
References 291

10 Generalisation in the Context of Schematised Maps 299
William Mackaness and Andreas Reimer
10.1 The Nature of Schematised Maps 299
10.2 A Definition of Schematisation 301
10.3 A Classification of Schematised Maps 302
10.4 Methods of Schematisation Production. 304
10.5 Schematisation Metaphors in Interactive Environments 306
10.6 Case Study I: Schematisation of Transportation Networks 307
10.7 Case Study II: Chorematic Diagrams 314
10.8 Case Study III: Schematised Maps for Multi Modal Travel . . . 319
10.9 Conclusions and Challenges 325
References 325

11 Generalisation in Practice Within National Mapping Agencies. . . . 329
Cécile Duchêne, Blanca Baella, Cynthia A. Brewer, Dirk Burghardt, Barbara P. Buttenfield, Julien Gaffuri, Dominik Käuferle, François Lecordix, Emmanuel Maugeais, Ron Nijhuis, Maria Pla, Marc Post, Nicolas Regnauld, Lawrence V. Stanislawski, Jantien Stoter, Katalin Tóth, Sabine Urbanke, Vincent van Altena and Antje Wiedemann
11.1 Introduction 331
11.2 Deriving Products Through Generalisation at the Institut CartogrÃ fic de Catalunya 333
11.3 The New Base Map Project: A Semi-Automated Production Line for Topographic Maps at IGN-France. 339
11.4 Producing Digital Cartographic Models at Swisstopo. 346
11.5 Automatic Map Derivation at Ordnance Survey GB. 351
11.6 Generalisation Methods Used for the USGS National Map and National Atlas. 355
11.7 Generalisation in Production at Kadaster NL 362
11.8 AdV-Project “ATKIS: Generalisation” Map Production of DTK50 and DTK100 at LGL in Baden-Württemberg. 369

11.9 Multi-Scale Data in Spatial Data Infrastructures: Developments in INSPIRE at the JRC 373
11.10 Synthesis: Recent Achievements and Future Challenges Regarding Generalisation in NMAs 378
References 387

12 Conclusion: Major Achievements and Research Challenges in Generalisation 393
Dirk Burghardt, Cécile Duchêne and William Mackaness
12.1 Major Achievements 393
12.2 Future Challenges 398
12.3 In Conclusion 402
References 403

Index 405

Chapter 1
Map Generalisation: Fundamental to the Modelling and Understanding of Geographic Space

William Mackaness, Dirk Burghardt and Cécile Duchêne

Abstract It would be a mistake to see map generalisation merely as the automation of a set of cartographic practices. The process of representing various geographies at different levels of detail goes to the heart of geographical understanding. Comprehension and context comes from being able to examine information at multiple levels of detail. An automated environment that can support such interaction depends upon a rich understanding of the qualities, behaviours and relationships among the various geographic phenomena that are being mapped. In this chapter we seek to explore the complexity of map generalisation, reflecting on the impact of the changing ways in which we gather and interact with geographic information. This in turn provides a justification for the structure of the book which is then summarised in the second half of this chapter.

1.1 Map Generalisation: Why So Complex?

"Nothing is less real than realism. Details are confusing. It is only by selection, by elimination, by emphasis that we get to the real meaning of things." So said Georgia O'Keeffe (Stuhlman 2007, p. 22) in a 1922 interview with the New York Sun. She is considered to be one of the twentieth Century's greatest artists and her thinking reflects the idea of the close association between the process of

W. Mackaness (✉)
School of GeoSciences, The University of Edinburgh, Edinburgh, UK
e-mail: william.mackaness@ed.ac.uk

D. Burghardt
Dresden University of Technology, Dresden, Germany
e-mail: dirk.burghardt@tu-dresden.de

C. Duchêne
Laboratoire COGIT, IGN, Paris, France
e-mail: cecile.duchene@ign.fr

D. Burghardt et al. (eds.), *Abstracting Geographic Information in a Data Rich World*, Lecture Notes in Geoinformation and Cartography, DOI: 10.1007/978-3-319-00203-3_1,

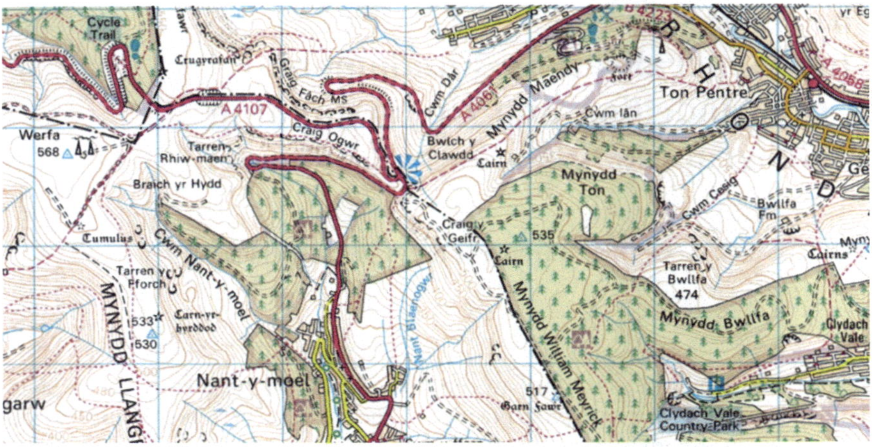

Fig. 1.1 Not lines, points, text and polygons but a 'landscape of relationships' (Crown Copyright: Ordnance Survey; OS/EDINA supplied service)

abstraction and the search for truth. A similar line of thinking is reflected in Krippendorff's assertion that design is about 'making sense of things' (1989); and more broadly it is worth noting Muller's (1991) argument that from an epistemological perspective, generalisation is 'a process which attempts to establish the universality of a statement' (p. 457). Maps in their incongruous forms (Ormeling 2013) reflect that search for truth—seeking to be 'faithful' in their conceptualisation of what is important and giving emphasis to a particular set of real world relationships. This is the foundation point of map generalisation.

When we come to explore the topic from a pragmatic perspective, we often present this process as being made up of a modelling component and a cartographic component. The complexities of map generalisation lie in modelling this abstractive process via these two components. The modelling side requires us to make explicit a subset of all the relations that might variously exist between one geographic phenomenon and another. The cartographic component is concerned with the rendering of that subset of relations through the symbolisation of various geographic entities. The aim is for the resulting map to convey those relationships elegantly and efficiently through their simple arrangement and juxtaposition. There is no need to write on the map that 'this is close to that', or 'this line connects these stations' because (hopefully) the creator and the user have a shared understanding of what those symbols represent and how they behave and interact with one another. At its simplest, we might say that the map is made up of a set of symbolised primitives (points, lines, areas and text), but this is not what the user 'sees'. What the user 'sees' is the geography of the world through that arrangement of simple data types. For example, the user does not see a 'line' but a twisted road climbing steeply over mountain passes connecting remote villages that are sometimes cut off in the winter (Fig. 1.1). Long ago, research into map generalisation accepted that high levels of automation could only be achieved if we

Fig. 1.2 Missing the context: A map that is correct, yet has little meaning

explicitly modelled such behaviours and relationships among such geographic entities. Long ago, researchers abandoned the idea that it was sufficient to see map generalisation as a set of geometric operations applied to a set of primitives.

Our capacity to interpret the landscape in which we find ourselves depends on a successful mapping between the real world and the abstracted view that we call 'the map'. This in turn requires that there is a shared understanding between the cartographer and the map user, of the symbology that is used. It is also a requirement that the cartographer preserves a sufficient proportion of those relationships that the map user is readily able to move between the real and abstracted view. In other words, that there is sufficient context to give meaning to that abstraction (Fig. 1.2). The complexity arises in finding a compromise between the choice of entity, their form and detail of representation, and the space available in which to display them (the scale of the map). It is this idea of trying to model 'compromise' that produces such a breadth of potential solutions.

1.2 The Map as a System of Relationships

A well designed map is a silent record of the many relationships that exist between the entities mapped. It is delicious to watch and hear map users give voice to those properties as they jab and finger the surface of a map! A well-designed map enables us to comprehend the rich and intricate properties of that place; these are semantic properties that are metric, topological and Gestalt[1] in form. From looking at a map we can describe various relationships between entities in terms of alignments, clusters, distances, angles, and extents. And we can use topological descriptors to describe their connectivity, adjacency and containment relationships (for example). Thus when thinking about the map generalisation process, it is

[1] Gestaltism is the idea that the human eye sees a collection of objects in their entirety before perceiving their individual parts.

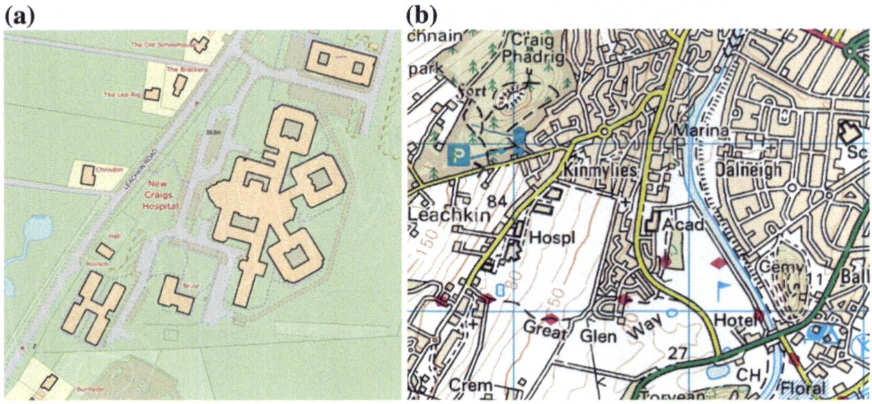

Fig. 1.3 A hospital in Inverness: Different scales in order to discern *different* patterns, *different* relationships, among *different* entities (Crown Copyright: Ordnance Survey; OS/EDINA supplied service)

useful to think of a map as a 'system of relationships' (rather than view the problem as one of object manipulation) and for map generalisation to be thought of as the process by which we manipulate those relationships. This is precisely why there is so much interest in ontological modelling among map generalisation researchers. Couclelis (2010) presents ideas of ontologies of geographic information sufficient to allow us to represent entities as hierarchical composites, thus reflecting the connections among the functionalities and scale interdependencies of geographic entities. Such ontologies provide a framework by which we can construct and manipulate these relationships; it also enables us to 'build a bridge between spatial reasoning and spatial database approaches' (Masolo and Vieu 1999, p. 235). Couclelis goes on to argue that we need a conceptual vocabulary linking purpose, function and granularity. It is exciting to observe that increasingly these ideas are being explored within the map generalisation research community because what complicates (and thus distinguishes) map generalisation from other types of modelling is the fact that it must deal with these conceptual transitions across scale, and must also seek to link 'purpose, function and granularity' in the construction of the map. Muller referred to these conceptual transitions when he wrote about 'crossing conceptual cusps' as we move from the very detailed (large scale mapping) to the very coarse or granular (small scale) mapping. As we change scale, we can no longer preserve the detail; and so by definition neither can we continue to include all the relations that were conveyed at the large scale. As one set of relations 'disappear' from the map, so a new set of relations take their place; a set of relations only revealed at the smaller scale.

For example at one scale we might convey the curvilinear arrangement of a set of buildings alongside a road, but at a smaller scale, this pattern is subsumed by the more dominant relationship between, say, the building blocks and the road network, or the suburb's relation to the city centre (Fig. 1.3b). Indeed this is precisely why we change scale—so that we can see *different* patterns, *different* relationships, between *different* entities. This is the ambition of the map generalisation process.

We observe three things from this discussion: first, that the relationships we choose to convey are scale and theme dependent. For example in Fig. 1.3a, the thematic choice, in this instance, gives emphasis to buildings and roads, but equally it could have given emphasis to the function of each building. Secondly, that the smaller scale features are forged from the more detailed objects. For example, we readily understand that a hospital is made up of a close arrangement of car parks, specialist facilities, and wards. It is this understanding that enables us to simplify that arrangement down to a simple form with the word 'Hospl' next to it (Fig. 1.3b). Third, that there are a very large number of ways by which we might model and describe the relationship between those entities both within and between scales. Returning to Fig. 1.3, we can see that there are many ways by which we might describe the building's relationship with the town. For example, it might be described in terms of its function ('it's the town's only hospital'), its broader significance ('it's the only hospital in that region'), in terms of the topological network of roads by which we might travel to the hospital, or its positional relationship with respect to the town ('it's on the west side of town') (as conveyed in Fig. 1.3b). For the driver racing to take advantage of its services, all these visualisations at their various scales are relevant (hence the importance of an interactive zoom facility!)

Of course, there are many, many other ways in which we might convey to the user, the relationship between this one entity and all the other map entities. Our choice of which relationships are pertinent are tightly coupled to the intended use. In the context of the example above, are we giving spoken directions to a pedestrian, or describing its importance to a regional planner, or is it for display on an in-car SatNav? One can easily envisage more complex relationships. For example, how might we model (and thus describe) the relationships between the entities that describe the hydrology of a landscape (how rivers, lakes, and canals are networked together)? What are the relationships between the hydrology and other entities? (between say, rivers and bridges, or rivers and topography). The question then becomes: which of these relationships do we need to make explicit? For any given use, and scale, which ones are to be preserved and how best are they conveyed? What can be left 'out' in order to give emphasis to what is left 'in'? These are very complex questions that are scale, and thematic dependent. To try and model many of these relationships in anticipation of all the possible types of maps that might be requested is a daunting prospect. It becomes even more challenging when we include feature classes associated with different thematic maps (geological, climatic, epidemiological, cadastral, etc.), or consider how we might also model the relations between feature classes (such as the partonomic relationship between a city, its suburbs, and the buildings within). We argue that such a goal is justified if (1) we are to achieve high levels of automation, (2) we are to conceal this complexity from the unsuspecting user, (3) we wish to work across all types of maps, and (4) provide functionality that is instant, tailored, and web enabled.

1.3 The Importance of Data Enrichment

It is the ambition to make explicit these relationships that has led to a significant focus on 'data enrichment'. Data enrichment is the process by which we make explicit those relationships implicit in the patterns and associations among features represented on the map. The requirement for data enrichment is reflected in the huge library of cartometric techniques that have emerged. Such techniques are applied to model clusters of discrete objects and continuous fields (e.g. minimum spanning trees, Delaunay triangulation), to model topological properties (e.g. planar graphs, Voronoi diagrams), or measure the shape characteristics of objects (e.g. the Hausdorff distance)—to name but a few. These techniques have come to exist in response to limits arising from trying to build automated systems based on simple data models and databases of points, lines, polygons and text. In order to reason about these primitives, it has been necessary to build enrichment techniques that reveal the 'geography' inherent in the data. Such information can either be stored explicitly within the database or calculated on-the-fly. Either way, without this data enrichment stage, relations are mismanaged—and the map starts to fall apart; it begins to lose its credibility and indeed its truth. If we consider a map to be a system of relationships, then generalisation is all about the modelling, filtering, and visualisation of those relationships. Hence the need to make them explicit within automated generalisation systems. A nice example of this mismanagement of relations is the case of the river that follows the point of steepest descent down a hill. In the absence of an understanding of this dependency, the generalisation process may simplify the contours in a way that no longer preserves or conveys the relationship between the river and the underlying topography (Fig. 1.4).

If we are to ensure that these generalised forms remain faithful to their various relationship then either we need the tools to make explicit the geography that is implicit in the data, or we need data models that explicitly store that geography (Couclelis 2010). Where we have the capacity to model the geographic, we can link cartographic techniques to the behaviours and interactions of those entities. Indeed that was the ambition of the AGENT project in which agent based modelling techniques were used to reason about the cartographic process.

Over the years, sophisticated cartometric techniques have been developed to automatically create such geographic models. But three research questions remain unanswered: (1) what is the relationship between scale (the level of detail) and the choice of relations that we wish to show? (2) what proportion of those relations is it necessary to make explicit within automated generalisation systems? and (3) how do we manage changes in the detail and emphasis given to the representation of those relations and properties? The various attempts to measure line distortion when reducing the number of points used to represent a line, is testament to this third question! Various cartometric techniques have been used to identify characteristic points, and thus minimise loss of geographic information whilst simplifying

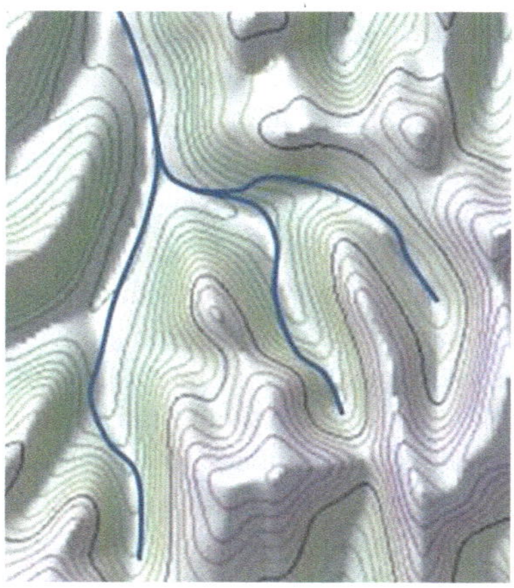

Fig. 1.4 A loss of belief, a loss of 'truth', a mismatch in expectation

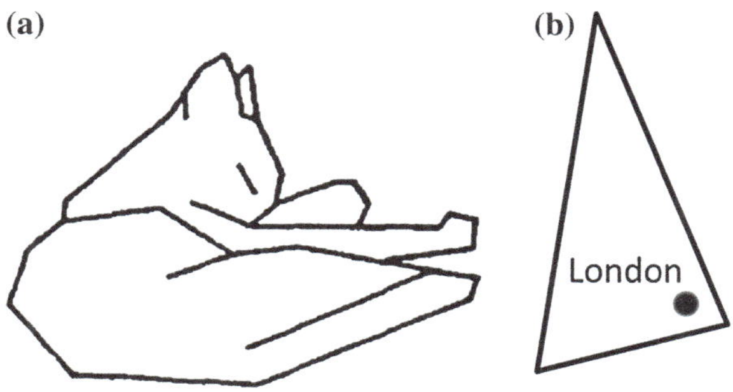

Fig. 1.5 **a** 38 points defining Attneave's cat (Attneave 1954), **b** 3 points defining a country; both are an exercise in generalisation

its form. Attneave had a similar ambition when he sought to identify the simplest silhouette of a sleeping cat (Fig. 1.5a) arguing that the goal of perception is to reduce this redundancy and to "encode incoming information in a form more economical than that in which it impinges on the receptors" (Attneave 1954, p. 189). What research in perceptual studies has shown is that the eye will tolerate huge shape distortion yet still recognise the entity being represented (Fig. 1.5b). Such research highlights the difficulties of measuring information content and calls into question the wisdom of applying geometric measures in the absence of knowing what it is you are trying to represent. Data enrichment is about providing that information necessary to support the map generalisation process. The challenge then becomes one of how best to utilise that information in the design of solutions.

1.4 Alternate Paradigms to Map Generalisation

So the argument is made that we must have a very rich geographic model in order to achieve high levels of automation in the map generalisation process. To date, those models have typically been developed within NMAs who sought to derive various topographic and thematic solutions from a single, highly detailed database (in order to minimise redundancy and facilitate the process of update). There are significant overheads in maintaining and keeping up to date, a system of high quality able to respond in real time, to multiple communities of users.

Such systems are resource intensive and are complex to implement and maintain. Not surprising then, that outside the institutional walls of National Mapping Agencies, more simplistic generalisation functionality is used. Web based mapping services such as Google maps, Bing maps and OpenStreetMap (OSM)[2] use scale dependent rendering—a simple filtering mechanism based on the level of detail and the entities' attributes. The map tiles are pre-generalised offline and cached at various levels of detail. At run-time, only those map tiles needed for a particular level of detail are retrieved from the cache. The capacity to 'zoom in' avoids the need for perfect legibility at each level. Indeed the illegible text at one level acts almost as a 'pointer'—indicating to the user that they need to zoom in, in order to read what is written! This simple utilisation of generalisation algorithms 'works' because the context is so different from that of NMAs. Different in the following ways: first, the task of generalisation is 'distributed'—effectively it is spread amongst those involved in the data gathering process (unbeknownst to the user, generalisation considerations are implicit in the detail with which they choose to capture a particular feature). Second, the storage capacity of the cloud is so great that high levels of data redundancy are acceptable—there is no imperative to maintain a single detailed database (neither is there much of an imperative to maintain an integrated one). The fact that the user can zoom in and out so easily enables them to resolve any ambiguity at the smaller scale. Third, user interaction is simplistic, and user expectations in terms of quality and completeness of data are low. Indeed this is a 'user beware' service—there are no legal or institutional imperatives governing its use. So when an island goes 'missing' (Fig. 1.6), there is little surprise in the community. Indeed users should not expect the levels of precision, completeness and accuracy typically associated with NMA data (Fairbairn and Al-Bakri 2013).

But even these differences are changing. The difference between 'formal and exhaustive' and 'informal-user generated' is becoming increasingly blurred with many NMAs gifting their data for free—OSM being the dominant recipient in this regard.[3] 'Partnerships' are forming between NMAs and the user generating community. An example of this, is where NMAs update their maps using crowd

2 http://www.openstreetmap.org/

3 http://wiki.openstreetmap.org/wiki/Potential_Datasources

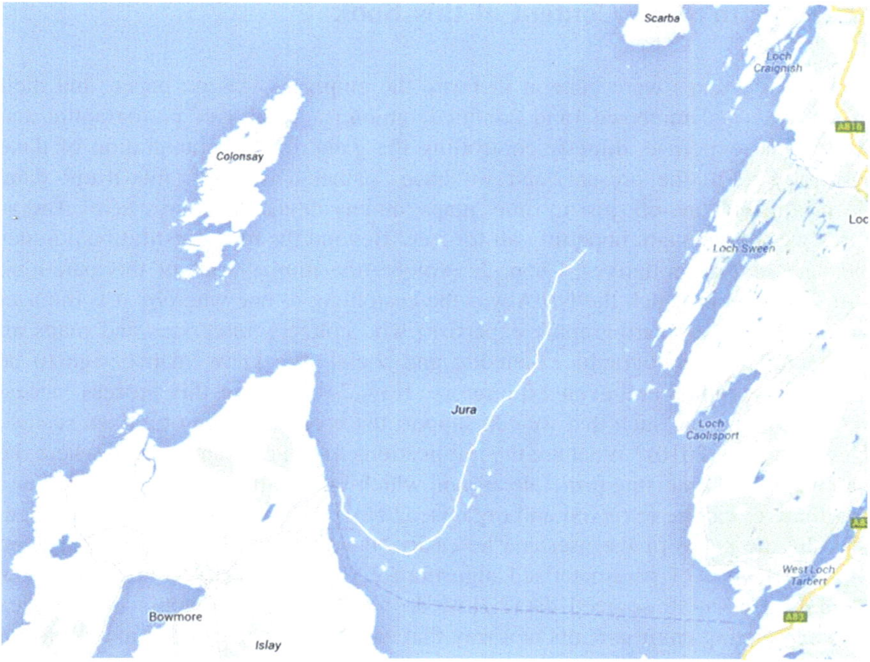

Fig. 1.6 Google 'apologised' when the Island of Jura, western coast of Scotland went 'missing' in July 2013 (http://www.maps.google.com—copyright: 2013 Google)

sourced data. The USGS is doing precisely this with its TNM project[4] in which individuals can collect a wide range of structural and natural features which are subsequently incorporated into 'The National Map'—precisely mirroring the OSM paradigm. In time, given (1) the range of scales of contributing data, (2) the variety of thematic and topographic data, and (3) the expectations of a growing community of users (particularly in their need for analysis), issues of data integration will become more pressing; there will also be a great need to manage volume, to generalise data to a shared scale prior to integration and to visualise it according to personal preference. Thus it is argued that the core objectives of generalisation will remain the same, however, future developments need to anticipate these changing contexts of use. Contexts that reflect a more distributed model, greater sharing of resource (reflecting the ambitions of the Open Geospatial Consortium[5]), fast, web-based, on-demand solutions, to a user community unknowing of the intricacies of map generalisation science. It was with these various ideas and changing contexts that the editors set about deciding on the structure and content of this book—one that built upon past knowledge, but responded to changing environments of use.

---

[4] http://nationalmap.gov/index.html

[5] http://www.opengeospatial.org

1.5 Structure and Content of this Book

Traditionally maps were general in form, the emphasis was on paper, and their production was immersed (and confined) among a wealth of cartographic and institutional expertise, directly controlling the gathering and integration of data. Fast forward to the present, and we have 'gather-it-yourself, mix-it-up' data, 'do-it-yourself, one-off, just-in-time' maps, on any device, and anywhere! And as if this was not enough, hopefully all for free! Beyond the many institutional issues and data sharing initiatives, Chap. 2, explores the implications of this paradigm shift from one in which the NMA was the custodian, to one whereby it is the user (with their limited cartographic expertise) who collects, integrates and maps all manner of data, topographic, thematic and social. We know map design to be inherently complex and even exploratory. How do we make this process 'accessible' to the citizen such that we can support the broader vision of citizen science (Newman et al. 2010)? What are the implications for map generalisation where the map forms an organising principle around which information (both qualitative and quantitative) can be retrieved and organised? How do we provide an environment whereby the needs of the user can be easily stated, then automatically translated into a sequence of parameterised algorithms? How is quality controlled in the integration of multi sourced data? How do we facilitate interaction in complex decision making environments in a way that caters to the growing communities of users, and an increasing diversity of maps? Are design specifications different for maps that are required for static display versus those for analysis and interaction? All of these questions have important ramifications for the underlying data models that must support these various activities, many of them perhaps necessarily hidden from the unknowing user.

Pragmatically then, the focus is on the creation of a map. What underpins that ambition is the fact that the power of maps lie in their abstraction and that the map is a model; a model that enables us to reason about the space around us. At different scales (levels of detail), we see different patterns of relationships and associations. This is why there is such a close coupling between the geographic and the cartographic. In Chap. 3, the authors explore the need for databases to be more 'geographic'—not just 'cartographic'. Can we list the types of relations that need to be stored explicitly, or determined through cartometric analysis? Is there an ontology of relations? What relations do we need to model beyond the metric, the topological, and the perceptual? And most critically, how do we model the changes in those relations as the features on a map 'transition', from those we see at the most detailed level (buildings, pavements, gardens), through to the very small scale map (capitals, countries, continents). For it is argued that in highly automated environments, and in the context of on-demand mapping, we need to make explicit these relations in order to produce meaningful maps—ones that give the correct emphasis and caricature among the multitude of relations that potentially describe the features on a map. In essence, the argument is made that the capacity to automatically reason about space is dependent upon an ontology of its relations.

In creating any map, it is important to acknowledge the impact that the web and mobile devices have had on how we interact with geographic information. Among the many anachronism associated with paper based cartography is the production of scale specific maps (1:10,000, 1:50,000, 1:250,000, etc.). From an historic context this was well justified but this idea has little purchase in the context of web based and mobile mapping, where the user wishes to 'pinch' and zoom, in a seamless and continuous manner, unconstrained by such 'stepped' scalings. Many research questions fall from such changing contexts: how do we ensure smooth and continuous zooming via remote devices? How do we prioritise data transfer in response to continuous panning and zooming, whilst minimising wait time? What classification mechanisms are required to control the merging and splitting of objects? These are some of the research questions explored in Chap. 4,—a chapter that acknowledges the rapidly growing importance of mobile map users and the consequent need for vario-scale data structures that support 'continuous' generalisation and thus provide a smooth transformation in the rendering of objects from the very detailed to the very coarse. How can we minimise data redundancy in such structures and how can we conserve topological properties during the generalisation process? How do we compress and partition such data in order to work within limited bandwidth and small screens? What are the partonomic structures that govern how objects are merged and split? Such research offers new ways of modelling the relationship between scale and space, and opens the door to new ways of visualising geographic data (such as mixed scale visualisations in which the same rendering contains features visualised at different levels of detail).

Technology has not only affected the delivery of geographic information, it has also opened the door to new ways of gathering such information both in a benign form (in which the user unwittingly becomes part of a sensor network) and proactively, in the form of volunteered geographic information (VGI). Both forms of content production can take many forms. Its strengths (and its weaknesses) lie in its unconventional, and instantaneous form; the notion that it is 'scale free' insofar as the user has little notion of the scale dependency of the data; VGI varies enourmously in quality and completeness, and rarely conforms to any standard or shared classification schema. How then do we integrate this 'informal' data with the formal? How do we convey the quality of that data? What role can generalisation methods play in reducing the data down to a shared denominator of scale? Chapter 5 examines these issues, demonstrating how a variety of methods (principally aggregation, selection, and dimensional change techniques) can be applied as a precursor to the application of matching techniques that enable the explicit joining together of features common across different datasets. The benefits arising from this are considerable; it is then possible to semantically link and integrate, not just conventional map data, but different types of social media (tweets, videos, imagery). In this sense the map becomes the organising principle underpinning the internet of things—a fusion of interconnected datasets. Fundamental to this ambition is the requirement that we achieve both a semantic and a geometric matching, and effective exploitation requires us to manage error handling and find ways of conforming to standards as part of the semantic sensor web. More broadly, this capacity is another paradigm shift in the way that we capture data and challenges the traditional authority of the NMAs.

It is against the backdrop of these changing contexts of data capture and visualisation, that Chap. 6 reviews developments in map generalisation techniques. It is these changing contexts that require operators to be more 'aware' of the fundamental relationships and properties of the phenomena represented on the map. Beyond the requirement of greater flexibility in their application (in concert with other generalisation operators) is the capacity to *reason* about map conflicts—to model the notion of compromise, to preserve various qualities, to know when a solution is out of reach and thus efficiently direct manual intervention. The requirement to model the underlying behaviour and to be responsive to various contexts leads inexorably to ideas of data enrichment and structural analysis. Generalisation Operators illustrates the maturing of robust generalisation techniques sufficient for their incorporation in commercial GIS. The case studies illustrate a growing sophistication in the cartographic modelling of geographic phenomena, rather than past approaches which saw the isolated application of individual techniques to primitive data types. That confidence is reflected in their broader application—broader both in the thematic data considered, and the wider range of scales over which they can be applied.

Whilst Chap. 6 reviewed advances in map generalisation operators, Chap. 7, considers ways in which map generalisation can be delivered as a service, over the web. What is required are ways of 'bundling' these operators together, incorporating sufficient 'know how' that they know when, in what order, where and by how much they need to be applied in order to create the required map. The research questions fall thick and fast from such an ambition: How do we formalise the procedural knowledge that enables us to chain these processes together? Is there an ontology to on-demand mapping? How do we manage this complexity (and conceal it from the user), how can we share and re-use these algorithms? What service oriented architectures best facilitate rapid processing of data? Do we transfer the data to the service, or the service to the data? How do we manage multiple formal and informal datasets or select those datasets most appropriate in providing a thematic context? How do we control the quality of the product, or assess the veracity of the solution (determining when the process can stop)? What role do machine learning or parallel processing have in such environments? These are the sorts of questions that have recently arisen from developments in web based services, fuelled by the expectations of users.

When it comes to map generalisation, there is something of a tendency to focus on anthropogenic features. Of course the location and extent of such features is hugely influenced by the shape of the underlying landscape morphology. A simple approach to the generalisation of the landscape might be to treat it as a simple manifold, a continuous surface that extends over the land and under the sea. But this is far removed from how we perceive and respond to such landscapes. We see valleys, mountain chains that divide, fenland at risk of flooding, and we fear seamounts that might gouge a hole in the hull of a ship! Chapter 8 examines the role of map generalisation in these contexts, arguing that we need to classify the landscape in order to apply operators that variously smooth and abstract such surfaces whilst preserving their most salient qualities (the highest point on the seamount in the case of our captain). Map generalisation is about conveying caricature whilst preserving truth in the relationship between features. So if for reasons of clarity the contours are slightly displaced, can we ensure that the river still follows the steepest path? Though the focus is on the third dimension, the same issues of data enrichment and behaviour modelling to support generalisation arise. We need a geographic understanding of their significance and relationship to other features if we are to achieve high levels of automation.

Any automated environment needs 'stopping criteria'—the point at which a generalised solution is deemed sufficient or cannot be improved upon without human intervention. Many factors govern that point: the interpretive skills of the user and their aesthetic preference, the context of use, not to mention the underlying complexity of the features being represented on the map. What is meant by a satisfactory solution? Is it still meaningful to use traditional paper maps as exemplars? Is the map legible? Does it have 'too much ink'? Does it effectively convey the essential qualities of the map or is it 'too busy'? What can we learn from human subject evaluation of maps? Chapter 9 examines such frameworks and the points in the generalisation process where evaluation techniques are necessary. Whilst there are many metrics by which we might quantitatively assess the correctness of a map (metric, topological, semantic and even Gestaltic properties), the fact remains that a map reflects the idea of compromise. The chapter therefore presents ways by which we might aggregate various evaluation criteria. Considerable challenges remain since the breadth of maps (and their context of use) extend beyond the traditional repertoire used to define 'good maps'. As generalisation is applied over larger and larger changes in scale, the challenges of defining these stopping criteria can surely only grow!

Increasingly generalisation techniques are being applied in thematic contexts and beyond. Can map generalisation techniques facilitate the design of metro maps for example? Can we apply 'strong' forms of generalisation that result in highly abstracted, 'essentialist' map forms—ones reduced down to their simplest form? Such abstractions work beyond traditional map constraints, sometimes heavily distorting the geographic space in the interests of simplicity and clarity—particularly pertinent in the context of low exposure times, where rapid comprehension is required. Chapter 10, explores these questions, examining the design of highly synoptic maps, such as those proposed by Brunet (Ormeling 1992), as well as a range of other maps that fall under the heading of 'schematised maps'. Are constraint based paradigms relevant in contexts where the emphasis might be on conveying causality, connectivity or distributive patterns among datasets? Does research in computational geometry and graph theory provide an extension to map generalisation methodologies? Automated rendering of highly abstracted forms are particularly pertinent to mobile devices (minimum data transfer and limited screen estate) and raise interesting questions about how we control generalisation techniques when applied in the creation of highly abstracted outputs.

For all of these many research questions, we might measure the competence and utility of our research in terms of its uptake in commercial systems and businesses. Chapter 11 demonstrates beyond doubt a maturing, robustness and sophistication that has resulted in the incorporation of many research outputs in the production environments of various NMAs (at least in Europe and the USA). NMAs remain the predominant users of generalisation methodologies. The eight case studies contained in this chapter reflect a growing acknowledgement of the need to provide their services in the context of on-demand mapping; in other words, the idea that their data will be used as a backdrop to third party data (user's own generated content). Tightly entwined in the ambitions and remits of NMAs is the need for low cost solutions (maximising levels of automation), that conform to measurable quality indicators, and work within frequent update cycles. In that sense generalisation techniques are not seen as some isolated end process but something that needs to be integrated throughout the data life cycle. The transition from research to commercial production environments has built on close cooperation between NMAs, vendors and research communities (both internal and external to the institution). The acceleration in adoption has enabled NMAs to think very differently about the information science associated with their products—enabling them to broaden the range of products, via different media, and in different contexts. The case studies in this chapter variously illustrate the broadening range of scales over which these techniques are applied, and also reflect a greater thematic variety.

Standing back from all these activities among researchers and practitioners, Chap. 12, lists those research questions currently uppermost in our minds. It attempts to distil the consequences of new, integrative, locational technologies, and the impact of new paradigms in the capturing and sharing of geographic information, upon the science of map generalisation. The ambition of map generalisation is many things; at its heart lies the challenge of viewing the earth at multiple levels of detail but it is also about understanding the needs of the user; it's about the efficiency with which they comprehend that geographic information; it's about modelling and abstracting space, it's about conveying the essence of place, revealing the patterns and associations among various geographic phenomena; ultimately it's about improving the quality of spatial decision making. Though the contexts in which maps are used may change, these ambitions remain the same; this book, like its predecessor, continues to reflect those ambitions.

References

Attneave F (1954) Some informational aspects of visual perception. Psychol Rev 61(3):183–193

Couclelis H (2010) Ontologies of geographic information. Int J Geogr Inf Sci 24(11–12):1785–1810

Fairbairn D, Al-Bakri M (2013) Using geometric properties to evaluate possible integration of authoritative and volunteered geographic information. ISPRS Int J Geo-Inf 2(2):349–370

Krippendorff K (1989) On the essential contexts of artifacts or on the proposition that 'design is making sense (of things)'. Design Issues 5(2):9–39

Masolo C, Vieu L (1999) Atomicity vs. infinite divisibility of space. In: Freksa C, Mark D (eds) Spatial information theory. Proceedings of COSIT'99. Springer, Berlin, pp 235–250

Muller JC (1991) Generalisation of spatial databases. In: Goodchild M, Rhind D, Maguire DJ (eds) Geographical information systems. Longman Scientific, London, pp 457–475

Newman G, Zimmerman DE, Crall A, Laituri M, Graham J, Stapel L (2010) User friendly web mapping: lessons from a citizen science website. Int J Geogr Inf Sci 24(11–12):1851–1870

Ormeling FJ (2013) Cartography as a window to the world. Cartographic J 50(2):175–181

Ormeling FJ (1992) Brunet and the revival of French geography and cartography. Cartographic J 29(1):20–24

Stuhlman J (2007) Georgia O'Keeffe: circling around abstraction. Hudson Hills Press, Manchester, VT

Chapter 2
Map Specifications and User Requirements

Sandrine Balley, Blanca Baella, Sidonie Christophe, Maria Pla, Nicolas Regnauld and Jantien Stoter

Abstract In traditional generalisation flow lines, the target map is specified upstream, manually, by cartographers and is intended to answer generic, well-identified user needs. In the emerging context of on-demand mapping, maps have to be derived automatically for users whose requirements are not known in advance, and who may need to integrate their own data. The definition of suitable target map specifications thus becomes part of the service, which raises challenges that are explored in this chapter. The first challenge is to set up a formal map specifications model, rich enough to guide the whole map derivation process. The second challenge is to collect requirements and to assist the user, who is not supposed to be a map designer, in the specification of a map usable for their task and one that respects cartographic standards.

S. Balley (✉) · N. Regnauld
Ordnance Survey Great Britain, Southampton, UK
e-mail: Sandrine.Balley@gmail.com

N. Regnauld
e-mail: Nicolas.Regnauld@1spatial.com

B. Baella · M. Pla
Institut Cartogràfic de Catalunya, Barcelona, Catalonia, Spain
e-mail: Blanca.Baella@icc.cat

M. Pla
e-mail: Maria.Pla@icc.cat

S. Christophe
Laboratoire COGIT, IGN, Saint-Mandé, France
e-mail: Sidonie.Christophe@ign.fr

J. Stoter
Delft University of Technology and Kadaster NL, Delft, The Netherlands
e-mail: j.e.stoter@tudelft.nl

D. Burghardt et al. (eds.), *Abstracting Geographic Information in a Data Rich World*, Lecture Notes in Geoinformation and Cartography, DOI: 10.1007/978-3-319-00203-3_2,

2.1 Introduction

Using current generalisation solutions, it is possible to derive a considerable range of different maps and datasets from the same database. This potential is still under-exploited (Mackaness et al. 2007). If generalisation could be more flexible it could adapt to various needs depending on the context of use.

The idea of adaptive generalisation was first studied to enable advanced location-based services (Reichenbacher 2004). Here, maps have had to adapt to new, but well-defined parameters defining user requirements: taking account of the user's location and mobility, the screen size and resolution, and the user's task (Sarjakoski and Nivala 2005). A demonstration of this was done by the EU-funded GiMoDig project (Sarjakoski and Sarjakoski 2007), with the first prototype of adaptive generalisation in a mobile context. Beyond location-based services, in the wider domain of web cartography, user requirements are more diverse and their description is a key research topic, as illustrated by the working groups on usability, map use and user issues (at Agile[1] from 2001 to 2006, at ICA[2] since 2005).

On-demand mapping is the research domain that seeks to derive automatically maps tailored according to expressed user requirements. Although some aspects of on-demand mapping have been studied and some prototypes designed (Cecconi 2003; Jabeur 2006; Sarjakoski and Sarjakoski 2007; Foerster et al. 2010; Christophe 2011; Balley et al. 2012), the whole issue remains a to-be-solved, cross-domain puzzle (Balley and Regnauld 2011). This chapter focuses on one of the challenges of on-demand mapping, which is the acquisition and interpretation of the user's needs in order to specify a usable target map. Another challenge lies in the automatic design and orchestration of the target map's derivation process, in particular the generalisation part of it (Chap. 7).

On-demand mapping can provide an answer to an increasingly common activity, namely creating a cartographic mash-up. A mash-up is a map combining cartographic layers from different data sources. As in Fig. 2.1, mash-ups are often composed of a background topographic layer (accessed through the API of a major provider such as a national mapping agency, Google or OpenStreetMap), and of a user-generated, thematic layer in the foreground. Mash-ups are prone to legibility issues for two main reasons. Firstly, since the data currently available online to the general public is not customisable, the content and legend of the background layer cannot be adapted to the foreground layer. In the example in Fig. 2.1, this results in a too dense road network and unnecessary topographic features (such as the details associated with the parks). Secondly, mash-ups often fail to depict the spatial and semantic relationships between themes (Jaara et al. 2012). In our example, the map does not clearly indicate that cycle facilities *follow* roads and that some cycle support services are *associated with* the roads. Instead, cycle

[1] http://www.agile-online.org/, accessed 28 May 2013

[2] http://icaci.org/commissions/, accessed 28 May 2013

Fig. 2.1 Extract of a mash-up featuring cycling facilities (OpenCycleMap). As argued by Gaffuri (2011) and others, generalisation should dramatically improve the quality of such internet maps in the future

facilities appear as standalone entities and over-dominate the roads. Highlighting relationships between themes requires conflation tools and map design skills that cannot be expected from every mash-up author. The idea is that such tools could be embedded in an on-demand mapping system in the future. One of the challenges for this system would be to take account of the user's inputs in terms of their requirements (e.g. the intended map use, the nature of the thematic data and the user's preferences), to identify the most relevant background content and inter-theme relationships that it was important to show on the map.

This chapter focuses on the translation of expressed user requirements into map specifications, i.e. into the formal description of a target map in terms of content and representation. Since the state of the art is still limited, this translation is explored not only with respect to generalisation but more generally with respect to the whole mapping process. How to adapt the map making process to reach these specifications is out of the scope of this chapter. Section 2.2 describes the translation process and defines the needs, requirements and specifications. Section 2.3 focuses on user's requirements, how they influence map specifications and how they can be automatically interpreted. Section 2.4 focuses on the acquisition of user requirements. Three Case studies are then presented. Section 2.5 presents an initiative to collect and formalise the cartographic constraints of national mapping agencies for a European generalisation test bed. Section 2.6 describes a research model for map specifications dedicated to on-demand mapping including user data at Ordnance Survey Great Britain (OSGB). Section 2.7 presents a dialogue-based prototype from the COGIT laboratory (IGN France) that enables the automated creation of personalised legends. Section 2.8 is an overview of the issues and identifies future research challenges.

2.2 Key Concepts: Needs, Requirements and Specifications

2.2.1 Deriving a Custom Map: Process and Definitions

While on-demand mapping is a long-term research goal, mapping agencies do currently deliver customised maps. Automatic services are available, but the customisation is still limited to the spatial extent, scale, title and cover of the map.

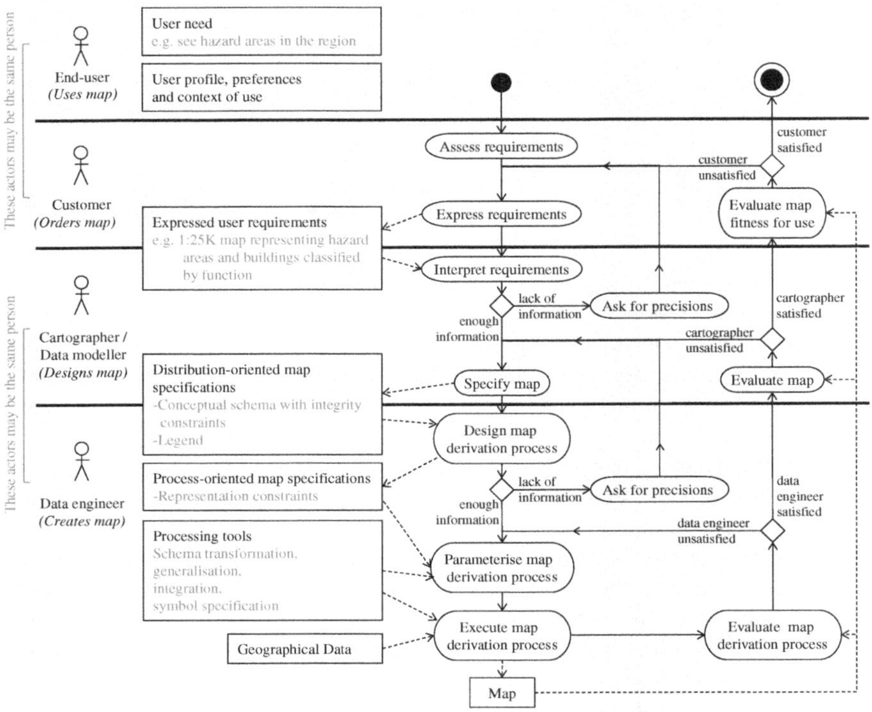

Fig. 2.2 The manual process of deriving a customised map. On-demand mapping is an automatic version of this process

With the increasing use of multiple-representation databases, these services will soon be able to deliver instant maps with different, predefined levels of detail. More flexibility can be achieved through a manual approach, represented in Fig. 2.2. Since this process is iterative and involves several actors, it is very costly and affordable to professional users only. Going through this process, this section introduces the concepts used throughout this chapter.

At the root of the process, the *end-user* (Fig. 2.2) has a *need* (e.g. to plan a trip or to carry out an analysis task) and it is assumed that they will require a map in order to answer it. The end-user also has a *profile* (i.e. a few personal characteristics such as age and nationality), some *preferences* regarding the map, and a *context* in which the map will be used (at some place, using some device—e.g. an emergency situation). The combination of these parameters is called *user requirements.* Unfortunately, the end-user is not always given a chance to express their requirements. Instead, they are assessed by the *customer* who will order the map (the user and customer may be the same person). The customer expresses the assessed requirements to the *cartographer*, whose role will be to design a product that best fits those requirements. The expression and interpretation of requirements is typically an iterative process. The cartographer progressively translates the interpreted requirements into *map specifications*, i.e. a description of the target map in terms of information content and information representation. At this stage, these are distributed-oriented specifications: they are human-readable and informative. They are typically composed of a conceptual schema, potentially enriched with some integrity constraints that are specific to the user's need, and a legend. The role of the *data engineer*, who may be the same person as the cartographer, is to create and run a map derivation process including the choice of source data, content selection, restructuration, generalisation, integration and symbolisation. By examining the specifications, and requesting clarifications from the cartographer if required, the data engineer creates *process-oriented map specifications* which will be used to parameterise each sub-process. Unlike distribution-oriented specifications, process-oriented ones must be machine-readable and detailed enough to guide the automated parts of the derivation process. The cartographic constraints used in automated generalisation are examples of process-oriented map specifications. The map derivation is followed by an iterative evaluation process. The map is reviewed by the data engineer (who can select other tools or modify their parameters to make the output closer to the specifications), by the cartographer (who can change or tighten the specifications) and by the customer (who can express new requirements). The process terminates when the customer is satisfied.

On-demand mapping is an automatic version of the process represented in Fig. 2.2. The cartographer and data engineer are replaced by an expert system interacting directly with end-users. No distribution-oriented map specifications are required but only process-oriented ones, which play a crucial role since the map derivation process is fully automated. The next section deals with the content and formalisation of these specifications.

2.2.2 Process-Oriented Map Specifications

From this point, we will use "map specifications" to refer to "process-oriented map specifications". The next section details the content of map specifications. The following one clarifies the relation between map specifications and the cartographic constraints used in the generalisation process.

2.2.2.1 Content of Process-Oriented Map Specifications

This section lists the required content of process-oriented map specifications. It relies on proposals issued not only for generalisation, but also for data production (ISO 2005; Inspire 2008), data integration (Gesbert 2005) and legend design (Bucher et al. 2007; Christophe 2009).

Map specifications must describe the *information content* of the target map. This includes the spatial extent, the geographic concepts represented (e.g. roads and buildings), the scale (for paper maps) or scale range (for digital maps), and the overall level of detail. In addition, for each concept represented on the map, the specifications must make explicit the entities selected for representation (e.g. for a map of car accidents: only important roads, plus roads involving a high number of accidents). This specification element provides clues as to the choice of source data, the feature selection and the generalisation process.

Map specifications must also describe the expected *geometric modelling* of concepts: (1) the primitives used, (2) the conditions under which entities must be represented as aggregates (e.g. buildings are combined if they are within 50 m of each other and collectively cover at least 1,000 m^2), and (3) the spatial relations that should occur between themes (e.g. cycling facilities *follow* roads). This information will guide the generalisation steps (e.g. collapse, aggregation) and the integration of user's data.

The *thematic modelling* of each concept must be specified, i.e. the represented properties and their domain of values (e.g. the "importance" of roads, with an enumerated domain of values: "low", "medium" and "high"). This information influences the choice of source data and may lead to some data restructuration. This information can govern a specific option for data integration, where the user's data is not displayed geometrically but is projected as new properties on the referential features (as exemplified in Sect. 2.6).

If the target product is a vector map, map specifications may include information about the expected *data schema*, i.e. the organisation of entities and their properties into feature types, attribute types and association types. This information drives possible schema transformations (Balley 2007; Letho 2007) which are part of model generalisation (Foerster et al. 2010).

The last specification element describes how represented entities should be *symbolised* on the map. The styles must be defined, possibly supported by the explicit mention of relations between themes (association, opposition or order) as defined in the theory of semiology (Bertin 1967).

From this discussion we make three observations. Firstly, every specification element may involve relationships between objects and between themes that will guide some step of the map derivation process. Spatial relations (e.g. cycling facilities *follow* roads) may strongly influence data integration and generalisation; while semantic relations (e.g. cycling facilities *are attached to* roads) may influence schema transformation and symbol specification. The modelling of spatial relations in automated environments is the topic of Chap. 3.

Secondly, map specifications must be a description of how the final map should be, and not of how it can be generated. Consequently, map specifications should not refer to any data source, derivation algorithm or implemented tool. This ensures the sustainability of map specifications independently of the evolution of data models and tools, which inevitably occurs in map production environments. This also enables reuse of specifications in different environments, (e.g. being able to create the same type of map in different countries).

Thirdly, for the same independence purpose as above, map specifications should not refer to implemented feature types but to constructs from higher abstraction levels. In standardised contexts such as the INSPIRE infrastructure, these constructs can be feature concepts taken from a feature concept dictionary (ISO 2009). Feature concepts can be seen as interfaces for feature types proposing a meaning and a set of properties. In non-standardised contexts, where no assumption can be made on the schemas of available datasets, specifications can rely on generic feature types pointing to geographic concepts (e.g. road or city) from an ontology. Defined by Gruber (1993) as "the specification of a *conceptualisation*", ontologies of real world concepts were introduced in GIS by Fonsesca (2001) and have been used more and more intensively since then for various applications (Klien et al. 2006; Lemmens 2006; Regnauld 2007; Touya and Duchêne 2011; Gould and Chaudhry 2012; Abadie et al. 2010; Dominguès et al. 2009). The specification models proposed in Gesbert (2005), Touya and Duchêne (2011), Balley et al. (2012) refer to ontologies of topographic concepts. Because map specifications strongly rely on spatial relations, an ontology of spatial relations was proposed by Touya et al. (2012) (Chap. 3).

2.2.2.2 The Relationship Between Process-Oriented Map Specifications and Cartographic Constraints

Map specifications are a set of conditions that the on-demand map should satisfy. In the context of generalisation, such conditions have long been represented through cartographic constraints (Beard 1991; Ruas and Plazanet 1996). This section clarifies to what extent process-oriented map specifications as defined in the previous section can be formalised by cartographic constraints.

The reader should refer to Zhang (2012) and Touya (2011) for an up-to-date review of developments in cartographic constraint methodologies. Several classifications of constraints have been proposed (Harrie and Weibel 2007). Following the EuroSDR testbed of generalisation software (Sect. 2.5), Burghardt et al. (2007) produced a fine-grained classification of cartographic constraints (Fig. 2.3). Constraints are first categorised according to the generalisation goal: improving the map legibility or preserving relevant visual characteristics. They are then classified by the constrained property (e.g. minimal dimensions, topology, position, shape, etc.), by the number of objects involved, and by the geographic concept affected.

Cartographic constraints (as presented in Fig. 2.3) can express some of the specification elements listed in Sect. 2.2.2:

- The information content can be partly specified. Minimal dimensions constraints can help delimit the set of represented entities. The concepts that should not appear can be specified by removal constraints (e.g. "if the use of the building equals 'habitation', then the building should not be represented"). However, the entities that should appear cannot be specified. This is due to the fact that, in generalisation, the information content of the target map is implicitly determined by the information content of the original map.
- The geometric modelling can be partly specified by topological constraints (e.g. "no intersection between roads and buildings") and position constraints

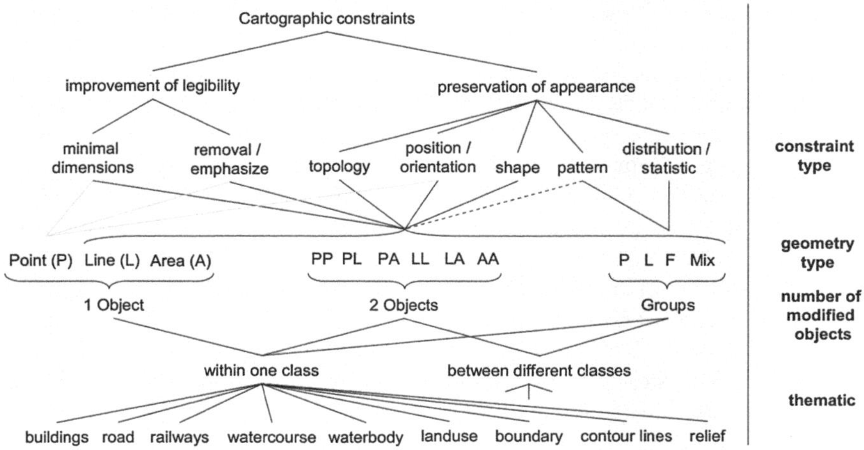

Fig. 2.3 Typology of cartographic constraints for automated generalisation (Burghardt et al. 2007)

(e.g. "cycle itineraries follow roads"). However, in their generalisation form, many of these constraints are expressed through a maximum degree of variation between original and final states. This may not be adapted to on-demand mapping, where the source data is not supposed to be known in advance.

It appears that more categories of constraints—they could be generically called representation constraints—are needed to express the other specification elements (thematic modelling, data schema and symbol specification) in order to support more of the steps in the map derivation process. The research models now presented could help fill the gaps.

Gesbert (2005) designed a model to formalise database specifications and applied it to products of IGN France. The model consists of linking database feature types with topographic concepts via four types of representation rules. Selection rules are used to determine whether or not a real world object must be represented in the database. Cutting rules state whether a real world object is represented individually or by its parts. Aggregation rules state whether a real world object is represented individually or in a grouping of similar objects. Instantiation rules indicate in which class and with which attributes a real world object and its properties are represented. The entities affected by a rule are delimited by constraints related to their nature, properties and relations to other entities. The model is used at IGN France for schema matching (Abadie 2009) and database discovery (Mechouche et al. 2013).

Bucher et al. (2007) proposed a map specification model intended for legend design that enabled the definition of relationships between themes (order, association or difference) and communication levels (e.g. a group of elements that need to be given visual emphasis). Christophe (2011) proposed a version of this model especially enriched for the description of colours. This model is exploited in the dialogue-based system for legend design presented in Sect. 2.7.

Building upon the contributions described in this section, a map specification model based on representation constraints is being experimented with for on-demand mapping at OSGB (detailed in Sect. 2.6).

2.3 Inferring Map Specifications from User Requirements

This section deals with the instantiation of process-oriented map specifications inferred from user requirements. It investigates how to automate the activity "design map" traditionally performed by the cartographer (Fig. 2.2). Here, "user" refers to the end-user and not the customer. Section 2.3.1 analyses how the user requirements may influence map specifications. Section 2.3.2 reviews knowledge-based mechanisms that enable inference of map specifications to be made. Section 2.3.3 focuses on the knowledge required by these mechanisms.

2.3.1 User Requirements and Their Influence on Map Specifications

As defined in Sect. 2.2.1, user requirements are a combination of user needs, user profile, preferences and a context of use. We propose to further decompose these elements into "user variables". Figure 2.4 synthesises common user variables and states again, the map specification elements described in Sect. 2.2.2.

The needs of the user are determined by the task of the map user (e.g. "hiking" or "plan an itinerary"). This task determines the map topic, which in turns influences the represented concepts and their assigned symbols. For instance, a topographic map depicts the nature of the terrain features with a "neutral" legend, while a navigation map focuses on and emphasises communication networks and landmarks. The needs "visualise hazard zones" and "perform a risk analysis" call for the same map topic but for different generalisation levels (potentially going up to schematic maps) and legend choices. To perform this analysis, professional users need task-specific concepts, that sometimes do not exist generically and need to be derived (Harding 2011). They also need a suitable data schema, (e.g. one with explicit network structures).

There are as many classifications of user profiles (Fig. 2.4) as applications exploiting them. Users can be categorised according to their familiarity with maps (professional users, non-professional users, non-professional users who are novices in the use of maps), their age, gender and nationality. The user profile influences the level of detail of the map. For example, children or sight-deficient readers may need maps representing a limited number of concepts, a limited amount of objects with a high level of generalisation and/or large symbols.

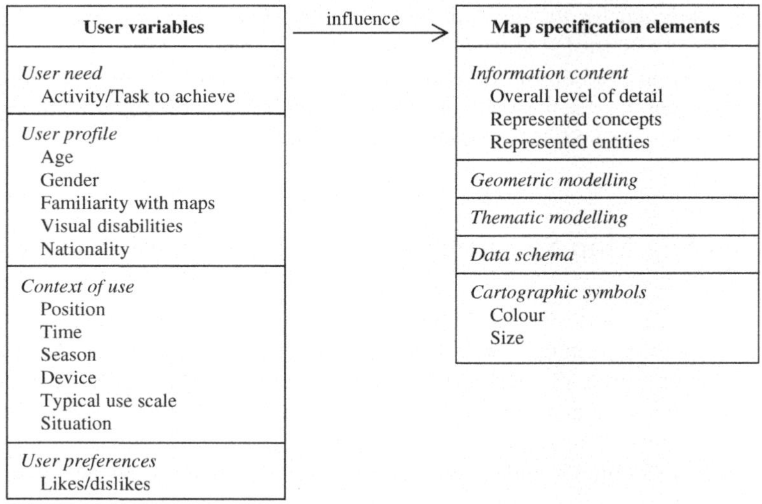

Fig. 2.4 User variables influencing map specification elements

However, the definition of what constitutes an "expert" and its influence over the ability to interpret maps efficiently is a controversial issue (Ooms et al. 2011). Non-professional map users may be accustomed to maps displaying many different concepts (e.g. tourist maps). The profile also influences the choice of colours. For instance, as detailed in Christophe (2009), children like primary colours and European people tend to prefer different hues from North-American people. Colour-blind users need specific colour arrangements to enhance the perceived contrast (Brewer 1997; Dhée 2011).

The context of use (Fig. 2.4) refers to the place, time and situation in which the activity is carried out. If the map is needed to situate the user in a mobile context, the user's position, and more specifically the type of place they are located in (e.g. in the underground, in the city or on the mountain), influences the choice of represented concepts. The type of media (paper or digital map, screen size, battery life), time and season (for contrast reasons), influences the generalisation level and choice of colours, as studied by projects on ubiquitous mapping (Sarjakoski and Nivala 2005; Hoarau 2011). The situation variable may refer to "on-field study" versus "office study", "real time" versus "post-analysis", "standard" versus "emergency". As analysed by Duchêne et al. (2011), an emergency situation generates specific constraints on the map content—the message must be as simple as possible—and on the generalisation process itself—the map must be delivered quickly.

User preferences are the last set of variables. They enable the end-user to express direct constraints on any component of the map specifications. They must be used with caution: as pointed out by Harding et al. (2009), it is easier for users to describe what they want to do (i.e. their task or activity) than to define what geographical information they need. In addition, privileging user preferences over the expertise of the map designer may put at risk the map's efficiency and consequently, the user's satisfaction. However, as the manual process represented in Fig. 2.2 shows, user's preferences about maps derived on-demand can help refine the system's proposition (Sect. 2.4).

2.3.2 Knowledge-Based Mechanisms Used to Infer Map Specifications

This section focuses on systems that infer personalised map specifications from user requirements. Interfaces to collect requirements are discussed in Sect. 2.4.

One approach is to first set the requirements and to follow a static reasoning process using rules. Forrest (1999) designed an expert mapping system based on this approach. The targeted maps were small scale thematic maps. The initial user inputs consisted of a map topic selected from a list (e.g. geological, industrial, topographic, etc.), a map purpose (detailed study or general overview), a user profile (naive or expert) and an output medium. Using rules and importance scores defined by the cartographer, the system proposed background themes for the user to choose from, while limiting the amount of information on the map. The system proposed a legend (number of subclasses and symbols) by taking into account the map topic, scale and level of detail. The GiMoDig system used a similar approach, but added a form of machine learning (Sarjakoski and Sarjakoski 2005): map specification templates were manually created to suit a few predefined combinations of context parameters. These templates constituted the initial knowledge base of the system. For each new combination, an inference engine was used to select and refine the template that best matched the user's parameters. In both approaches, the user was allowed to refine the system proposal, e.g. to add or remove a theme.

More sophisticated systems rely on negotiation: user's preferences and cartographic knowledge are represented as constraints that the system tries to reconcile. The process is iterative: the user can react to proposed map specifications by expressing new preferences. This approach was proposed by Hubert (2003) to translate user preferences into generalisation parameters and by Christophe (2011) to collaboratively design legends. The latter system is presented in Sect. 2.7.

2.3.3 Cartographic Design Knowledge

We call cartographic design knowledge the knowledge that enables the system to infer suitable map specifications from user requirements. This section focuses on the acquisition and formalisation of that knowledge. Expert cartographers are the primary source of this knowledge. They describe the map specifications they judge usable for certain user requirements, but also compliant with the standards of cartographic design. Usability surveys are the second source of knowledge.

They must be carefully interpreted: as pointed out by Harding et al. (2009), data usability is sometimes difficult to distinguish from the usability of systems and interfaces used to manipulate the data (Haklay 2010).

2.3.3.1 Knowledge on Information Content and Structure

Different approaches have been used to describe the ideal information content of a map or a dataset. A pragmatic option is to use templates. With ScaleMaster, (Brewer and Buttenfield 2010) provide formal guidelines for the information content of multi-scale topographic maps, and "recipes" to derive them with a minimal workload. A library of styles is defined for each scale range. When the scale is decreasing, style modifications are preferred to geometric transformations, which only occur over significant changes in scale. The ScaleMaster model was implemented by Touya and Girres (2013) to monitor automatic multi-scale generalisation. Templates can also be used to record the knowledge on data schemas (Gnägi et al. 2006; Balley 2007): data providers, using internal knowledge and users' feedback, can suggest schema profiles, i.e. useful feature types which are initially implicit, but can be derived at no cost from their products. A second option, more flexible than using templates, is to represent the knowledge in the form of an ontology. In their ontological approach to data enrichment, Lüscher et al. (2007) relate spatial and temporal context information (e.g. "urban area", "Victorian period") with potentially relevant high-level geographic structures. The system relies on ontologies of geographic and GIS concepts, and application ontologies provided by domain experts.

Only a few projects have attempted to collect knowledge directly from the end-users. Usability researchers at OSGB conducted a survey of 56 professional users (Harding 2011) and captured the information content and structure needed for their tasks. Building on the interview results and on space descriptions from non-professional volunteers, Battye (2010) tried to assess which feature types are mentally encompassed by place types at various scales (from "locality" to "country"). The extracted knowledge resulted in guidelines for the design of usable products.

2.3.3.2 Knowledge Related to Legends

Most of the knowledge currently used in legend design was enunciated by semiologists such as Bertin (1967) and Robinson (1952) amongst others. The ColorBrewer system (Brewer 2003) proposes predefined, harmonious ranges of colours suitable for thematic maps with qualitative, sequential or diverging classifications. The suitability of each colour range to different devices or users (printed maps, projector, colour-blind people, etc.) is indicated. In his experiments on European legends, Renard (2008) asked cartographers to characterise 19 legends applied to the same three sample datasets. The result was a knowledge base where each legend is labelled with the themes it best suits (e.g. roads or habitation), the kind of zone it best represents (e.g. urban, rural or mountainous area), and also the general feeling it conveys. This study was building upon a corpus of emotional terms (e.g. rich, luminous, sober or happy) created through a user survey (Domingùes and Bucher 2006). It was used to characterise a collection of legend samples based on different colour schemes. The knowledge base and legend samples were used in an on-demand mapping prototype described by Jolivet (2009).

2.4 Collecting User Requirements

The previous section presented some solutions to the challenge of inferring map specifications from user requirements. This section discusses interfaces suitable to acquire these requirements. It focuses on the activity "express requirements" represented in Fig. 2.2, traditionally performed by the customer at the beginning of the process and during the evaluation phase. On-demand mapping would enable any end-user to express their own requirements, but it would require more support than currently provided. As a matter of fact, most map design wizards are essentially collecting specifications. This is feasible for simple maps if the background data is predefined and requires limited processing. Examples are provided in Sect. 2.4.1. Section 2.4.2 focuses on more sophisticated systems that require detailed process-oriented map specifications. It is important that intuitive solutions can be found that readily allow the user to influence these specifications.

2.4.1 Specifying a Thematic Map Through Contextual Menus

More and more services enable users to map thematic data over contextual data. Two types of integration services are available.

In the first type, some thematic data are projected on referential objects through a gazetteer. With a minimum of guidance, users can choose the background themes, select thematic data or upload their own data. Textual menus and cursor selection can be used to select which attributes and classes of values will guide the thematic classification. GeoCommons[3] and Geoclip[4] are examples of such services.

In the second type, the thematic data are simply overlaid on the contextual background. If no geometric adjustment is performed, users can simply set some of the user variables listed in Fig. 2.4 through menus and checkboxes, and let the system do the rest. If a geometric adjustment was performed, interfaces would be

[3] http://geocommons.com, accessed 28 May 2013.

[4] http://www.geoclip.net, accessed 28 May 2013.

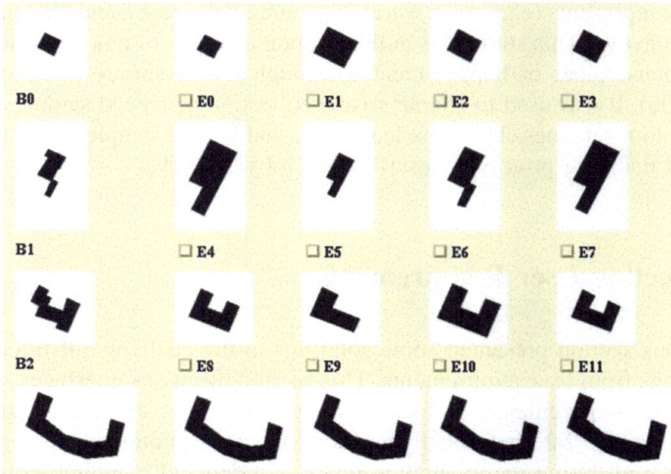

Fig. 2.5 Extract of a dialogue interface to adapt generalisation constraints to user preferences (Hubert and Ruas 2003)

required to let the user choose their reference objects and the spatial relations that must be satisfied. Such services are being investigated (Jaara et al. 2012) but no interface has yet been fully specified.

2.4.2 Collecting Preferences Through Map Samples

Unlike other user variables, user preferences are difficult to collect through contextual menus because they may relate to specification elements that are difficult to conceptualise. Being able to visualise the influence of these elements on a map is much more intuitive and easy to comprehend. This section reports on proposals whereby user preferences are collected through map samples, in order to help the user see what the resulting map would look like.

Hubert et al. (2003) designed a dialogue engine enabling non-generalisers—but still experienced map makers—to influence the cartographic constraints used by a generalisation process, and consequently the geometric modelling of entities on the map. In the implemented case study, the system and the user tried to reach an acceptable solution for the generalisation of buildings. In this dialogue, the user and the system were communicating through map samples and simple natural language statements:

- The system proposed a list of map samples representing an initial building set and their generalised counterparts (Fig. 2.5). Each generalised sample was internally associated with the parameters of the cartographic constraints used to generalise it.

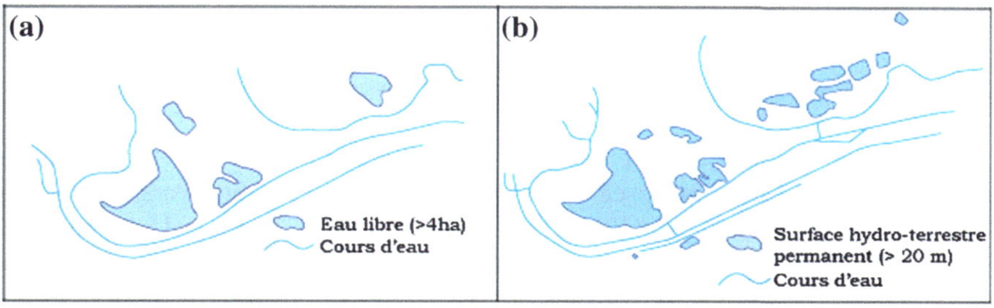

Fig. 2.6 A web interface using feature examples to explore databases specifications (Gődér 2003). IGN France BDTopo® (**a**) and former BDCarto® (**b**) represent hydrographic features with different selection criteria

- The user was reacting either by selecting one or several map samples, or by commenting on them (e.g. "too big", "too square"). These comments led the system to choose other combinations of parameters and present the associated map samples.

Another dialogue engine based on map samples for legend design, and more specifically for the choice of colours, is described in Sect. 2.7 (Christophe 2011).

The shape of objects and the colour of symbols are only part of the specifications required for on-demand mapping. More example-based solutions need to be imagined to allow users to express preferences influencing the information content, geometric and thematic modelling, data schema and other legend variables. For example, in Fig. 2.6, data samples are used to illustrate differences in the selection of entities between two databases (Gődér 2003).

2.5 Case Study I: Specifying Generic and Specific Map Specifications—A EuroSDR Case Study

Jantien Stoter, Blanca Baella and Maria Pla

The previous sections explained the process of deriving map specifications for automated generalisation from user requirements. This section explains how this was realised within a EuroSDR project.

The EuroSDR project on the state-of-the-art of automated generalisation software was a cooperative project of Universities, NMAs and private industry across Europe (Stoter et al. 2009a, 2010). The main objective of the project was to study the state-of-the art of currently available commercial generalisation software packages, examining their capabilities at generalising a complete map. The map specifications were defined by the NMAs and harmonised before use. Additionally, the project sought to gain insight into other issues, such as how to specify requirements for generalisation, how do generalisation processes work, how to set up a case to study the-state-of-the-art in generalisation, and how to perform evaluation of generalisation output. Evaluation aspects are covered in detail in Chap. 9.

Between June 2007 and Spring 2008 tests were performed by project team members (from NMAs and research institutes) on out-of-the-box versions of four generalisation systems: ArcGIS (ESRI), Change/Push/Typify (University of Hanover), Radius Clarity (1Spatial) and axpand (Axes Systems). At the same time the vendors carried out tests with the same test cases with improved and/or customised versions of their systems.

This section presents the approach employed in the EuroSDR project to define a shared model of generalisation constraints. This requirements analysis (carried out in 2006/2007), resulting in generic and NMA-specific map specifications, consisted of four steps:

- Selection of test cases representative of typical generalisation problems (Sect. 2.5.1).
- Formalisation of NMA map specifications for automated generalisation (Sect. 2.5.2).
- Harmonisation of the specifications resulting in one generic set of NMA map specifications within the context of the study (Sect. 2.5.3).
- Analyses of the defined specifications to learn more about similarities and differences between map specifications of NMAs (Sect. 2.5.4).

2.5.1 Selecting the Test Cases

The first step in the requirement analysis was the selection of test cases representing problems for automated map generalisation. To meet this objective, the EuroSDR project team generated a list of outstanding map generalisation problems based on the OEEPE research (Ruas 2001) complemented with the research team's own experiences. Examples of these problems are: building generalisation in urban zones, mountain road generalisation, solving overlapping conflicts in locally dense networks, pruning of artificial networks, and ensuring consistency between themes in particular areas such as coastal zones. Some of these problems have been tackled in research, resulting in at least partial solutions. However, the EuroSDR project wished to evaluate complete solutions in commercial systems, and, therefore, these problems were also identified as representative map generalisation problems. Four test cases were selected that included all these problems (Table 2.1). The test cases were provided by Ordnance Survey Great Britain (OSGB), Institut Géographique National (IGN France), The Netherlands' Kadaster (Kadaster) and Institut Cartogràfic de Catalunya (ICC).

Table 2.1 Test cases selected for the EuroSDR research

Area type	Source dataset (k)	Target dataset (k)	Provided by	No. of feature classes	Main feature classes
Urban area	1:1250	1:25	OS Great Britain	37	Buildings, roads, river, relief
Mountainous area	1:10	1:50	IGN France	23	Village, river, land use
Rural area	1:10	1:50	Kadaster, NL	29	Small town, land use, planar partition
Coastal area	1:25	1:50	ICC Catalonia	74	Village, land use (not mosaic), hydrography

The NMAs modified their test datasets to prepare them as input for the generalisation tests. For example, details such as rich classifications were removed from the datasets and the datasets were translated into English. In addition, to be able to define specifications of the output maps with respect to symbolised objects and to assure uniform outputs, the four NMAs defined symbols for the output maps. Figure 2.7 shows samples of the source datasets. These inputs differ slightly with the original datasets and symbols as used in production, since they were adjusted (i.e. simplified) for the project.

2.5.2 Formalisation of NMA Map Specifications for Automated Generalisation

In the task of formalising map specifications for automated generalisation, the EuroSDR project distinguishes between two stages. The first stage is to describe the specifications in a way that the users (in this case the testers of the systems) fully understand what they should try to obtain from the system. The second stage is to translate these specifications in a format understandable by the generalisation system. The first stage was completed by means of cycles between the data providers and the research team. The testers completed the second stage during the test process.

To implement research theories, NMA map specifications were defined as a set of cartographic constraints that had to be respected. As mentioned in the theoretical part of this chapter, the use of constraints is a common method for defining specifications and to control and evaluate the automated generalisation process. Constraints express how generalisation output should look without addressing the means—for example the sequence of operations—by which this result should be achieved.

A template was developed for a uniform definition of the constraints in the four test cases. In the template, specific properties of the constraint can be defined such as conditions that must be respected and the geometry type and feature class(es) to which the constraint applies (Table 2.1). The template distinguishes between

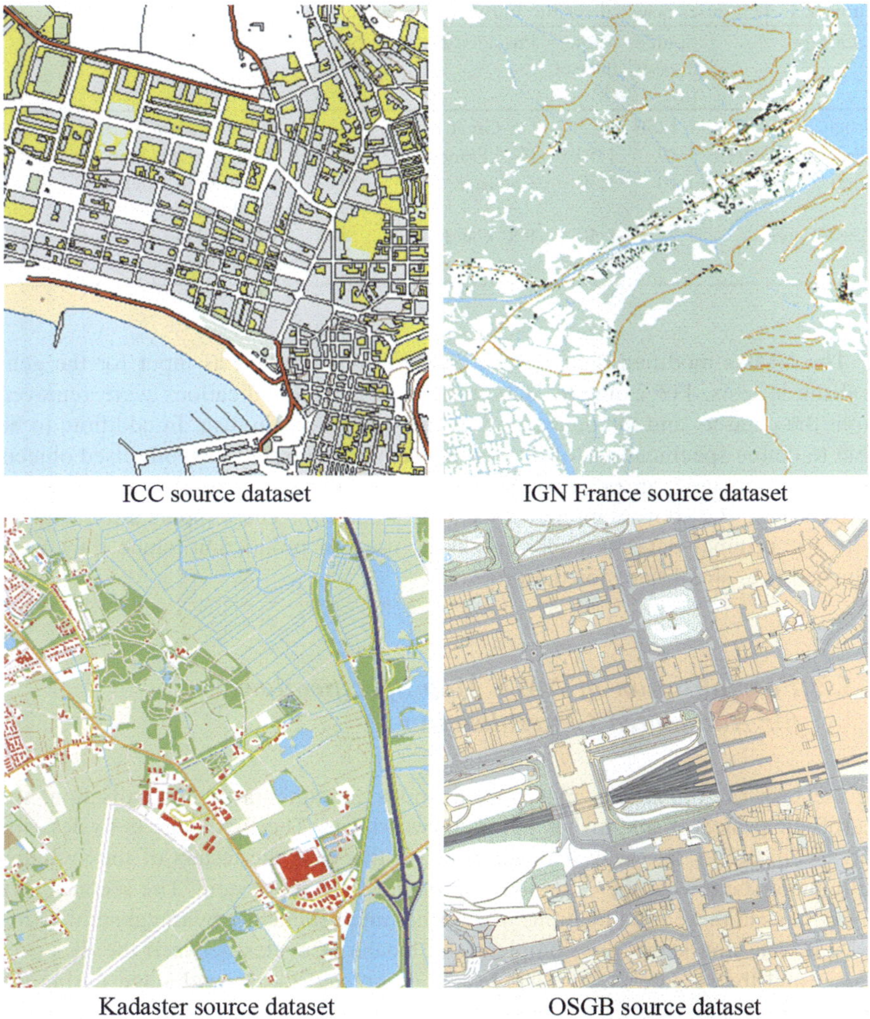

Fig. 2.7 Samples of source datasets in the EuroSDR generalisation study (maps are reduced in size)

constraints on one object, on two objects, and on groups of objects. An importance value indicates the importance of satisfying the specific constraint in the final output. This value does not indicate in what sequence the constraints should be solved (Ruas 1999). Satisfying less important constraints first may be necessary to satisfy more important constraints later. For example, generalisation of buildings should start by reducing density before trying to cope with overlaps, even though non-overlapping constraints are more important than density constraints. NMAs could also propose an action to support the tester in finding the most desired generalisation solution. This is because in some cases NMAs know what action should be taken to meet the constraint optimally. For example the constraint "minimal depth of protrusion of a building" can be solved by the two actions "exaggerate detail" or "eliminate detail" but each provides very different results. Touya et al. (2010) elaborated on this outcome of the EuroSDR project afterwards, by proposing a model mixing constraints and rules to define generalisation specifications (see also Case study I, Chap. 7).

2.5.3 Harmonising NMA Map Specifications for Automated Generalisation

NMAs defined their map specifications for automated generalisation in the developed template by analysing text-based map specifications, mapping applications and cartographers' knowledge. Initially a large number of constraints were defined for the four test cases (about 250), but which often covered similar situations.

In the next step, the EuroSDR research team harmonised the constraints. The aim was to identify constraints that are similar in the specifications provided by different NMAs, and replace them with a single one that can be tuned. This was needed for two reasons. Firstly, to simplify the tests; once a tester had defined the constraint within the software, they could perform the same actions to express a similar constraint for a second test case. Secondly, harmonisation allowed comparison of results for similar constraints across the test cases. An additional reason for this harmonisation process was that it provided important knowledge on how similar the constraints (and hence the specifications) were among different NMAs.

For the harmonisation process, similar constraints across the four test cases were identified by carefully comparing the four constraint sets. The harmonisation resulted in a list of generic constraints. A few constraints were so specific that they remained as specific constraints. Examples are OSGB constraints addressing how buildings should be aggregated depending on the initial pattern. Table 2.2 shows examples of generic constraints on one object, two objects, and a group.

Once all four NMAs had agreed on the harmonised constraints, they redefined their initial constraints in terms of the generic constraints with their own feature classes, thresholds, parameter values, and preferred actions. Table 2.3 is an example from the ICC (all NMA specific information is indicated in bold, italic). This resulted in about 300 NMA specific constraints, i.e. 50 more than initially expected. This is because the harmonisation process looked at the specifications across NMAs and pointed at specifications that were identified by one NMA but were missing in the other set. It should be emphasised that these constraints do not completely resemble the NMAs' map specifications, since they have been altered in order to meet the needs of the project.

Table 2.2 Examples of generic constraints

Constraint type	Property	Formal generic constraint
Constraints on one object		
Minimal dimensions	Area	Target area > x map mm^2; target area = initial area ± x %
	Width of any part	Target width > x map mm
	Area of protrusion/recess	Target area > x map mm^2
	Length of an edge/line	Target length > x map mm
Shape	General shape	Target shape should be similar to initial shape
	Squareness	[initial value of angle = 90° (tolerance = ± x°)] target angles = 90°
	Elongation	Target elongation = initial elongation ± x %
Topology	Self-intersection	[initially, no self-intersection] no self-intersection must be created
	Coalescence	Coalescence must be avoided
Positon/Orientation	General orientation	Target orientation = initial orientation ± x %
	Positional accuracy	Target absolute position = initial absolute position ± x map mm
Constraints on two objects		
Minimal dimensions	Minimal distance	Target distance > x map mm
Topology	Connectivity	[initially connected] target connectivity = initial connectivity
Position	Relative position	Target relative position = initial relative position
Constraints on a group of objects		
Shape	Alignment	Initial alignment should be kept
Distribution and Statistics	Distribution of characteristics	Target distribution should be similar to initial distribution
	Density of buildings (black/white)	Target density should be equal to initial density ± x %

Table 2.3 Example of ICC map specifications defined as constraints that extend the EuroSDR harmonised constraints

Item in constraint template	Example on one object	Example on two objects	Example on group of objects
Constraint ID	ICC-1-22	ICC-2-21	ICC-3-18
Generic-Constraint-ID	EuroSDR-1-8	EuroSDR-2-3	EuroSDR-3-13
Geometry type	Polygon	Polygon—line	Polygons
Feature class 1	Quay_adjacent_to_sea	Building	Building
Condition for object being concerned with this constraint (for class 1 if there are two objects)	Depth of protrusion > 1 map mm	Distance between building and road < 0.5 map mm	
Constrained property	Width of protrusion/recess	Orientation	Density of buildings (black/white ratio)
Condition depends on initial value?	No	Yes	Yes
Condition to be respected	Target width > 0.2 map mm	Building must be parallel to road	Target density should be equal to initial density ±20 %
Action	Collapse to a line		
Importance of constraint (1 to 5, 1 is less important)	3	3	3
Exception			
Schema to illustrate if needed			
Additional for constraints on two objects			
Feature class 2		Road	
Condition for both objects being concerned with this constraint		Objects are parallel (±15°)	
Additional for constraints on group of objects			
Kind of group			Urban block
Kind of objects of the initial data composing the group			Buildings surrounded by minimal cycle of roads (in urban areas)

2.5.4 Analysing the Test Cases

To obtain more in-depth knowledge on NMA requirements for automated map generalisation, the final step of the requirements analysis was the comparison of constraints across the four test cases. One should note that the constraint sets do not reflect all the generalisation problems of NMAs, because the NMAs had to limit their constraints to those describing the main problems in the test area and to constraints that were more or less straightforward to formalise.

For the comparison of constraints among the four test cases, three criteria were considered: (1) the number of objects taken into account in the constraints, (2) the type of constraints, and (3) the feature class for which the constraints were defined.

The following observations were made. Firstly, most constraints are defined for one object in all four cases, whereas the fewest constraints are defined for groups of objects. This is probably because it was difficult to formalise constraints on groups of objects. In addition, constraints for ensuring minimal dimensions are important in all four test cases, probably because it is straightforward to define this type of constraints. Another observation is that position and orientation constraints are rarely specified by all NMAs, and they refer only to buildings. A final conclusion of this analysis concerns the feature classes that were included in the constraint definitions. All four test cases contain many constraints on buildings, land use, and roads. The reason for the importance of these classes in the constraint sets is most likely because these are the most frequently occurring objects and the most significant for users of the map and therefore most (interactive) generalisation is applied to these objects.

In conclusion, the EuroSDR project was significant in examining generalisation from a cross-countries perspective. Therefore it was possible to deliver a harmonised model of generic and NMA-specific map specifications. The harmonised constraint set, although not complete, serves as a common base for better understanding of the complexity of the generalisation process, for developing and improving automated generalisation solutions as well as for automated evaluation of automated generalisation (further discussed in Chap. 9).

2.6 Case Study II: A Map Specifications Model for On-Demand Mapping at Ordnance Survey

Sandrine Balley and Nicolas Regnauld

Section 2.2.2 pointed out the difficulty of representing map specifications precisely enough to drive a complete map-making process. This section describes an experimental model designed at OSGB for on-demand mapping. We particularly focus on how this model supports data integration.

2.6.1 Context

OSGB has designed and is populating a new multiple-resolution database which will feed the production systems of existing products and provide a flexible environment to create new products (Regnauld 2007). However, with the current design, components will be chosen and combined manually. Thus, creating new products will still be relatively slow and expensive and custom products will only be viable for a few large customers. An exploratory project launched in 2010 investigated a more advanced use of this database in order to bring the benefits of custom products to a wider range of customers (Balley and Regnauld 2011). In this approach, the geographic components are picked and mixed to create products on demand automatically, therefore at low cost. This opens the door to products that support integration of customer data with OSGB reference data providing a contextual background.

A high-level distributed architecture was proposed for this on-demand mapping system (Chap. 7, Fig. 7.1). Starting from the target product specifications, a derivation engine utilises its internal knowledge, the descriptions of available data and the descriptions of available services to build a map derivation workflow. All these resources can interoperate by using shared concepts from a semantic referential (Kuhn 2003) composed of an ontology of geographic concepts (e.g. "road", "itinerary"), an ontology of GIS concept (e.g. "polyline", "feature type", "colour") and an ontology of operations (e.g. "filter", "reclassify").

We now present the map specifications model and how it can represent data integration requirements for the derivation engine.

2.6.2 The Map Specifications Model

The model adopted for the project relies on representation constraints, enclosing and extending the concept of cartographic constraints used for generalisation. Types of representation constraints were inspired by the model of Gesbert (2005) concerning dataset specifications. These constraints are expressed over Mapped Concepts (i.e. geographic concepts from the semantic referential) and their properties, as described by Touya and Duchêne (2011) or Zhang (2012). Figure 2.8 lists the different types of representation constraints together with examples. The model allows redundancies and correlated information: for instance, the fact that "Land cover" is a background mapped concept and the fact that it is represented in a faded colour can both be represented in the specifications (under symbolisation constraint), even if the latter has (or could have) been inferred from the former. The integration need is more specifically expressed through modelling and symbolisation constraints, as detailed in the next section.

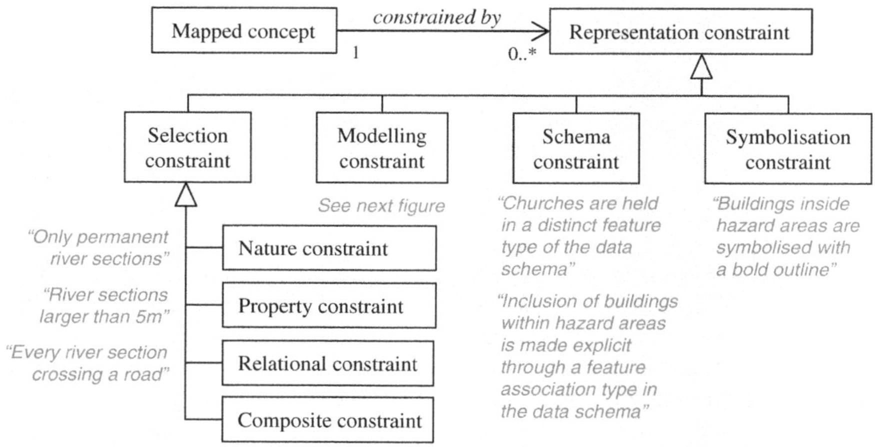

Fig. 2.8 Representation constraints of the map specifications model

2.6.3 Representing Integration Requirements Using the Specifications Model

The integration requirement is twofold: putting user's data into context, and making user's data geometrically consistent with the referential. For each facet, we list the required specification elements and their representation in our model.

2.6.3.1 Putting User's Data into Context

Putting data into context consists of providing a map background with relevant themes. In that purpose, the user must inform the system of the meaning of their data, by "joining" the semantic referential at some level (e.g. "cycle routes" or "accidents" or, at a higher abstraction level, "itineraries" or "punctual events"). As in other expert mapping systems (Forrest 1999; Sarjakoski and Sarjakoski 2005), rules or specification templates can be used by the system to identify contextual mapped concepts (e.g. "roads" and "landmarks"). For a finer selection, the system needs clues about the thematic and spatial relationships the user wants to emphasise (e.g. "the sections of itinerary equipped with cycle lanes" or "the accidents occurring at crossroads"). Using this information, the system will organise themes into different reading levels, from the "first-sight" themes to the background themes (Bucher et al. 2007), which is formalised through symbolisation constraints (Fig. 2.9) and will govern the choice of styles. The choice of symbols for the map foreground will also be guided by the geometric modelling

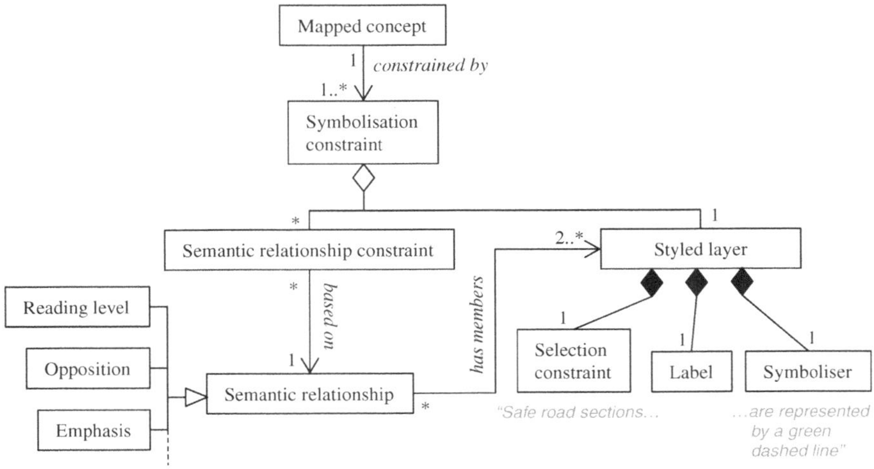

Fig. 2.9 Symbolisation constraints (detail of the map specifications model)

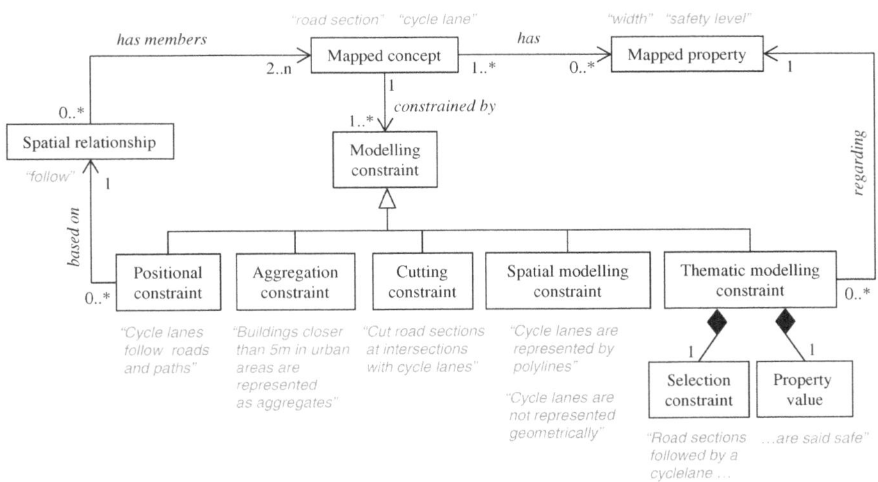

Fig. 2.10 Modelling constraints (detail of the map specifications model)

(see Spatial modelling constraint, Fig. 2.10) of features constituting the map background, and vice versa. For example, it is better not to use red hues for the background data if the user's data deals with risk (to avoid disturbing the cartographic message), or if the user has expressed a preference for displaying their own data in red (to ensure a good contrast) (Chesneau 2007).

2.6.3.2 Making User's Data Geometrically Consistent with the Referential

Ensuring geometric consistency between data sources is the second integration requirement: it makes the map accurate and readable, and emphasises spatial relations that are relevant to the user (e.g. the fact that cycle routes follow roads). Depending on the map use and theme, three integration options are possible.

The first option consists of overlaying and geometrically aligning user's objects on the referential background. This makes sense if the target product is for visualisation purpose only, and if the user's objects represent physical features that are not already represented in the reference data (e.g. public phones). This choice can be expressed through a positional constraint (Fig. 2.10).

The second option consists of discarding user's object geometries and projecting their properties on the referential objects as new thematic attributes. This makes sense if the product is meant to be used for thematic analysis, if the user's data represents non-physical entities (e.g. statistical data or itineraries), or if it represents physical entities that already are in the referential with less thematic attributes but better geometric accuracy. This can be expressed through a thematic modelling constraint (Fig. 2.10).

The third option consists of keeping user's object geometries, aligning them and adapting the referential topology accordingly. Depending on the user's need, it may for instance be useful to reorganise the road network using the user's objects as new nodes. This requirement can be expressed through a positional constraint and a cutting constraint (Fig. 2.10).

2.6.4 Conclusions

The map specifications model presented in this Case study extends the principle of cartographic constraints to support not only generalisation, but also other processes required by on-demand mapping, notably data integration. The specification model was instantiated for a use case where user-generated cycle routes are integrated with a reference road network. It was implemented and utilised by an on-demand mapping prototype to demonstrate the proposed high-level architecture (Balley et al. 2012). The model now needs to be tested against various scenarios. In the current stage of the project, map specifications (and not user requirements) are the input of the on-demand mapping system. Collecting the user requirements and setting the map specifications dynamically will be investigated in later phases.

2.7 Case Study III: COLorLEGend—Design of Personalised and Original Maps

Sidonie Christophe

Selecting colours to symbolise features on a map is still a complex problem. Though cartographic theory provides some recommendations about colours (Bertin 1967; Robinson et al. 1995), there is no generic method for handling colours in a map. The following question remains: "How to select and combine colours to render data, according to the message that the map is supposed to convey?". The problem has no unique or optimal solution: only the cartographer is able to validate a solution judged satisfactory. In the context of on-demand map design, the problem may be mostly stressed by the level of cartographic expertise of the user (i.e. the map maker). The user may have no skills to correctly manipulate the colours of cartographic objects. The suitability of colour choice requires basic knowledge that may not be provided by cartographic tools.

The Case study presented here is the COLorLEGend (COLLEG) system, a cooperative method to help users to make personalised and creative colour choices (Christophe 2009). This application has been implemented in the IGN-France production services. The project consisted of providing both knowledge and methods to help users choose colours to render their geographical data. The principle of COLLEG is a dialog engine interacting with the user and relying on a knowledge base on colours. Cartographic knowledge has been previously identified, acquired and stored to be used as a knowledge base in the COLLEG system (Christophe 2011). A specific approach to collecting user's preferences is proposed through *inspiration sources* (Sect. 2.7.1). Then, user's preferences, considered as constraints on the map legend, are validated against cartographic expertise in order to infer some cartographic solutions that are acceptable to the user (Sect. 2.7.2). The challenge is to find suitable cartographic solutions in cases where user's preferences and cartographic knowledge on colours may a priori appear to be incompatible.

2.7.1 Collecting User's Preferences

It is not easy for users to express their colour choices. Colours preferences may be difficult to specify on a blank page, whether it be by writing a text or by clicking on colour squares: the set of possibilities is too numerous for the user to manage. Once chosen, the user may change their preferences once they have seen the consequences of their choice. So the user should be able to modify or refine them anytime during the cartographic process. They should be allowed to express their feelings on colours regarding the overall map or in greater detail, e.g. regarding a single geographical theme.

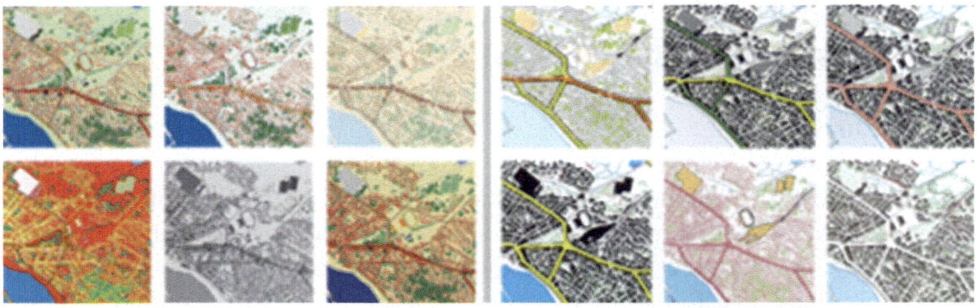

Fig. 2.11 Inspiration sources 1 (extract)—harmonic (Dominguès and Bucher 2006) (*left*) and European (Christophe 2009) (*right*) map samples

The proposition of collecting user's preferences is inspired by the work of Hubert and Ruas (2003) where a dialogue engine presents to the user some examples of generalisation results that can be selected to help them to parameterise generalisation processes (Sect. 2.4.2). This analogy between examples and related parameterisation of a process is reused in the present context: examples are presented to facilitate the selection of colours and of how they are used on the map. Users can draw from two types of *inspiration sources:* map examples and famous paintings.

2.7.1.1 First Inspirational Sources: Cartographic Examples

Previous research has provided a database of map samples: several colour schemes were drawn from a book about colour harmony (Sawahata 2001) and applied on the same geographical dataset with the same legend structure (Dominguès and Bucher 2006). This database has been extended with new colour schemes coming from European topographic map legends (Renard 2008; Christophe 2009) (Fig. 2.11). It was considered that these inspirational sources may be useful in helping the user to pick satisfactory colours schemes for a map, or satisfactory colours independently for a geographical theme.

2.7.1.2 Second Inspirational Sources: Famous Paintings

In order to encourage creativity and to open the user's mind in terms of choices of colours, inspirational sources drawn from another graphical domain were investigated. A few famous paintings were selected, together with their associated colour palette and colour patterns (general surfaces, repartitioned in spots or area, neighbouring colours) (Christophe 2011) (Fig. 2.12). It was considered that these inspirational sources may be useful in helping the user to pick harmonious combinations of colours, i.e. not only a colour palette, but also a specific colour composition. The user may prefer a colour palette, some colours in a palette or a colour for a geographical theme.

Fig. 2.12 *Inspiration sources* two paintings and related colour palettes

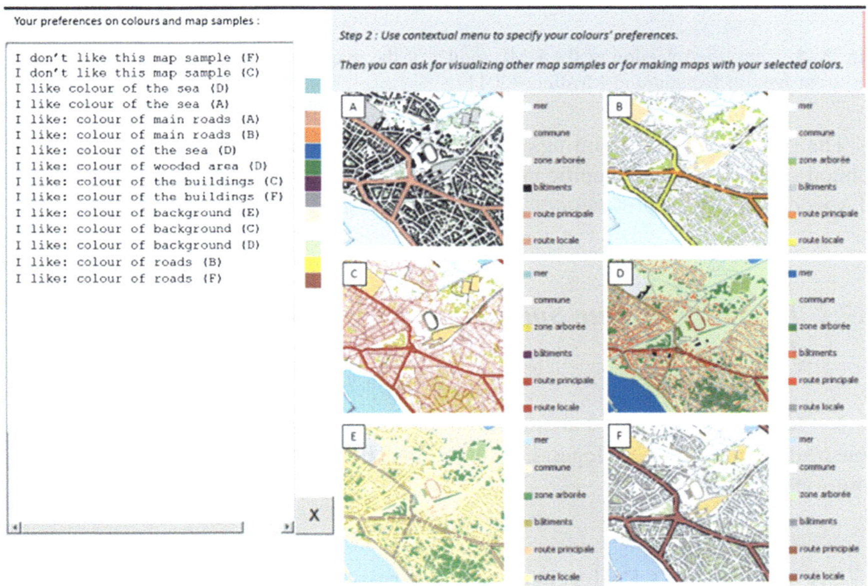

Fig. 2.13 User's preferences on colours on map samples

2.7.1.3 Specifying Preferences on Colours

With the help of the COLLEG system, the user starts the map design process by selecting one type of inspirational source (map samples or paintings). Then they pick their preferred colours in the proposed inspirational sources.

In the map samples strategy, the process consists of presenting a set of six samples that a user may annotate (by their likes/dislikes for the maps and their colours). According to these initial preferences, a new set of six samples are proposed again to collect more preferences. This continues until the user decides to launch the making of maps. In the example shown in Fig. 2.13, the user "disliked" two map samples and "liked" thirteen colours for a theme. The purpose is to explore a large amount of possibilities and then to converge towards satisfactory colours: this process relies on a classification of map samples (Christophe 2009).

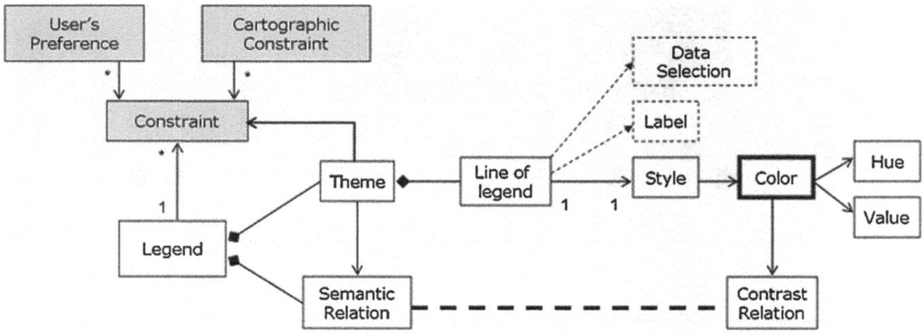

Fig. 2.14 Simplified version of the UML model of the map legend Christophe (2009) extended and improved in Hoarau and Mustiere (2011)

The principle of the paintings strategy is to let the user select a painting and its related palette or some specific colours in this palette. Using this strategy, the step of specifying user's preferences is relatively simple.

2.7.2 *Inferring Colour Specifications from User's Preferences*

Once the user's preferences are acquired, the COLLEG system translates them into constraints on the map legend. Thus COLLEG is able to infer various map solutions.

2.7.2.1 From User's Preferences to Constraints on the Map Legend

COLLEG manages a model of the map legend structured in themes rendered by a specific symbolisation. At this stage, COLLEG translates the user's preferences, i.e. some likes and dislikes of colours, into constraints on the legend. The user's preferences are divided into two types of constraints: a colour may be applied to any theme of the legend, or, a colour should be applied to the specific theme specified by the user. Thus the user's constraints are managed as objects impacting the map legend and the theme objects (Fig. 2.14).

2.7.2.2 From Constraints on Map Legend to Various Map Solutions

COLLEG relies on a constraint satisfaction problem: each geographic theme may be rendered by any user's colour satisfying constraints, resulting in various map solutions. As detailed in Christophe (2011), the constraints reflect a cartographic knowledge of colours:

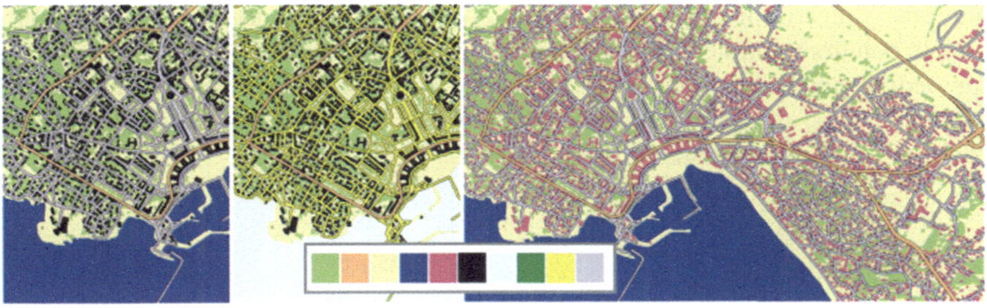

Fig. 2.15 Three possible maps coming from user's preferences on map samples

Fig. 2.16 Various maps coming from Derain's palette and user's preferences on colour uses

- Semantic relationships between themes (association, difference, order) should be rendered by specific contrasts between colours of the related themes.
- Background colour should be high contrasted in value with other themes.
- Conventional colours should be used to render sea and wooded area.

Figure 2.15 shows examples of maps resulting from the user's preferences based on the samples presented in Fig. 2.13: semantic relationships between road themes and conventional colours are preserved. Figure 2.16 shows various maps coming from a painting, according to cartographic or artistic constraints: not only cartographic conventions, but also artistic colour compositions of the painter may be enforced or relaxed.

As a final step, COLLEG proposes a refining tool, similar to an elaborated colour picker, in order to improve application of colours in the given map solutions: the refining tool is adapted to current maps and therefore proposes suitable colours to refine those maps according to existing cartographic and artistic knowledge. Some flexibility in interactions with COLLEG enables the user to rework the map design process at any stage, or to change initial preferences in order to make more satisfactory maps.

2.7.3 Conclusions

A usability test of the COLLEG system has been implemented and reviewed (Christophe 2009). According to participants, the system was clearly helpful in designing creative maps and was deemed efficient at proposing suitable map solutions, and exploring the space of possibilities bounded by cartographic knowledge and user's preferences. Two strategies were investigated to collect and manage user's preferences on colours; while the strategy based on map samples collects preferences upstream, the strategy based on paintings collects preferences directly on resulting maps and through the refining tool. In the future learning techniques will be utilised to improve the collection of preferences, in order to propose suitable strategies and to make the process faster according to user's profiles. Bigger databases of inspirational sources may be also considered: associated automated tools to extract required information and thus better interpret user's preferences are also worth developing (Christophe et al. 2013).

2.8 Conclusions

Generalisation is not restricted to the production of predefined map series anymore. The current challenge is to automatically adapt to changing user requirements. This chapter presented the issues related to the creation of formal map specifications resulting from user requirements. The objectives of this research are:

- to enable on-demand generalisation and on-demand mapping, resulting in good quality, usable maps that support integration of user's data,
- to make advanced mapping processes available to those users who do not have the skills to create map specifications.

This chapter has shown that map usability is receiving more and more attention as the ranges of map users and map uses grow. Applications delivering maps adapted to predefined user profiles have emerged, especially in the domain of mobile maps. In parallel, map and dataset specification models have become more expressive and more interoperable.

However, no global specification model to date is able to drive the entire map-making process. The first reason lies in the fact that on-demand mapping encompasses many processing steps. Specification models—and associated user profiles—relevant for different steps of the map-making process need to be integrated, which requires re-examination of the very concepts underlying these models. The second reason is that some of the cartographic knowledge still cannot be—and might not be in the medium term—formalised as map specifications (Stoter et al. 2009b). This should not be an obstacle to on-demand mapping. We need to consider again what we mean by map quality, in order to deliver maps that are not as "good" as maps involving manual editing, but are usable for a task and given context.

The emergence of such specification models will lead to the issue of their instantiation, first by cartographers, and then by the end-users of the service. More formalised knowledge on map design will be required to assist these users (Ory et al. 2013). Reusable map specifications might be useful. This would assume that the specification model is shared, that specification templates are proposed, and that "map provider profiles" are formalised, enabling mapping agencies to retain their own trademark.

The collection of user requirements and their automatic interpretation has not been sufficiently explored. Innovative interfaces based on map samples, and their associated interpretation mechanisms, have been designed for a few steps of on-demand mapping. Could the approach be extended to other steps such as content selection and user data integration? As the optimal map specifications depend on the task the user wants to achieve, task-oriented interfaces could be envisaged, as well as interfaces dedicated to users who are getting more accustomed to map-making and need to interact at different expertise levels. At each expertise level, we must decide how, and by how much, cartographic standards can incorporate user preferences.

References

Abadie N (2009) Formal specifications to automatically identify heterogeneities. In: Proceedings of the 12th international conference on geographic information science AGILE, pre-conference workshop "challenges in spatial data harmonisation", Hannover, 2–5 June 2009

Abadie N, Mechouche A, Mustière S (2010) OWL-based formalisation of geographic databases specifications. In: Proceedings of the 17th international conference on knowledge engineering and knowledge management, Lisbon, 11–15 Oct 2010

Balley S (2007) Aide a la restructuration de donnees geographiques sur le Web—Vers la diffusion a la carte d'information geographique. PhD Thesis, Université Paris Est

Balley S, Regnauld N (2011) Collaborating for better on-demand mapping. In: Proceedings of the 14th ICA workshop on generalisation and multiple representation, jointly organised with ISPRS commission II/2 working group on multiscale representation of spatial data, Paris, France, 2011

Balley S, Jaara K, Regnauld N (2012) Towards a prototype for deriving custom maps from multisource data. In: Proceedings of the 15th ICA workshop on generalisation and multiple representation, jointly organised with EuroSDR commission on data specifications, Istanbul, Turkey, 2012

Battye E (2010) Conceptualising space project: final report. Ordnance Survey Great Britain Research, unpublished

Beard K (1991) Constraints on rule formation. In: McMaster R, Buttenfield B (eds) Map generalization: making rules for knowledge representation. Longman Group, UK, pp 121–135

Bertin J (1967) Sémiologie Graphique. Gauthier-Villars, Paris. English edition: Bertin J (1983) Semiology of Graphics (trans: William JB). University of Wisconsin Press,Wisconsin

Brewer CA (1997) Spectral schemes: controversial color use on maps. Cartography Geogr Inf Syst 24(4):203–220

Brewer CA (2003) A transition in improving maps: the Colorbrewer example. Cartography Geogr Inf Sci 30(2):159–162

Brewer CA, Buttenfield BP (2010) Mastering map scale: balancing workloads using display and geometry change in multi-scale mapping. Geoinformatica 14(2):221–239

Bucher B, Buard E, Jolivet L, Ruas A (2007) The need for web legend services. In: Ware JM, Taylor GE (eds) Web and wireless geographical information systems. 7th International Symposium W2GIS, Cardiff, 28–29 (2007) Lecture Notes in Computer Science, vol 4857. Springer, Heidelberg, pp 44–60

Burghardt D, Schmid S, Stoter J (2007) Investigations on cartographic constraint formalisation. In: Proceedings of the 11th ICA Workshop on Generalisation and Multiple Representation, Moscow, Russia, 2007

Cecconi A (2003) Integration of generalisation and multi-scale databases for enhanced web mapping. PhD Thesis, University of Zurich

Chesneau E (2007) Improvement of colour contrasts in maps: application to risk maps. In: Proceedings of the 10th International Conference on Geographic Information Science AGILE, Aalborg, 8–11 May 2007

Christophe S (2009) Aide à la conception de légendes personnalisées et originales: Proposition d'une méthode coopérative pour le choix des couleurs. PhD Thesis, Université Paris Est

Christophe S (2011) Creative colours specification based on knowledge (ColorLegend System). Special issue: ICC Paris 2011. Cartographic J 48(2):138–145

Christophe S, Perret J, Hoarau C (2013) Extraction de palettes de couleurs pour l'aide à la conception cartographique. Revue des Sciences et Technologies de l'Information (in press)

Dhée F (2011) Amélioration des cartes topographiques pour les daltoniens. In: Proceedings of the 25th international cartographic conference ICC, Paris, 3–8 July 2011

Dominguès C, Bucher B (2006) Legend design based on map samples. In: Proceedings of the 4th international conference on geographic information science GIScience, Münster, 20–23 Sep 2006, pp 55–59

Dominguès C, Christophe S, Jolivet L (2009) Connaissances opérationnelles pour la conception automatique de légendes de cartes. In: Actes des 20èmes Journées Francophones d'Ingénierie des Connaissances, Hamamet, 25–29 May 2009

Duchêne C, Christophe S, Ruas A (2011) Generalisation, symbol specification and map evaluation: feedback from research done at COGIT laboratory, IGN France. Special issue: validation of EO-derived information for crisis management: a digital earth perspective in the Valgeo expert community. Int J Digital Earth 4(1):25–41

Foerster T, Lehto L, Sarjakoski T, Sarjakoski T, Stoter J (2010) Map generalization and schema transformation of geospatial data combined in a Web Service context. Comput Environ Urban Syst 34(1):79–88

Fonsesca FT (2001) Ontology-Driven Geographic Information Systems. PhD Thesis, University of Maine

Forrest D (1999) Developing rules for map design: a functional specificationfor a cartographic-design expert system. Cartographica 36(3):31–52

Gaffuri J (2011) Improving Web mapping with generalization. Special issue: internet mapping: selected papers from the 25th conference of the international cartographic association, Paris, 3–8 July 2011. Cartographica 46(2):83–91

Gesbert N (2005) Étude de la formalisation des spécifications de bases de données géographiques en vue de leur intégration. PhD Thesis, Université Paris Est

Gnägi HR, Morf A, Staub P (2006) Semantic interoperability through the definition of conceptual model transformations. In: Proceedings of the 9th international conference on geographic information science AGILE, Visegrad, 20–22 Apr 2006

Gődér G (2003) Représentation comparée de schémas et de spécifications de contenu. Mémoire de DESS de Cartographie et d'Information Géographique, Université Paris Est
Gould NM, Chaudhry O (2012) An ontological approach to on-demand mapping. In: Proceedings of the 15th ICA Workshop on generalisation and multiple representation, jointly organised with EuroSDR commission on data specifications, Istanbul, Turkey, 2012
Gruber T (1993) A Translation Approach to Portable Ontologies. Knowl Acquisition 5(2):199–220
Haklay M (ed) (2010) Interacting with geospatial technologies. Wiley, Chichester
Harding J (2011) Usability of geographic information: factors identified from task-focused user interviews. In: Proceedings of the 25th international cartographic conference, Paris, 3–8 July 2011
Harding J, Sharples S, Hakla M, Burnett G, Dadashi Y, Forrest D, Maguire M, Parker CJ, Ratcliffe L (2009) Usable geographic information—what does it mean to users? In: Proceedings of AGI Geo-community, Stratford Upon Avon, 23–24 Sept 2009
Harrie L, Weibel R (2007) Modelling the overall process of generalisation. In: Ruas A, Sarjakoski LT (eds) Generalisation of geographic information: cartographic modelling and application. Elsevier, Amsterdam, pp 67–87
Hoarau C (2011) Reaching a compromise between contextual constraints and cartographic rules: application to sustainable maps. Cartography Geogr Inf Soc J 38(2):79–88
Hoarau C, Mustière S (2011) GeOxygene, 'Semiology' Plugin—Base model for works on map legends in GeOxygene. http://oxygene-project.sourceforge.net/doc/geoxygene-semiology-0.4.pdf. Accessed 29 May 2013
Hubert F, Ruas A (2003) A method based on samples to capture user needs for generalisation. In: Proceedings of the 7th ICA workshop on progress in automated map generalization, Paris, France, 2003
Inspire Drafting Team on Data Specifications (2008) Deliverable D2.6: methodology for the development of data specifications. http://inspire.jrc.ec.europa.eu/reports/-ImplementingRules/inspireDataspecD2_6v2.0.pdf. Accessed 29 May 2013
ISO TC211 (2005) 19131 Geographic information: data product specification. International Standard
ISO TC211 (2009) 19123 Geographic information: Feature concept dictionaries and registers. International Standard
Jaara K, Duchêne C, Ruas A (2012) A model for preserving the consistency between topographic and thematic layers throughout data migration. In: Proceedings of the 15th international symposium on spatial data handling sDH, Bonn, 22–24 Aug 2012
Jabeur N (2006) A multi-agent system for on-the-fly web map generation and spatial conflicts resolution. PhD Thesis, Université Laval
Jolivet L (2009). Characterizing maps to improve on-demand cartography—the example of European topographic maps. In: Proceedings of the 17th annual conference on GIS reasearch GISRUK, Durham, 1–3 Apr 2009, pp 345–348
Klien E, Lutz M, Kuhn W (2006) Ontology-based discovery of geographic information services—an application in disaster management. Comput Environ Urban Syst 30(1):102–123
Kuhn W (2003) Semantic Reference Systems. Int J Geogr Inf Sci 17(5):405–409
Lemmens R (2006) Semantic interoperability of distributed geo-services. PhD Thesis, ITC, University of Twente
Letho L (2007) Real-time content transformations in a web service-based delivery architecture for geographic information. PhD Thesis, Helsinki University of Technology
Lüscher P, Burghardt D, Weibel R (2007) Ontology-driven enrichment of spatial databases. In: Proceedings of the 11th ICA workshop on generalisation and multiple representation, Moscow, Russia, 2007
Mackaness WA, Ruas A, Sarjakoski LT (2007) Observations and Research Challenges in Map Generalisation and Mutiple Representation. In: Mackaness WA, Ruas A, Sarjakoski LT (eds) Generalisation of Geographic Information: Cartographic Modelling and Applications, Elsevier, pp 315–323
Mechouche A, Abadie N, Prouteau E, Mustière S (2013) Ontology-based formal specifications for user-friendly geospatial data discovery. In: Guillet F, Pinaud B, Zighed DA, Venturini G (eds)

Advances in knowledge discovery and management, studies in computational intelligence 471. Springer, Heidelberg, pp 151–176

Ooms K, De Maeyer P, Fack V (2011) Can experts interpret a map's content more efficiently? In: Proceedings of the 25th international cartographic conference, Paris, 3–8 Jul 2011

Ory J, Christophe S, Fabrikant SI (2013) Identification of styles in topographic maps. In: 16th international cartographic conference, Dresden, 25–30 Aug 2013

Regnauld N (2007) A distributed system architecture to provide on-demand mapping. In: Proceedings of the 11th international cartographic conference, Moscow, 4–10 Aug 2007

Reichenbacher T (2004) Mobile cartography: adaptative visualisation of geographic information on mobile devices. PhD Thesis, Technische Universität München

Renard J (2008) Analyse et caractérisation de légendes européennes. Mémoire de stage, Ecole Nationale des Sciences Géographiques

Robinson A (1952) The looks of maps. University of Wisconsin Press, Madison

Robinson A, Morrison JL, Muehrcke PC, Kimerling AJ, Guptill SC (1995) Elements of cartography (6th edn). Wiley, Chichester

Ruas A (1999) Modèle de généralisation de données géographiques à base de contraintes et d'autonomie. PhD Thesis, Université Paris Est

Ruas A (2001) Automatic generalisation project: learning process from interactive generalisation. OEEPE official publication 39

Ruas A, Plazanet C (1996) Strategies for automated generalisation. In: Proceedings of the 7th international symposium on spatial data handling SDH, Delft, 12–16 Aug 1996

Sarjakoski LT, Nivala A (2005) Adaptation to context—a way to improve the usability of mobile maps. In: Meng L, Zipf A, Reichenbacher T (eds) Map-based mobile services: theories, methods and implementations. Springer, Heidelberg, pp 107–123

Sarjakoski T, Sarjakoski LT (2005) The GiMoDig public final report. Public EC report, GiMoDig project, IST-2000-30090

Sarjakoski T, Sarjakoski LT (2007) A real-time generalisation and map adaptation approach for location-based services. In: Mackannes WA, Ruas A, Sarjakoski LT (eds) Generalisation of geographic information: cartographic modelling and applications. Elsevier, Amsterdam, pp 137–159

Sawahata L (2001) Color harmony workbook. Rockport Publishers

Stoter J, Burghardt D, Duchêne C, Baella B, Bakker N, Blok C, Pla M, Regnauld N, Touya G, Schmid S (2009a) Methodology for evaluating automated map generalization in commercial software. Comput Environ Urban Syst 33(5):311–324

Stoter J, van Smaalen J, Bakker N, Hardy P (2009b) Specifying map requirements for automated generalization of topographic data. Cartographic J 46(3):214–227

Stoter J, Baella B, Blok C, Burghardt D, Duchêne C, Pla M et al (2010) State-of-the art of automated generalisation in commercial software. EuroSDR Official Publication

Touya G (2011) Orchestration d'un processus multi-modèles de généralisation d'espaces géographiques hétérogènes. PhD Thesis, Universite Paris Est

Touya G, Duchêne C (2011) Collagen: collaboration between automatic cartographic generalisation processes. In: Ruas A (eds) Advances in cartography and GIScience vol 1, Selection from ICC 2011 Paris. Lecture notes in geoinformation and cartography, Springer, Heidelberg

Touya G, Girres J (2013) ScaleMaster2.0: a scalemaster extension to monitor automatic multi-scales generalisations. CaGIS—special Issue ICC 2013 (in press)

Touya G, Duchêne C, Ruas A (2010) Collaborative generalisation: formalisation of generalisation knowledge to orchestrate different cartographic generalisation processes. In: Proceedings of the 6th international conference GIScience 2010, Zurich, 14–17 Sep 2010

Touya G, Balley S, Duchêne C, Jaara K, Regnauld N, Gould NM (2012) Towards an ontology of generalisation constraints and spatial relations. In: Proceedings of the 15th ica workshop on generalisation and multiple representation, jointly organised with EuroSDR commission on data specifications, Istanbul, Turkey, 2012

Zhang X (2012) Automated evaluation of generalised topographic maps. PhD Thesis, ITC, University of Twente

Chapter 3
Modelling Geographic Relationships in Automated Environments

Guillaume Touya, Bénédicte Bucher, Gilles Falquet, Kusay Jaara and Stefan Steiniger

Abstract Automated processes such as cartographic generalisation require formal abstraction of the geographic space in order to analyse, process and transform it. Spatial relations are key to understanding geographic space and their modelling is a critical issue. This chapter reports on existing classifications and modelling frameworks for spatial relations. A generic model is proposed for building an ontology of spatial relations for automatic processes such as generalisation or on-demand mapping, with a focus on so-called multiple representation relations. Propositions to use such ontology in an automated environment are reported. The three use cases of the chapter describe recent research that uses relations modelling. The first use case is the extension of CityGML with relations for 3D city models. The second use case presents the use of spatial relations for automatic spatial analysis, and particularly the grouping of natural features such as lakes or islands. Finally, the third use case is a data migration model guided by relations that govern the positioning of thematic data upon changing reference data.

G. Touya (✉) · B. Bucher · K. Jaara
Laboratoire COGIT, IGN, Saint-Mandé, France
e-mail: guillaume.touya@ign.fr

B. Bucher
e-mail: benedicte.bucher@ign.fr

K. Jaara
e-mail: kusay.jaara@ign.fr

G. Falquet
Centre for Computer Science, University of Geneva, Geneva, Switzerland
e-mail: gilles.falquet@unige.ch

S. Steiniger
Geomatics Engineering, University of Calgary, Calgary, Canada
e-mail: ssteinig@ucalgary.ca

D. Burghardt et al. (eds.), *Abstracting Geographic Information in a Data Rich World*, Lecture Notes in Geoinformation and Cartography, DOI: 10.1007/978-3-319-00203-3_3,

3.1 Introduction

Map generalisation is not simply the simplification of geographic features for legibility purposes but more generally a holistic abstraction process that seeks to represent geographic features at a given scale for a given purpose. Patterns and in particular spatial relations are key to the understanding of geographic space (Mackaness and Edwards 2002). For instance, Fig. 3.1a shows a town symbol placed over a road symbol, which can be expressed as the spatial relation "the town symbol *is on* the road". This relation conveys the information to the map reader that the road crosses the town. In contrast, the relation "the town symbol is *near* the road" (Fig. 3.1b) conveys the information that the road passes by the town. This simple example illustrates the importance of understanding spatial relations in the map generalisation processes. In Fig. 3.1c, the spatial relation "the cycle path is along the road" has to be explicit to adapt the road symbol with an additional dotted centreline. More generally, the generalisation process should as much as possible preserve the spatial relations existing in initial data. For example if a building is within a forest clearing, it should still be there after generalisation (Fig. 3.1d)!

The detection of spatial relations in order to guide generalisation has long been identified as one of the challenges for automation (McMaster and Shea 1988; Brassel and Weibel 1988). To achieve this goal, Mackaness and Edwards (2002) suggested reifying both spatial relations and patterns in order to ensure their preservation during generalisation. In the CartACom model (Ruas and Duchêne 2007; Duchêne et al. 2012), relations are additional objects on which constraints are defined, which guide the generalisation of geographical features.

As a prerequisite we need to formalise spatial relations in order to develop such relation-driven generalisation processes and also to facilitate relational constraint modelling (Burghardt et al. 2007) and to monitor such processes. A better formalisation of spatial relations would also improve process interoperability (Chap. 7) and help users define their needs (Chap. 2) (Touya et al. 2012).

This chapter presents recent research on modelling spatial relations for automatic mapping environments. Its second part describes related work on spatial relations classifications. The third part presents a proposal for a spatial relations ontology and the fourth part explains how such an ontology can be used in automatic environments. Section 3.5 presents a first Case study on spatial relations for 3D city models. Section 3.6 describes the second Case study on relations based on spatial analysis. Section 3.7 details the third use case on data migration guided by explicit relations. Finally, conclusions are drawn on some research perspectives on spatial relations for generalisation.

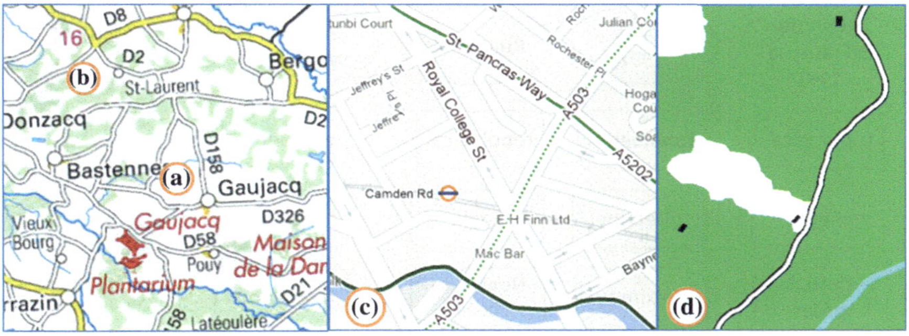

Fig. 3.1 The town symbol is on the road when it actually crosses the town (**a**) or near the road when it passes by the town (**b**) (©IGN). **c** The relation "the cycle path *follows* the road" facilitates the symbolising of the road with the *dotted line*. **d** The relation "the building *is inside* a clearing" should be preserved by generalisation

3.2 Spatial Relations Classification

This section mainly focuses on spatial relations between pairs of geographical objects, but the relations between more than two objects are briefly discussed in Sect. 3.2.3.

3.2.1 Classification and Formalisation of Spatial Relations

Over the past thirty years, several spatial relations classifications have been proposed (Egenhofer and Franzosa 1991; Jones 1997; Ruas 1999; Steiniger and Weibel 2007; Burghardt et al. 2010). They were designed from different points of view, and some were specifically dedicated to map generalisation (Fig. 3.2). We note that most classifications distinguish between topological and geometric relations, the focus being mainly on the former. Steiniger and Weibel (2007) add more detail by subdividing their geometric relations classification into into size, position, shape and orientation relations while semantic relations are divided into similarity, priority, resistance/attraction and causal/logic relations.

In respect of the topological relations, models were developed to enable automatic reasoning, such as the 4-intersection model (4IM) (*included, includes, covered by, covers, overlaps, equals, meets, disjoint*) that manages topological relations between polygons. The most commonly used topological model is the 9-intersection model (Egenhofer and Franzosa 1991), adopted by the OGC, and the Region Connection Calculus (Randell et al. 1992)—somewhat similar to the former.

There are many ways to define a set of spatial relations for an automatic application such as map generalisation. The question then becomes: which one has the greatest utility? Cohn and Hazarika (2001) claim that a set of spatial relations

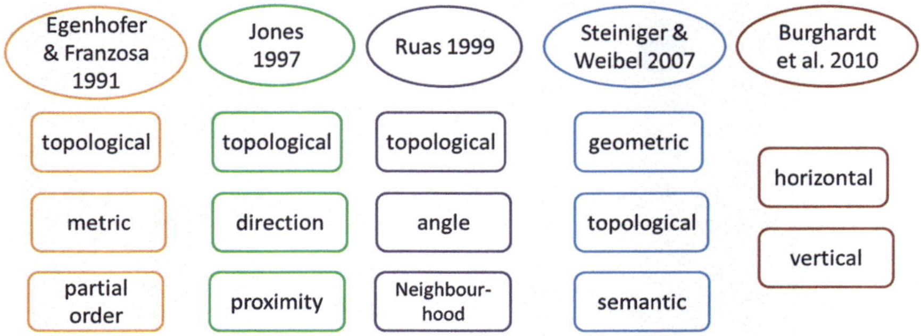

Fig. 3.2 Several spatial relations classifications, the last three being dedicated to generalisation

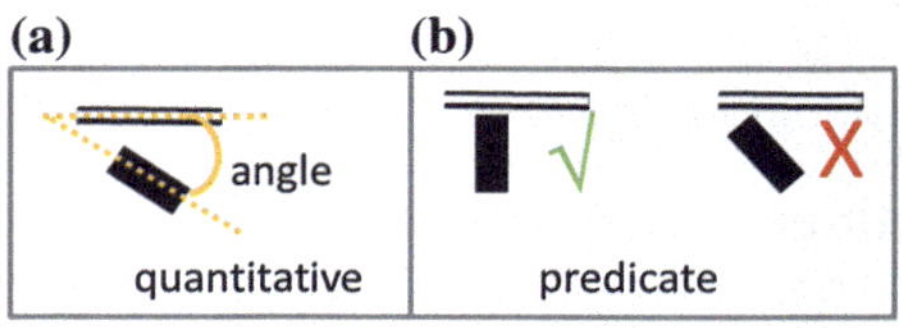

Fig. 3.3 **a** A relative orientation quantitatively described by an angle. **b** An orthogonal relation described by a predicate

has to be relevant to the task being performed, and that none is universal. Following the same idea, Clementini (2010) divides relations models into three levels: geometric (e.g. *touch*), computation (e.g. line *touch* polygon) and user level (e.g. dam *touch* lake) where application related relations are defined (further discussed in Sect. 3.5.3). In Sect. 3.3, a user level relation model is proposed.

3.2.2 Quantitative Versus Binary Relations

In the literature, two ways of describing a relation between a pair of objects can be found (Touya et al. 2012). The first one quantitatively considers a spatial relation with a measure: e.g. a proximity relation is characterised by a distance, or a relative orientation relation is characterised by an angle (Fig. 3.3a). This is a convenient model for the "metric relations" of Egenhofer and Franzosa (1991). The second way of describing a relation considers a predicate that can be true if the relation exists, or false. For instance, Fig. 3.3b shows a spatial relation described by the predicate *is orthogonal to*. Only crisp relations are discussed here, fuzzy relations being discussed later in the section.

Some authors have defined families of predicates that cover all possible relations between a pair of objects from a particular point of view. In this case, a pair of objects only meets one predicate of the family, e.g. the 4IM or the 9-intersection

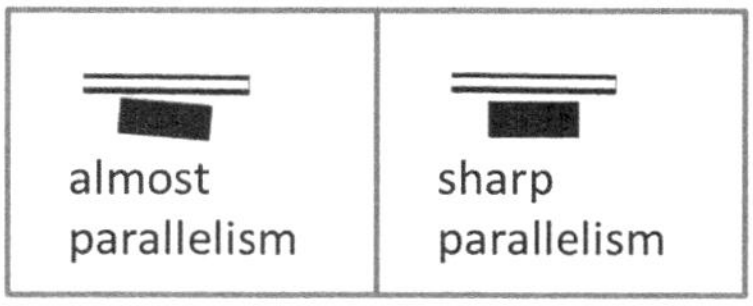

Fig. 3.4 A sharp relation and its fuzzy version considered for cognitive perception limits

model (9IM) (Egenhofer and Franzosa 1991). Other families may not follow this principle: for example a family made of two predicates *is orthogonalto* and *is parallel to* does not cover all the types of relative orientation.

Whether a model of spatial relations considers one relation (e.g. relative orientation) and several associated predicates (*parallel*, *orthogonal*), or several relations (*parallelism*, *orthogonality*) is an arbitrary choice. We choose the second, because, in the context of map generalisation, the qualitative evaluation of the preservation or transformation of the relation is easier considering several relations. As a consequence, it is possible to distinguish between *quantitative relations* (relations described by a quantitative measure) and *binary relations* (relations described by a predicate, which are present, or not, between a pair of objects). Binary does not refer, here, to the mathematical definition that means that a pair of objects is related.

Now, a particular situation with respect to binary relations is the situation where the relation is not completely present, but almost (Duchêne et al. 2012). For instance, in Fig. 3.4, the building is not strictly parallel to the road but almost parallel: it is not clear if it should read as parallel to the road or not, particularly if scale is reduced. This situation is tricky in the context of generalisation, which seeks to avoid the fuzziness that blurs legibility, and replace it by a sharp relation through caricature. So, for each sharp binary relation we propose to add an associated fuzzy relation in our ontology corresponding to the case where the sharp relation is "almost" present, such as *near parallelism*.

Fuzzy topological relationships have been studied, [for example by Winter (2000) and Bejaoui et al. (2009)] in the context of objects with fuzzy limits. Here, in the context of generalisation, the same models can be applied, in which the fuzziness is a 'perceived fuzziness' due to the 'noise' that distracts perception.

To summarise, *quantitative* and *binary* relations are distinguished, and among binary relations, *sharp* and *fuzzy* relations are further distinguished.

3.2.3 Spatial Relations Between More than Two Features

Up to now, we mainly considered spatial relations between pairs of objects, what Mustière and Moulin (2002) call non-hierarchical relations, in opposition to hierarchical 1-to-n (or n-to-m) relations. Mustière and Moulin (2002) distinguish objects being part of a group [like the components of the meso objects from Ruas

Fig. 3.5 The toponym, buildings, sports fields and footpaths are all functionally related to the complex object school (outlined with *dashes*)

and Duchêne (2007)], and objects being inside an area (e.g. a mountain road is inside a mountainous area). For their part, Mackaness and Edwards (2002) state that relations related to patterns or structure can be divided in two categories: *taxonomies* and *partonomies*. Taxonomies refer to categorisation hierarchies, e.g. an orchard *is a* forest. Partonomies refer more to a conceptual and geometrical division of space, e.g. buildings *are part of* cities (Chaudhry and Mackaness 2007). Mackaness and Chaudhry (2011) propose methods to automatically retrieve urban functional partonomies such as schools (Fig. 3.5) or retail areas from their sub-components.

Steiniger and Weibel (2007) proposed a vast classification of spatial relations between more than two objects, divided into two main categories: statistical and density relations and structural relations which cover the partonomies and functional relations. Ternary relations such as *above* or *left*, with a reference as third object, are also worth noting (Borrmann and Rank 2009).

3.3 An Ontology of Spatial Relations

In light of the literature previously presented, this section describes a spatial relation ontology dedicated to generalisation and on-demand mapping, first proposed by Touya et al. (2012). The first subsection presents the general model of the ontology, the second proposes a taxonomy integrated to the ontology, and the third focuses on multiple representations relations.

3.3.1 Modelling Spatial Relations

This section describes the formalisation of spatial relations, proposed by Touya et al. (2012), as the upper concepts of an ontology that would contain the spatial

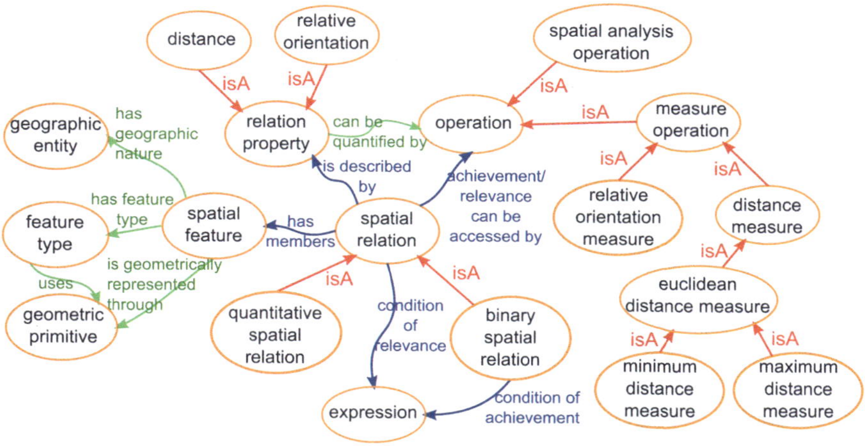

Fig. 3.6 An OWL model of spatial relations (Touya et al. 2012)

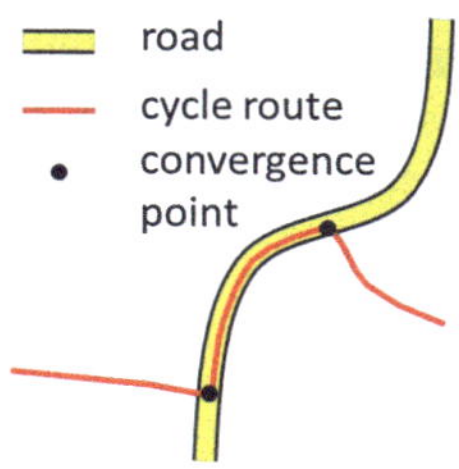

Fig. 3.7 A cycle route that *follows* a road

relations relevant to generalisation or on-demand mapping (Fig. 3.6). Figure 3.7 shows an instance of the "cycle route *follows* the road" relation that will be used to explain the features of the proposed model.

A spatial relation has two members denoted as spatial features (e.g. linear road). Each of these spatial features is described by a geographic entity (e.g. road), its geometric primitive (e.g. line) and possibly its feature type (e.g. roads in the INSPIRE schema). As relations are not all symmetrical (e.g. Fig. 3.9e), the property 'has members' is divided into 'has member 1' and 'has member 2'.

The 'condition of achievement' property describes the configurations where the spatial relation holds. It only applies to binary relations. For instance, a cycle route *follows* a road when the distance between a part of the route and a part of the road is small. In some geographic contexts, a spatial relation may become irrelevant, which can be described by the 'condition of relevance' property of the ontology. For instance, there is no *proximity* relation between two close buildings that are separated by a river. A spatial relation may be described by several properties. For instance, the *follow* relation is described by the convergence points (two per

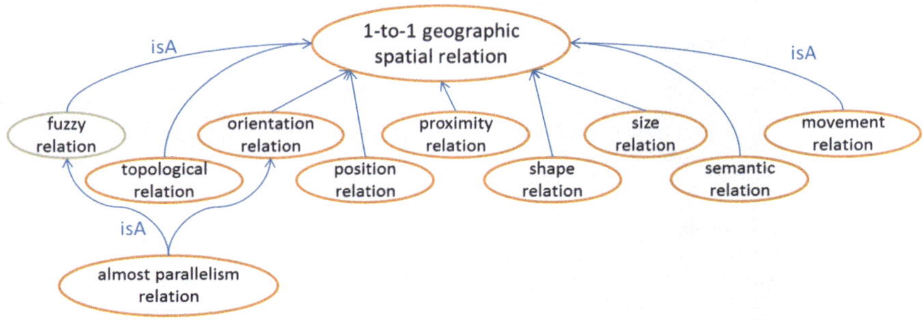

Fig. 3.8 Proposed taxonomy for spatial relations between two geographic entities

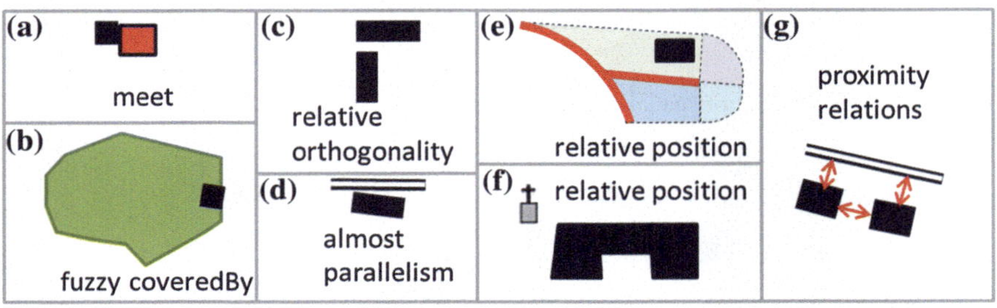

Fig. 3.9 Examples of spatial relations

convergence section) on the road (Fig. 3.7) and the distance between lines. Finally, relevance and achievement can be assessed by an operation, e.g. the achievement of the *follow* relation is assessed by a network matching operation (Chap. 5).

3.3.2 A Taxonomy of Spatial Relations for Generalisation

Touya et al. (2012) proposed to fill the ontology with spatial relations following a taxonomy, dedicated to generalisation and on-demand mapping, which combines various classifications presented in the literature (Sect. 3.2.1). A spatial relation belongs to one of the eight types, and can be sharp or fuzzy (Fig. 3.8).

Figures 3.9 and 3.10 show examples of the different relation types of the taxonomy. Topological relations include the 9IM relations (e.g. meet Fig. 3.9a) including as well their fuzzy counterpart (Fig. 3.9b). *Relative orthogonality* (Fig. 3.9c) *and almost parallelism* (Fig. 3.9d) are examples of orientation relations. Position relations contain *relative position relations* (Papadias and Theodoridis 1997; Matsakis et al. 2008) (Fig. 3.9f), that can be specific such as the *relative position of buildings with dead ends* proposed by Duchêne et al. (2012) (Fig. 3.9e). *Proximity relations* represent objects close to each other (Fig. 3.9g).

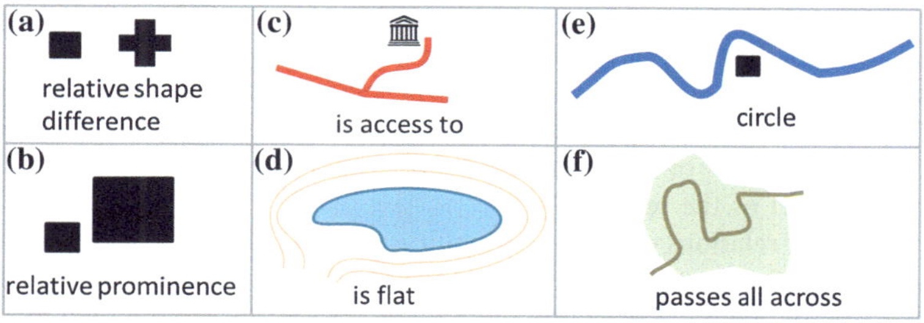

Fig. 3.10 Further examples of spatial relations

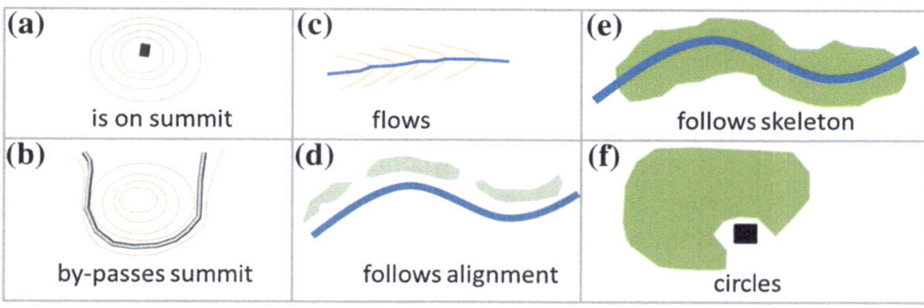

Fig. 3.11 Spatial relations between a feature and a part (or an implicit structure) of a feature

Figure 3.10a, b present shape and size relations that generalisation typically seeks to preserve between buildings. The semantic relations can be illustrated by "the road *is access to* the touristic site" relation (Fig. 3.10c) or by the "lakes *are flat*" relation with respect to relief, visualised in Fig. 3.10d through contour lines. Finally, so-called movement relations are spatial relations that can be named by a movement verb (Mathet 2000) such as "the river *circles* the building" (Fig. 3.10e) or "the path *passes across* the forest" (Fig. 3.10f).

3.3.3 Multiple Representations Relations

Different abstractions of real world features lead to different possible representations for the same features in geographical databases. Mathet (2000) states that some features are polymorph, and can be seen either as polygons, or as lines, or points, in spatial relations. This can be reformulated as spatial relations that may occur between implicit alternative representations of features. For instance, Fig. 3.11e shows a *follows* spatial relation between a river and the implicit linear representation of a forest area.

Mustière and Moulin (2002) claim that spatial relations are a scale-dependant notion: spatial relations may occur between representations of features dedicated to different scales. For instance in Fig. 3.11d, the three forest patches can be seen as a line that is a small scale representation of a row of trees along the river.

Finally, some features may be in relation to some small part of a feature and not to the complete feature. For instance, the limit of the forest locally circles the building in Fig. 3.11f, or the building is on a summit in Fig. 3.11a, which is a characteristic part of the relief.

Such kinds of relations can be called multi-representations relations and require an adjustment of the spatial relations model of Fig. 3.6 to be included in the ontology. Corcoran et al. (2012) propose a method to identify and model the multi-representations relation "a set of roads *is access to* a housing estate".

3.4 Spatial Relations Ontology to Support Automatic Processes

This sections presents research where the spatial relations ontology could be used in an automatic mapping environment. This can be accomplished by deriving constraints from the ontology or by improving interoperability of automatic processes.

3.4.1 Relational Constraints to Monitor Generalisation

Chapter 2 showed that user specifications for a map generalisation process are commonly expressed by generalisation constraints. Relational constraints are constraints on spatial relations that need to be preserved or caricatured (Duchêne et al. 2012). The spatial relations ontology proposed in the previous section helps to define a relational constraints ontology that would help users define their specifications regarding spatial relations (Fig. 3.12). The deeper the user goes into the ontology hierarchy, the less they have to specify constraint details, when defining their specifications.

The relations ontology helps defining a taxonomy of relational constraints for generalisation (Fig. 3.13). Four types of relational constraints are identified: (1) the relational preservation constraints that monitor the preservation of salient relations during generalisation; (2) the relation caricature constraints that monitor only fuzzy relations in order to make them sharp (e.g. caricature 'almost parallel' relations into 'parallel' relations); (3) the relation transformation constraints that change a relation into another during generalisation (e.g. transform road/building parallelism into adjacency); finally, (4) non-creation constraints that prevent non-existing relations from being created by generalisation transformations (e.g. no relative prominence relation between buildings are allowed to be created by building enlargement algorithms).

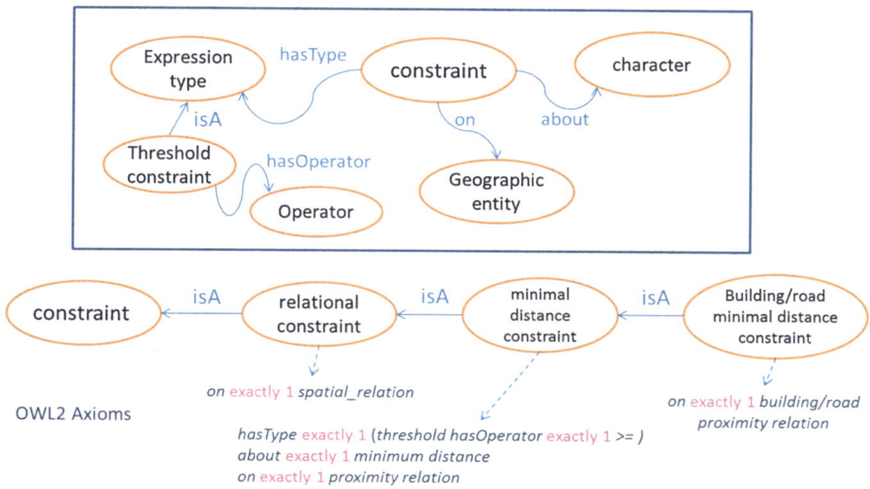

Fig. 3.12 The constraints ontology derived from Touya et al. (2010) and an example on how the axioms restrict the variables for a user to define their constraints

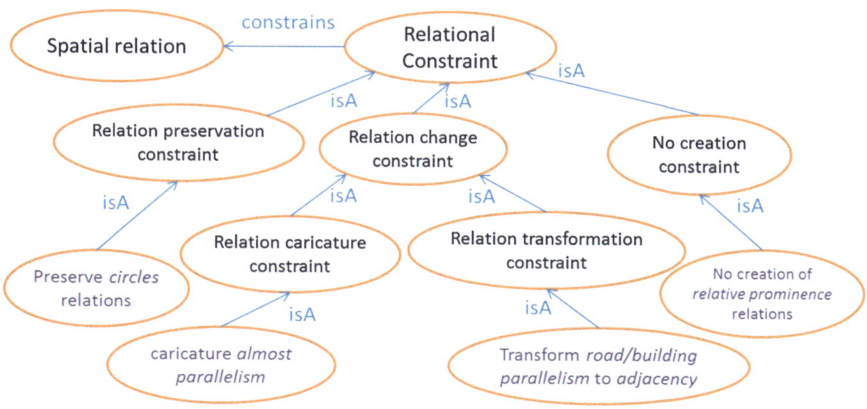

Fig. 3.13 Four types of relational constraints (Touya et al. 2012)

3.4.2 From Ontology to Algorithm

A challenge in defining spatial relations in automatic environments such as generalisation is to finding algorithms that assess these relations, i.e. defining the 'operation' in the model of Fig. 3.6. It is the same challenge that Steiniger and Weibel (2007) stated as finding measures to identify relations. With respect to possible algorithms, Mathet (2000) proposed geometrical methods to measuring movement relations such as *circles* or *passes through*, while Duchêne et al. (2012) proposed algorithms for the relations handled by the CartACom generalisation process.

To create the link between relations in an ontology and the algorithms for assessing them, several approaches have been proposed. A simple but non automatic solution is to semantically annotate the services that encapsulate the algorithms with the relations from the ontology, as proposed in Touya et al. (2010). When an ontology of the measures/algorithms is available, one has only to fill the 'achievement can be assessed by' property of the relations ontology (Balley et al. 2012). Finally, Gould and Chaudhry (2012) propose an automatic matching between relations and an algorithm ontology, provided the algorithm capabilities are described in the ontology.

3.5 Case Study I: Spatial Relations for Urban 3D Models

Bénédicte Bucher and Gilles Falquet

This section illustrates the relevance of an ontology of spatial relations in the context of applications based on city models. The first subsection describes the context of this use case and the requirements for the ontology. The second subsection describes how CityGML partially meets these requirements and the last subsection details design choices for extending the ontology.

3.5.1 Ontological Requirements for Urban Applications

With the growing availability of city models, more and more applications have begun exploiting possibilities offered by these models to support not only visualisation but also automated operations and analysis. Below are examples of such applications.

Wind comfort for pedestrians in an urban environment is a function of wind velocity. Therefore, the comfort of an urban area can be estimated from the average wind velocity at 1.5 m above ground. In wind comfort simulation applications, such as (Amorim et al. 2012), a 3D city model is needed to provide the geometry (terrain, buildings, vegetation) of the space in which the fluid dynamics simulator must operate. To visualise and exploit the simulation results, the 3D city model must support the representation of air flows and some of their characteristics, such as average speed, location of vortices, etc. It must also represent the relationship between the air flows direction, open spaces, and meteorological conditions.

Visual openness and visual exposure are important characteristics that determine the quality of life in a dwelling area (Fisher-Gewirtzman 2012). Visual exposure measures the number of points from which the interior of an apartment is visible. It can depend on the number of openings in other buildings that are in an intervisibility relationship with one or more openings of the considered apartment.

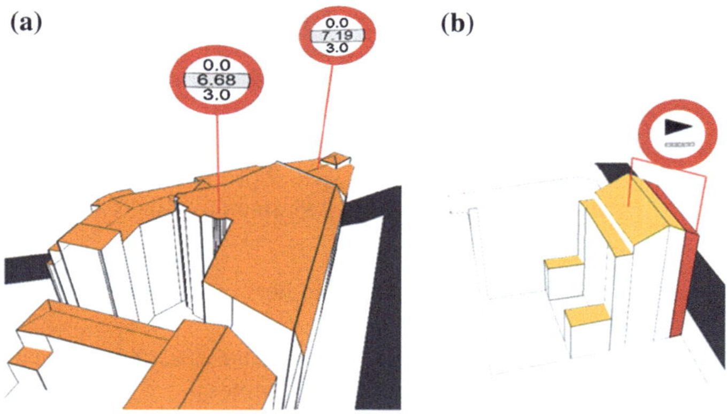

Fig. 3.14 Spatial properties and relations useful in urban planning. **a** Floor area ratio. **b** Distance between features

Intervisibility is the combination of a metric relation, the *distance* between the openings must be sufficiently short, and a projective relation, no opaque object must stand *between* the two openings.

Urban planning and civil engineering heavily rely on the formulation and evaluation of spatial properties and relations such as solar availability (of a parcel, of a room, of a terrace), intervisibility, minimal distance (between two buildings), walking distance, and accessibility. Moreover design can be seen as the activity of using primitives that can be combined as functional units to compose buildings. (Caneporo et al. 2007) proposed an ontology to support the design of new buildings based on a set of elementary components and relevant properties and relations that act as constraints on the possible combination of these primitives. More recently, Brasebin et al. (2011) have proposed an implementation in a 3D GIS of operations to support the automatic evaluation of 3D spatial relations and properties relevant to urban rules evaluation (Fig. 3.14).

The first 3D city models were merely dedicated to visualisation and entailed information about the terrain shape together with appearance (textures). More advanced applications call for a more structured model of information. The wind comfort application will require summarising the city in terms of canyons and will need to deliver its result not only as a 3D coverage but also as features related to the cityscape. The visual exposure application will require the evaluation of intervisibility relations. Civil engineering and urban planning will also require computing intervisibility and other spatial relations and properties. Sometimes end users do not have the geocomputational expertise to select the relevant type of data set and derive the necessary information. Furthermore, it should be noted that some relations such as "topological relations" or "touches" do not have the same meaning depending on the understanding of the author. Topology may refer to the fact that at the data level a network is correctly connected. It may also refer to the ability for a driver to go from one street to another at a cross road. While relations between pure geometric entities (points, lines, planes …) are well established, relations between application/urban objects (on the other side, close to, salient …) are more elusive.

To improve the usage of existing city data and the development and sharing of useful software we propose an ontology of spatial properties and relations that are meaningful to users and define their possible computation on available data, 2D or 3D (Bucher et al. 2012):

- It should serve as a vocabulary for researchers and application designers to avoid ambiguities.
- It should support the indexation of algorithms and assist an application developer to identify contributions from other disciplines.
- It could serve as a starting point for defining data types or database schemas for urban applications. Such schemas would be designed based on commonly used algorithms and useful datatypes to support them, such as the topological map datatypes to support routing of network data.
- The formal definition of relations, coupled with automated reasoning, can be used to automatically infer relations, and, to a certain extent, automatically validate relation computation algorithms.

3.5.2 CityGML

Soon after the *explosion* of virtual globes, Kolbe et al. (2005) proposed the CityGML model to enrich such terrain models with object, semantics and with more structured geometric information to support automated calculus. This model has been adopted as an OGC standard under continuous revision (OGC 2012). In CityGML, the city is modelled through city objects that have a geometric representation and a thematic representation. Class definitions are proposed for the most important city features: relief, buildings, city furniture, tunnels, bridges, water bodies, transportation (roads and railways), and vegetation (Fig. 3.15). The first introduced relation was the aggregation of objects. There is now the concept of CityObjectGroup to attach properties to an aggregation of objects and to specify the role of each component within the group.

An important feature of CityGML is to propose a scale to reference meaningful scales (or levels of detail) in cities: LOD0 to LOD4. The same object can have five different such representation (thematic and geometric) corresponding to the different levels in this scale and generalisation relations can also exist between aggregated objects to support the browsing of the city from one LODi to another.

CityGML objects can also have appearances features. The model also supports the explicit representation of topological relation between features and the terrain (or the water) firstly through a property "relativeToTerrain" (resp. "relativeToWater")—the values ranging from "entirelyAboveTerrain" to "entirelyBelowTerrain" and

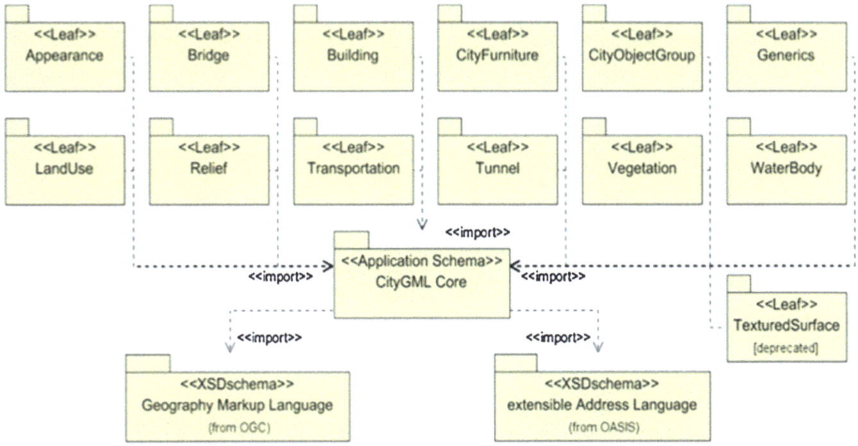

Fig. 3.15 CityGML packages [after (OGC 2012)]

secondly, for buildings, through a property terrainIntersection whose value is a Multicurve. The geometry used in city models is a profile of the GML3 geometry model. It contains 3D primitives and different kinds of combinations of primitives: composites (combination of elements of the same dimension with topological connections), aggregates (combination with a topological structure) and complex (free combination).

To conclude, CityGML is a big step forward in terms of an ontology of spatial relations and properties but does not fully meet our requirements since it contains a limited set of relations: aggregation, combination of geometries, topological relation with the terrain or water, and a generalisation relation between groups of objects.

3.5.3 Improving CityGML with an Ontology of Spatial Relations

We therefore, propose to extend CityGML with ontological items that will meet the requirements listed above (Fig. 3.16). Some items have already been introduced in this chapter and others are defined in the following section.

Several classification schemes for spatial relations have been mentioned in the beginning of this chapter. Clementini (2010) proposal is especially interesting since it provides a mapping between unambiguous relations and properties expressed in an application context and the possible evaluation of these relations and properties based on automated manipulation of city data. It has been extended in Bucher et al. (2012) to better fit urban applications requirements. Spatial relations may be:

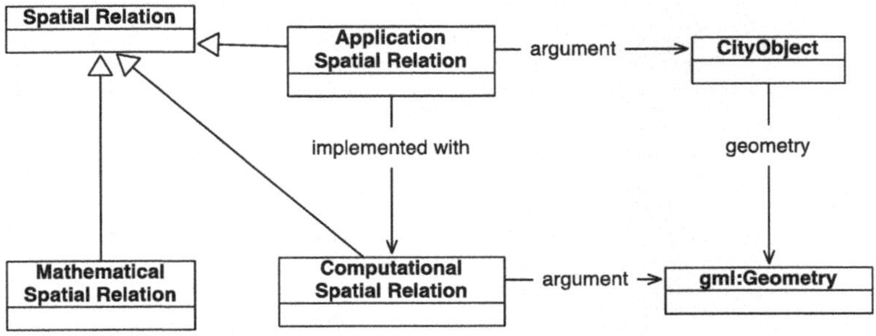

Fig. 3.16 Spatial relations separated into three levels: application, computational and mathematical

Mathematical relations, i.e. relations that exist in the scientific domain of mathematics. These can be geometric relations (distance, symmetry) or topological relations not always supported by the geometry.

Computational relations, i.e. relations that can be instantiated based on data and *Operations* (presented earlier in this chapter).

Application relation, i.e. relations that appear in the user's expression of needs and which reflects their experience of reality and background. Importantly, this level contains many common concepts, properties and relations across applications such as salience, visibility, and shape. As mentioned earlier, application-level relations often bear the same names as geometric relations: touches, between, closer, however, their exact meaning is much more difficult to define. Other application-level relations, such as intervisibility or accessibility, refer to application objects, must satisfy complex conditions, and depend on other objects that form an evaluation context. For application level relations, it is important to specify the *Context* which describes in what universe the relation holds. For example the intervisibility relation between two windows may suppose that there is neither fog nor truck passing by. This means that for an application, computing intervisibility may require data in addition to the city model, e.g. hypothesis on weather conditions.

Importantly, two important observations were made by Bucher et al. (2012) in the context of an ontology of spatial relations for urban applications.

The first is the property *FrameOfReference* of a spatial relation that refers to the point of view from which a spatial relation is observed (Trinh et al. 2011). It can be deictic such as in the relations "the street to the right of the church" where the meaning of the relations depends on the location of an observer. It can be intrinsic when it is attached to an object orientation such as "at the front of the car" (a car has a front and a back). In the case where it does not depend on an observer or on an object but depends on an absolute reference system such as "north of" it is called extrinsic. It is true for relations experienced in reality and for relations experienced in the representation: there are several ways for a user to interact with a 3D model of a city: they can manipulate the model as a single object (a digital mockup) or focus on an individual object (e.g. a building) or navigate within the model to simulate the experience of a human being walking in the city. The 3D model may also be used in an augmented reality application on a tablet or smartphone. These interaction patterns correspond to different frames of reference (absolute, object-centered, user-centered, etc.) and hence to different ways of understanding and computing relations such as left of, behind, etc. Frames of reference can be absolute—independent from the observer- or relative to an observer point of view (first person or third person).

The second observation relates to non binary relations that occur in many simulation applications: objects are interfaces between two other objects (e.g. for a car, a street network is the interface between an area to downtown). This is an extension from Billen et al. (2012) who suggested that buildings be described as a set of interfaces between outer empty space and indoor empty space with windows being the connections for specific agents (such as air). This proposal sounds very promising and could be extended to other kinds of interface (such as streets, bridges). This would be a vector counterpart to a raster view where a city model is decomposed into a mesh to run a simulation algorithm.

As 3D generalisation is still in its infancy with algorithms for individual features, such an extension of CityGML with relations information would greatly to step further to complex processes such as the ones presented in this book for two dimensions features.

3.6 Case Study II: Relations for the Extraction of Groups of Objects

Stefan Steiniger

This second Case study illustrates how spatial analysis can be used to model spatial relations. More specifically, we will model relations for the extraction of groups of objects. One Case study for this is the generalisation of naturally formed objects such as lakes and islands. Here, the preservation—or typification—of object groups during generalisation is important as patterns will often relate to the process by which they were formed (e.g. glacial processes). An example of the generalisation of a lake district, near Lyon in France, is for instance, given in Bertin (1983). Elements of the process he describes are: (1) the recognition of the overall shape of the lake group, (2) the identification of the individual lake shapes (since they may form a directional pattern), (3) the identification of a structural, visual skeleton within the group of lakes, and (4) the visual grouping of large lakes in the lake district.

An important part of Bertin's generalisation process description is the identification of structures, i.e. groups of objects, which are formed only by looking at them, i.e. through visual perception and cognition. To detect these groups they need to be described. For a group of lakes or islands, such description can be achieved by formalising the relations between the objects of the group, for instance by formalising the relation between the islands within one group and between groups. If these relations are sufficiently formalised, then we can use this description to extract the island groups from a dataset.

In the following two subsections, we describe a particular form of object groups, so-called "similarity groups". These groups emerge from perceptual similarities in shape and size of group individuals, (islands in our Case study). In the second subsection, we formulate relations for these types of groupings using the presented modelling approach.

3.6.1 Perceptual Grouping of Islands by Similarity

In the seminal work by Wertheimer (1938) on the "*laws of organization in perceptual forms*", several principles are described under which single objects, e.g. points and lines, are perceived as a group. These principles describe for instance inter-object characteristics such as object proximity, object similarity, good continuation (e.g. a sequence of line segments), and the observers past experience.

Different authors have noted the importance of the similarity principle in cartography e.g. the work of Bertin (1983) or work on building group detection for map generalisation by Li et al. (2004). Steiniger et al. (2006) and Steiniger and Hay (2008) describe an experiment in which participants were asked to group islands, lakes, and triangles. In that experiment participants seem to have applied the principle of similarity to group islands. Three examples for similarity groups, taken from that experiment, are shown in Fig. 3.17. An analysis of such similarity groups (i.e. not only those three groups) reveals that the following conditions seem to hold: (1) the members of the group are close to each other (proximity principle), (2) the members of the group have a similar size, (3) they have a similar shape, and (4) for the two groups, SG2 and SG3, the orientation of the island shape seem to be similar [see also Williams and Wentz (2008)]. We now describe these similarity groups by using relations.

3.6.2 Similarity Relations Among Island Groups

We identified four member properties, or conditions, of similarity groups: proximity, size similarity, shape similarity, and orientation similarity. Each of these four properties should be modelled as relations. This will allow us to detect such groups for generalisation and monitor their changes during the generalisation process. In the modelling example we will use measures and thresholds that are "informed guesses" to illustrate our ideas, rather than using tested measures and values. Figure 3.18 shows how "similarity in size" can be modelled for the island

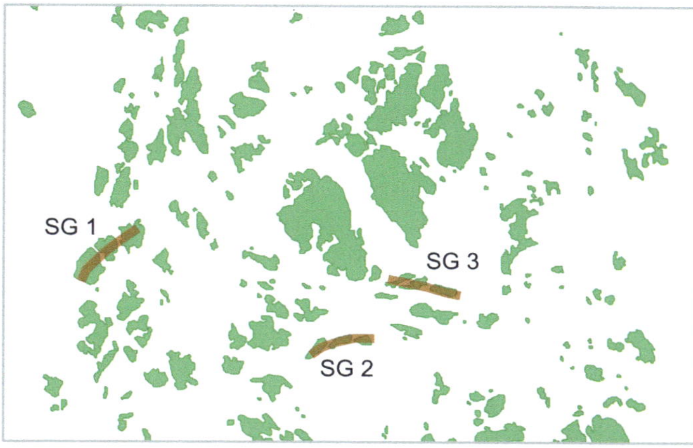

Fig. 3.17 Marked with *bold lines* are groups of islands delineated by participants of an experiment described in Steiniger and Hay (2008). These three groups are likely to be grouped due to similarities in size, shape and orientation of the islands

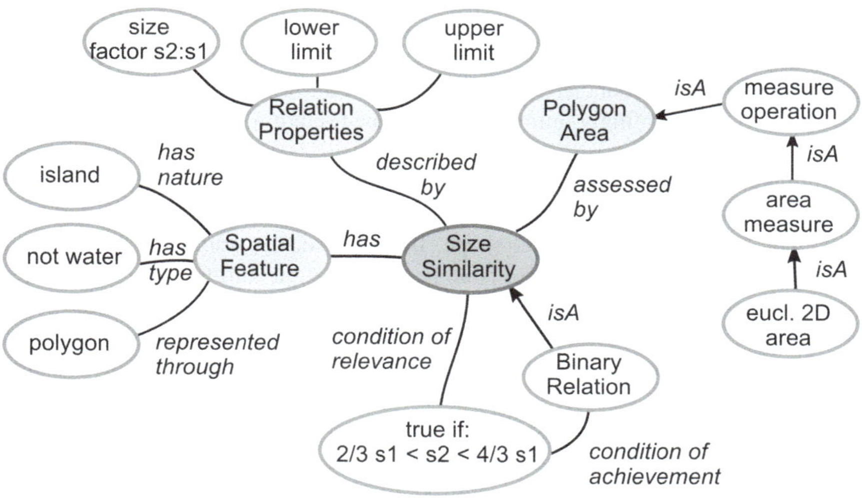

Fig. 3.18 Modelling of the size similarity relation for similarity groups

example by adopting the schema of Fig. 3.6. We assume that a relation is modelled only between two geographic features, i.e. island i_1 and island i_2.

In Fig. 3.18, the relation "*similarity in size*" is described by:

- the type of relation: *isA* binary relation,
- the expression: *true* if 2/3 size(i_1) < size(i_2) < 4/3 size(i_1),

- the members/spatial features: entity = *Island*, type != *water*, geometric primitive = *polygon*,
- the operation for the assessment: polygon area—*isA* measure operation, *isA* area measure, *isA* Euclidian 2D area measure,
- the properties: size-factor(i_2) = $size(i_2)/size(i_1)$, lower limit = 2/3, and upper limit = 4/3.

As we don't have sufficient space to describe the three remaining relations similar to the size similarity relation we have summarised this information in Table 3.1.

3.6.3 Open Modelling Challenges

There remain a number of open challenges that require further research. Three of these challenges are:

Challenge 1—In the applied modelling approach we have assumed that a relation is formed only between 2 geographic features. But a similarity group has at least 3 members. For the distance condition we need to describe at least two distances if three features exists, i.e. $d_{1,2}$ and $d_{1,3}$, assuming that $d_{2,3}$ may not be important to establish the existence of the group. While the modelling of only two distances avoids redundancies, this will does not allow us to monitor changes of $d_{2,3}$. So the third, seemingly unimportant, distance relation needs to be observed as well.

Challenge 2—When observing the two island groups that may be formed due to similarity principles in Fig. 3.17, we see that for the group *SG* 1 on the left side, the islands have a fairly compact shape. In this case, the orientation of the individuals is quite different, and there is no need to treat the similarity of orientations as a necessary relation. Thus the existence of an orientation similarity makes the classification and detection more reliable. Therefore, it would be good to have a general relation property that allows us to assign a weight or importance value to a relation.

Challenge 3—A broader issue remains: How do we connect the relations to each other and to a micro group object, which links to its individuals? One idea is to model the micro group object as a "similarity group relation". This relation could be a binary relation where the expression requires all other relations—i.e., proximity, shape similarity, size similarity, number of members, and eventually orientation similarity—to be evaluated as true. In that case the operation type would be a "spatial analysis operation" that links to the 3 (or 4) mentioned relations. This remains an idea in further need of development.

Table 3.1 Relations for micro groups formed by the similarity principle

Spatial relation	Relation type and expression	Spatial feature	Operation, type and measure	Properties
Proximity	*isA* binary relation, $d_{1,2} < d_{limit}$		Distance, isA measure operation, *isA* distance measure, *isA* eucl. dist., *isA* maximum distance	Distance $d_{1,2}$
Similarity of size	*isA* binary relation, $2/3\ s_1 < s_2 < 4/3\ s_1$	Entity: *island*, Feature type: *not water*, Geom: *polygon*	Polygon area, *isA* measure operation, *isA* absolute area measure, *isA* eucl. 2D area	Size factor s_2/s_1, lower limit, upper limit
Similarity of shape	*isA* binary relation, *diff* ($shape_1$, $shape_2$) < 0.2		Shumm Shape Index, *isA* measure operation, *isA* absolute shape measure	$Shape_1$, $shape_2$, difference limit
Similarity of orientation	*isA* binary relation, $d_{orient,1,2} < 25°$		Longest MBR axis orientation, *isA* measure operation, *isA* relative orientation measure	Orientation-difference $d_{orient,1,2}$, difference limit

3.7 Case Study III: Data Migration of User Data

Kusay Jaara

With the wide availability of topographic data for the general public from National Mapping Agencies and from collaborative geographic data resources (e.g. OpenStreetMap), users are able to create their own thematic geographic data using public topographic data as a reference. At a future point, users could have more current versions of their reference data, or they may want to change the reference data to use more or less detailed data. In order to obtain consistent data after the replacement of reference data, user data has to be processed. We call this processing *thematic data migration.*

If the relations between the users' thematic data and reference data are not taken into account during the thematic data migration, then errors can occur. For example, an accident (a user dataset) that took place in front of a bridge could be located behind the bridge after the migration process. For analysis purposes, such a change can be important. Moreover, users may want to emphasise some initial relations in order to make them visible. If we take the example of the accident near the bridge and the aim to obtain topographic data at a smaller scale, then the user may want to exaggerate, i.e. increase, the distance between the accident location and the bridge to clarify that the accident took place before the bridge and not on it.

A method of thematic data migration has been proposed in Jaara et al. (2013). It is an automatic process that takes into account geographic relations between a user's thematic data and reference data during data migration. The next sections describe the main principles of this method.

3.7.1 General Workflow of Thematic Data Migration

Relations between thematic and topographic data are not explicit in the initial database, so they have to be extracted and represented (Jaara et al. 2012). As the representation of the real world changes from one database to another, it is not always possible to maintain all initial relations. For instance, a roundabout can be represented as a set of road segments forming a round pattern in the initial topographic database and as a simple crossroad in the final topographic database. Relations in this case have to be modified according to the differences in representation of the same object in the topographic database. Relations may change even if the reference object is not modified. For example, a disjoint relation between a thematic region and a reference region might be discarded if the final reference data is presented at a smaller scale and the distance is small. In Fig. 3.19, the data migration workflow of Jaara et al. (2013) is presented.

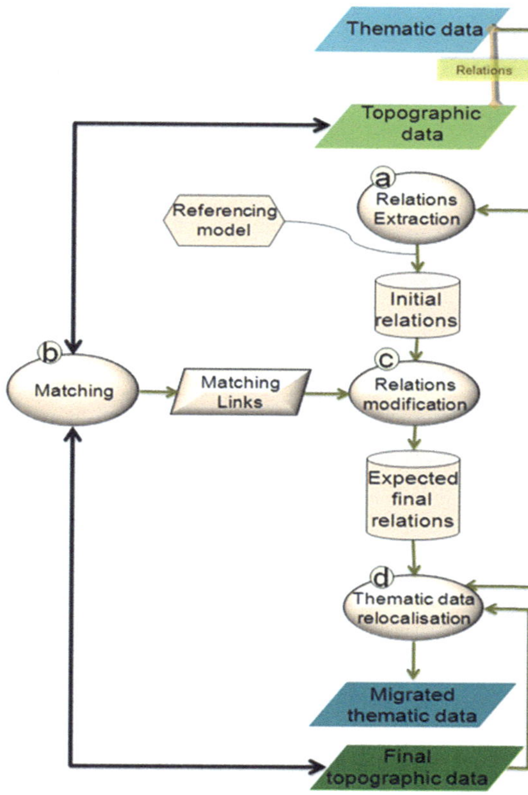

Fig. 3.19 Proposed workflow for the thematic data migration

According to the workflow, thematic data migration consists of:

(a) *Relation extraction and modelling*: significant types of relations are identified depending on the application case. Then relevant instances of these relations are extracted from the initial data (e.g. accident a1 is on road r1 and accident a2 is close to junction j1). These relations are represented using a referencing model (Sect. 3.7.2).
(b) *Matching*: initial and final topographic data are matched to detect the corresponding objects and changes on them (Chap. 5).
(c) *Relations modification*: expected relations within the final dataset are inferred, and are also modified when needed. The modification is based on matching links between the initial and final topographic data (Sect. 3.7.3).
(d) *Thematic data relocalisation*: the expected relations within the final dataset are used to control the spatial relocalisation process, i.e. the propagation of the reference data transformations on the thematic data. In some cases, we have to ignore one or more relations in order reach a solution (Sect. 3.7.4).

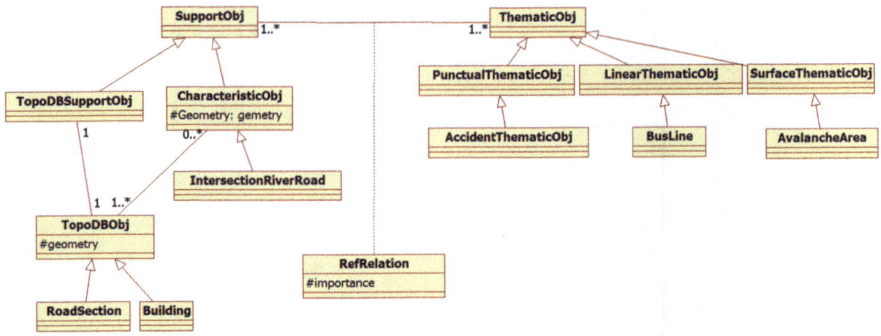

Fig. 3.20 Class diagram of the referencing model of Jaara et al. (2012)

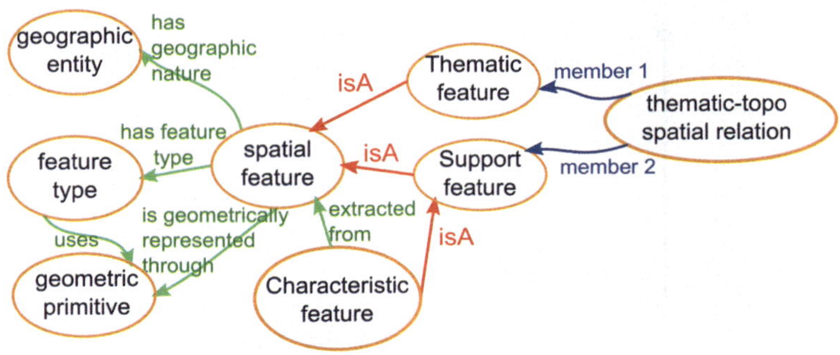

Fig. 3.21 Extension to the relations ontology of Touya et al. (2012) that includes spatial relations between thematic and topographic data

3.7.2 Referencing Model Between Thematic and Topographic Objects

The extracted relations have to be modelled in order to be stored and used in the subsequent stages. A model for relations between thematic and topographic data is presented in Jaara et al. (2012). In this model (Fig. 3.20), topographic objects are considered as 'support objects'. A relation connects one support object and one thematic object. Another type of support object are characteristic objects, which help to obtain a better description of the thematic data, such as roundabouts and road-river intersections. Characteristic objects are not explicitly represented in the initial data. Hence, they have to be extracted in a data enrichment stage. Relations can then be established for topographic objects and for characteristic objects.

The ontology of relations (Fig. 3.6) has been extended to include the thematic data referencing model—so that a spatial feature can either be characteristic, topographic or thematic. Every relation has two members: *Member*1 that is thematic and *Member*2 that is a support object. Figure 3.21 shows the resulting modifications.

3.7.3 Modification of Relations

Before the modification of relations step (Fig. 3.19), initial and final dataset have to be matched, the matching links being used by the modification of relations process.

The objective of this step is to modify the relations, if required, in order to reflect modifications in the topographic data (Jaara et al. 2013). Transition rules have to be defined at the beginning of the process. These rules define acceptable changes, necessary changes, and conditions of every transformation of relations. For example, a "*near parallelism*" relation, which is a fuzzy relation (Fig. 3.4), becomes a "*parallelism*" relation when the difference of scale between the original and the target scale exceeds a given (orientation angle) threshold. Rules can be used in order to make certain relations particularly visible (exaggeration). A transition rule for an "almost disjoint" relation between two regions can be defined in order to keep a visible distance between both regions in case a smaller scale map of the final topographic database is intended. The output of this processing stage is the expected relations between the final and initial topographic datasets.

3.7.4 Relocalisation Process

After identifying the expected relations in the final dataset (either preserving the initial or transforming them), the objective will be to find the position for the thematic object where these relations are best preserved or transformed. For this reason, we introduced the *relation satisfaction measure*. For a given position of thematic data, it measures how well the relation in the final reference data is preserved (or transformed). In the framework of the ontology of spatial relations (Fig. 3.6) proposed by Touya et al. (2012), the satisfaction measure is an operation that assesses the achievement (i.e. existence) of the spatial relation. The calculation of the satisfaction measure depends on the relation type and attributes. For quantitative relations, the satisfaction is equivalent to the difference of values before and after the migration (e.g. distances). For qualitative relations, neighbourhood graphs (Egenhofer and Franzosa 1991) could be used to extract the satisfaction, by measuring the distance between the expected relation and the actual relation, i.e. by evaluating the satisfaction.

The relocalisation can be treated as an optimisation problem: solutions are based on a local search near the initial position, where possible places for an object are calculated by adding regularly spaced vertices on the line that carries the object (the object can be moved to one of the vertices). Every possible place has a number of satisfaction values, one value for every expected relation. In some cases, certain relations have to be ignored in order to obtain a better result. In the accident example, in which the accident occurs next to the bridge and, additionally, in front of a building. If the building has a different position in the final

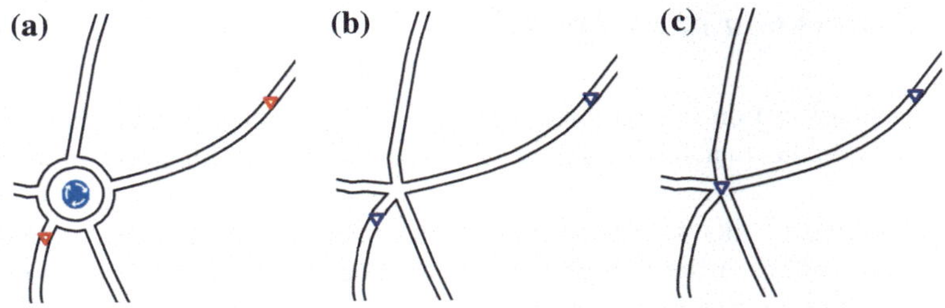

Fig. 3.22 Results of thematic data migration. **a** Initial combination of accidents with initial topographic database. **b** Final combination using curvilinear ratio. **c** Final combination using our methods

topographic database, we may prefer to ignore the relation with the building and leave the accident next to the bridge. It is more important to have an acceptable solution than to end up with a bad compromise. To solve the multi-criteria problem the PROMETHEE II approach (Brans and Mareschal 2005) has been used, because of its suitability and because it can be easily parameterised. It allows us to ignore some relations in instances of bad compromise. Solutions are scored and the one with the best scores are chosen.

3.7.5 Results

The model is illustrated using two examples. The first one illustrates the advantage of our thematic data migration method over using the simple curvilinear ratio method that could be the simple method one could use for data migration. With the curvilinear ratio, the thematic data is moved at a point that corresponds to the same curvilinear ratio on line length: for instance, if the thematic data is located at 50 m on the road curvilinear axis, with the road being 100 m long, then it is relocated at 45 m on the generalised road curvilinear axis that is 90 m long. The example shows how data migration is improved by the presented method (Fig. 3.22). The second example illustrates the usefulness of the chosen multicriteria system that favours some relations over others.

In the first example, two accidents located at different distances from a roundabout have to be migrated (Fig. 3.22); the final topographic database is a generalised version where the roundabout is changed into a crossroad.

The main relation that has to be preserved is the proximity relation to the roundabout. But as the roundabout is represented by a crossroad in the final database, the proximity relations to the roundabout is changed to a proximity relation to the crossroad. A transition rule is defined as follows: If an accident is close to a roundabout and if a roundabout is matched with a crossroad, then the accident has to be situated in the middle of the crossroad. The relocalisation is done based on the changed relations, which gives the result shown in Fig. 3.22.

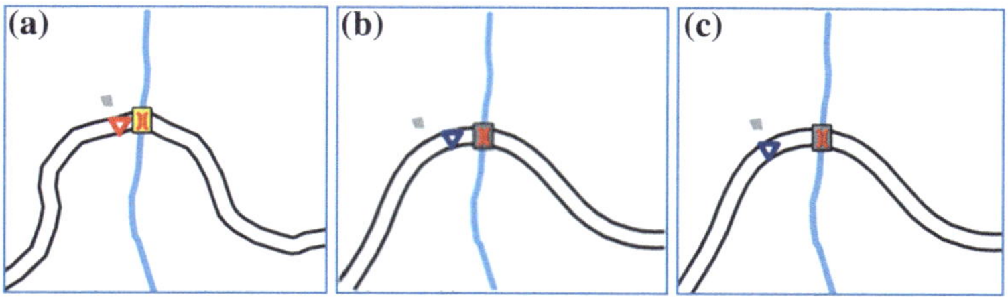

Fig. 3.23 **a** Initial data. **b** Migration with compromise between relations. **c** Migration using a multicriteria approach

In the second example (Fig. 3.23), an accident is situated next to the river-road intersection (proximity relation) and in front of a building (relative position relation). In the final topographic database, the road was generalised and the building has been displaced. If we try to keep both relations as much as possible by finding a compromise solution, the result will be a partial satisfaction of the two relations. The score of the best location while keeping the two relations is calculated. The cases when ignoring one of the two relations are also evaluated. The solution with the best evaluation is taken. Evaluations are based on relation importance, which is related to user needs or the nature of the thematic object.

3.8 Conclusions

Although research on spatial relations modelling began many years ago, their use in automated environments such as map generalisation is quite recent and research challenges remain. A model for a spatial relations ontology is proposed in this chapter but it requires further development to be useful for automatic processes. For instance, all relations identified by cartographers as important in the generalisation process should be included in the ontology. Then, it could be made freely available to spread this shared model in the generalisation research community and make it a standard resource of generalisation processes as well as of user requirement definition systems (see Chap. 2).

Moreover, the handling of relations during generalisation needs to be improved, particularly in the context of on-demand mapping with thematic data mapped onto topographic data. The migration use case raises the question of when to take relations into account during the generalisation process: before the process as parameters? Or afterwards with conflation or propagation techniques? Also, the handling and the definition of so-called multi-representations relations (Sect. 3.3.3) requires further research. For instance, how do we handle the relation of a building alignment *along* a dead end street when moving the street is required by the city generalisation process? By and large, a better management, in the generalisation process, of the interactions between geographic objects and their relations is required to improve generalisation automation.

References

Amorim JH, Valente J, Pimentel C, Miranda AI, Borrego C (2012) Detailed modelling of the wind comfort in a city avenue at the pedestrian level. Usage, usability, and utility of 3D city models, Nantes

Balley S, Jaara K, Regnauld N (2012) Towards a prototype for deriving custom maps from multisource data. In: Proceedings of the 15th ICA workshop on generalisation and multiple representation, jointly organised with EuroSDR commission on data specifications, Istanbul, Turkey

Bejaoui L, Pinet F, Bedard Y, Schneider M (2009) Qualified topological relations between spatial objects with possible vague shape. Int J Geogr Inf Sci 23(7):877–921

Bertin J (1983) Semiology of graphics: diagrams, networks, maps. University of Wisconsin Press, Madison

Billen R, Zaki CE, Servières M (2012) Developing an ontology of space: application to 3D city modeling. In: Leduc T, Moreau G, Billen R (eds) Usage, usability, and utility of 3D city models

Borrmann A, Rank E (2009) Specification and implementation of directional operators in a 3D spatial query language for building information models. Adv Eng Inform 23(1):32–44. doi:10.1016/j.aei.2008.06.005

Brando C, Bucher B, Abadie A (2011) Specifications for user generated spatial content. In: Geertman S, Reinhardt W, Toppen F (eds) Advancing geoinformation science for a changing world, vol 1. Springer, Heidelberg, pp 479–495

Brans JP, Mareschal B (2005) Promethee methods. In: Figueira J, Greco S, Ehrogott M (eds) Multiple criteria decision analysis: state of the art surveys. international series in operations research and management science, vol 78. Springer, New York, pp 163–186. doi:10.1007/0-387-23081-5_5

Brasebin M, Perret J, Haëck C (2011) Towards a 3D geographic information system for the exploration of urban rules: application to the French local urban planning schemes. In: 28th urban data management symposium (UDMS 2011)

Brassel KE, Weibel R (1988) A review and conceptual framework of automated map generalization. Int J Geogr Inf Syst 2(3):229–244. doi:10.1080/02693798808927898

Bucher B, Falquet G, Clementini E, Sester M (2012) Towards a typology of spatial relations and properties for urban applications. In: Leduc T, Moreau G, Billen R (eds) Proceedings of usage, usability and utility of 3D city models, COST TU0801 final conference, EDP Sciences, Nantes

Burghardt D, Schmid S, Stoter J (2007) Investigations on cartographic constraint formalisation. In: Proceedings of the 11th ICA workshop on generalisation and multiple representation, Moscow, Russia

Burghardt D, Petzold I, Bobzien M (2010) Relation modelling within multiple representation databases and generalisation services. Cartographic J 47(3):238–249. doi:10.1179/000870410X12699418769035

Caneparo L, Collo M, di Giannantonio D, Lombardo V, Montuori A, Pensa S (2007) Generating urban morphologies from ontologies. In: 2nd workshop COST action C21—towntology. http://www.towntology.net/Meetings/0710-Torino/articles/13Paper%20%28162-177%29.pdf

Chaudhry OZ, Mackaness WA (2007) Utilising partonomic information in the creation of hierarchical geographies. In: Proceedings of the 11th ICA workshop on generalisation and multiple representation, Moscow, Russia

Chaudhry OZ, Mackaness WA, Regnauld N (2009) A functional perspective on map generalisation. Comput Environ Urban Syst 33(5):349–362

Clementini E (2010) Ontological impedance in 3D semantic data modeling. In: Kolbe TH, König G, Nagel C (eds) Proceedings of 5th 3D geoinfo conference, Berlin, ISPRS, pp 97–100

Cohn AG, Hazarika SM (2001) Qualitative spatial representation and reasoning: an overview. Fundam Inf 46(1–2):1–29

Corcoran P, Mooney P, Bertolotto M (2012) Spatial relations using high level concepts. ISPRS Int J Geo-Inf 1(3):333–350. doi:10.3390/ijgi1030333

Duchêne C, Ruas A, Cambier C (2012) The cartACom model: transforming cartographic features into communicating agents for cartographic generalisation. Int J Geogr Inf Sci 26(9):1533–1562. doi:10.1080/13658816.2011.639302

Egenhofer MJ, Franzosa RD (1991) Point-set topological spatial relations. Int J Geogr Inf Syst 5(2):161–174. doi:10.1080/02693799108927841

Fisher-Gewirtzman D (2012) 3D models as a platform for urban analysis and studies on human perception of space. Usage, usability, and utility of 3D city models. Nantes

Gould N, Chaudhry O (2012) An ontological approach to on-demand mapping. In: Proceedings of the 15th ICA workshop on generalisation and multiple representation, jointly organised with EuroSDR commission on data specifications, Istanbul, Turkey

Jaara K, Duchêne C, Ruas A (2012) A model for preserving the consistency between topographic and thematic layers throughout data migration. In: Proceedings of 15th international symposium on spatial data handling (SDH'12), Bonn

Jaara K, Duchêne C, Ruas A (2013) Preservation and modification of relations between thematic and topographic data throughout thematic data migration process. In: Cartography from pole to pole: selected contributions to the 26th international cartographic conference 2013, Springer, Dresden

Jones C (1997) Geographical information systems and computer cartography. Prentice Hall, New Jeresy

Kolbe TH, Gröger G, Plümer L (2005) CityGML: interoperable access to 3D city models. In: van Oosterom P, Zlatanova S, Fendel EM (eds) Geo-information for disaster management. Springer, Berlin, pp 883–899

Li Z, Yan H, Ai T, Chen J (2004) Automated building generalization based on urban morphology and gestalt theory. Int J Geogr Inf Sci 18(5):513–534. doi:10.1080/13658810410001702021

Mackaness WA, Chaudhry OZ (2011) Automatic classification of retail spaces from a large scale topographic database. Trans GIS 15(3):291–307. doi:10.1111/j.1467-9671.2011.01259.x

Mackaness WA, Edwards G (2002) The importance of modelling pattern and structure in automated map generalisation. In: Proceedings of the joint ISPRS/ICA workshop on multi-scale representations of spatial data, pp 7–8

Mathet Y (2000) New paradigms in space and motion: a model and an experiment. In: Proceedings of the ECAI 2000 workshop on current issues in spatio-temporal reasoning, Berlin

Matsakis P, Wawrzyniak L, Ni J (2008) Relative positions in words: a system that builds descriptions around allen relations. Int J Geogr Inf Sci 24(1):1–23. doi: 10.1080/13658810802270587

McMaster RB, Shea KS (1988) Cartographic generalization in digital environment: a framework for implementation in a GIS. In: GIS/LIS'88, pp 240–249

Mustière S, Moulin B (2002) What is spatial context in cartographic generalisation? In: Joint International Symposium and Exhibition on Geospatial Theory, Processing and Applications. ISPRS & SIS, vol 34. pp 274–278

OGC (2012) OGC city geography markup language (CityGML) encoding standard, v2.0.0, OGC project document OGC12-019. In: Gröger G, Kolbe T, Nagel C, Häfele K-H (eds)

Papadias D, Theodoridis Y (1997) Spatial relations, minimum bounding rectangles, and spatial data structures. Int J Geogr Inf Sci 11(2):111–138. doi:10.1080/136588197242428

Randell DA, Cui Z, Cohn AG (1992) A spatial logic based on regions and connection. In: Proceedings of 3rd international conference on knowledge representation and reasoning, Morgan Kaufmann, Burlington, Massachusetts, USA

Ruas A (1999) Modèle de généralisation de données géographiques à base de contraintes et d'autonomie. Ph.D. dissertation, Université de Marne-la-Vallée

Ruas A, Duchêne D (2007) A prototype generalisation system based on the multi-agent system paradigm. In: Mackaness WA, Ruas A, Sarjakoski LT (eds) The generalisation of geographic information: models and applications, Elsevier, pp 269–284 (Chapter 14)

Steiniger S, Hay GJ (2008) An experiment to assess the perceptual organization of polygonal objects. In: Klippel A, Hirtle S (eds) You-are-here-maps—creating a sense of place through map-like representation, spatial cognition 2008, Freiburg, pp 38–44

Steiniger S, Burghart D, Weibel R (2006) Recognition of island structures for map generalization. In: GIS '06 Proceedings of the 14th annual ACM international symposium on advances in geographic information systems, Arlington, Virginia, pp 67–74. doi:10.1145/1183471.1183484

Steiniger S, Weibel R (2007) Relations among map objects in cartographic generalization. Cartography Geogr Inf Sci 34(3):175–197

Touya G, Duchêne C, Ruas A (2010) Collaborative generalisation: Formalisation of generalisation knowledge to orchestrate different cartographic generalisation processes. In: Fabrikant S, Reichenbacher T, van Kreveld M, Schlieder C (eds) Geographic information science. Lecture notes in computer science, vol 6292. Springer, Heidelberg, pp 264–278

Touya G, Balley S, Duchêne C, Jaara K, Regnauld N, Gould N (2012) Towards an ontology of generalisation constraints and spatial relations. In: Proceedings of the 15th ICA workshop on generalisation and multiple representation, jointly organised with EuroSDR commission on data specifications, Istanbul, Turkey

Trinh TH, Chevaillier P, Barange M, Soler J, Loor PD, Querrec R (2011) Integrating semantic directional relationships into virtual environments: a meta-modelling approach. In: Coquillart S, Steed A, Welch G (eds) JVRC11: joint virtual reality conference of EGVE—EuroVR, Nottingham, pp 67–74

Wertheimer M (1938) Laws of organization in perceptual forms. In: Ellis W (eds) A source book of Gestalt psychology, Routledge and Kegan Paul, London, pp 71–88 (English edn)

Williams EA, Wentz EA (2008) Pattern analysis based on type, orientation, size, and shape. Geogr Anal 40(2):97–122

Winter S (2000) Uncertain topological relations between imprecise regions. Int J Geogr Inf Sci 14(5):411–430

Chapter 4
Data Structures for Continuous Generalisation: tGAP and SSC

Peter van Oosterom, Martijn Meijers, Jantien Stoter and Radan Šuba

Abstract Spatial zoom and thematic navigation are indispensable functionalities for digital web and mobile maps. Therefore, recent map generalisation research has introduced the first truly smooth vario-scale structure (after several near vario-scale representations), which supports continuous or smooth zooming. In the implementation, the vario-scale representation of 2D geo-information can be stored as a single 3D (2D+scale) data structure. A single uniform scale map in 2D is then derived by computing a horizontal slice through the structure.

4.1 Introduction

In recent years, the Internet has become an important source of digital maps for mobile users. However, applications suffer from bandwidth limitations and restricted devices such as small displays. Sending a map with many details for each request is expensive and time consuming. From a user's perspective this is unsatisfactory when zooming or panning, for example, when first navigating to an area of interest. As this task does not require a map at the highest resolution, it is reasonable to send a less detailed map first. In order to define these representations such that characteristic features are preserved, automatic generalisation methods are needed.

P. van Oosterom (✉) · M. Meijers · J. Stoter · R. Šuba
Section GIS Technology, Department OTB, Faculty of Architecture and the Built Environment, Delft University of Technology, Delft, The Netherlands
e-mail: P.J.M.vanOosterom@TUDelft.nl

M. Meijers
e-mail: B.M.Meijers@TUDelft.nl

J. Stoter
e-mail: J.E.Stoter@TUDelft.nl

R. Šuba
e-mail: R.Suba@TUDelft.nl

D. Burghardt et al. (eds.), *Abstracting Geographic Information in a Data Rich World*, Lecture Notes in Geoinformation and Cartography, DOI: 10.1007/978-3-319-00203-3_4,

Technological advancements have led to maps being used virtually everywhere, for example map use in mobile smartphones. Map use is more interactive than ever before: users can zoom in, out and navigate via interactive maps. Therefore, recent map generalisation research has focused on continuous generalisation in contrast to 'traditional' generalisation applied to predefined scale-steps. Despite some useful efforts (Kreveld 2001; Sester and Brenner 2005; Cecconi and Galanda 2002), there is no optimal solution yet.

This chapter introduces the truly smooth vario-scale structure for geographic information: a small step in the scale dimension leads to a small change in representation of geographic features that are represented on the map. Continuous generalisations of real world features can be derived from a structure that can be used for presenting a smooth zoom action to the user. Furthermore, mixed-scale visualisations can be derived with more and less generalised features shown in one map in which the objects are consistent with each. Making such a transition area is one of the principal difficulties for 3D computer graphic solutions [e.g. using stitch strips based on triangles, such as used by Noguera et al. (2011)]. In addition, with the vario-scale approach there is no or minimal geometric data redundancy and there is no temporal delay between the availability of data sets at different map scales (as was and is the case with more traditional approaches).

Sections 4.2 and 4.3 provide the theoretical framework for vario-scale representations: the topological Generalised Area Partitioning-structure [tGAP, Sect. 4.2, largely based on van Oosterom (2005)] and the 3D SSC encoding of 2D vario-scale data [Space-Scale Cube, Sect. 4.3, largely based on van Oosterom and Meijers (2011b)]. After applying the theory to the two case studies in respectively Sect. 4.4 (constraint tGAP with independent constraint map), Sect. 4.5 (constraint tGAP with derived constraint map), the 3D approach of the SSC encoding of truly smooth tGAP is tested with Corine and ATKIS data in Sect. 4.6. Finally, Sect. 4.7 lists the main results and overview of future work.

4.2 Principles of the Generalised Area Partitioning Structure

This section presents the data structure for a variable scale representation of an area partitioning without redundancy of geometry, suitable for progressive transfer.

4.2.1 Introduction

There are a number of data structures available for multi-scale databases based on multiple representations, that is, the data are used for a fixed number of scale (or resolution) intervals. These multiple representation data structures attempt to explicitly relate objects at different scale levels, in order to offer consistency during the use of the data. The drawbacks of the multiple representations data structures are that they store redundant data (same coordinates, originating from the same source) and they support only a limited number of scale intervals. Most data structures are intended to be used during the pan and zoom (in and out) operations, and in that sense multi-scale data structures are already a serious improvement for interactive use as they do speed-up interaction and give reasonable representations for a given level of detail (or scale).

Need for Progressive Data Transfer: A drawback of multiple representation data structures is that they are not suitable for progressive data transfer, because each scale interval requires its own independent graphic representation to be transferred. Good examples of progressive data transfer are raster images, which can be presented relatively quickly in a coarse manner and then refined as the user waits a little longer. These raster structures can be based on simple (raster data pyramid) (Samet 1984) or more advanced (wavelet compression) principles (Lazaridis and Mehrotra 2001; Hildebrandt et al. 2000; Rosenbaum and Schumann 2004). For example, JPEG2000 (wavelet based) allows both compression and progressive data transfer from the server to the end-user. Also, some of the proprietary formats such as ECW from ER Mapper and MrSID from LizardTech are very efficient raster compression formats based on wavelets, offering multi-resolution access suitable for progressive data transfer. Similar effects are more difficult to obtain with vector data and require more advanced data structures, though a number of attempts have been made to develop such structures (Bertolotto and Egenhofer 2001; Buttenfield 2002; Jones et al. 2000; Zhou et al. 2004).

Multi-scale/Variable-scale Vector Data Structures: For single (line) objects, a number of multi-scale/variable-scale data structures have been proposed: Strip-tree (Ballard 1981), Multi-Scale Line tree (Jones and Abraham 1987), Arc-tree (Gunther 1988), and the Binary Line Generalisation tree (BLG-tree) (van Oosterom 1990). The Strip-tree and the Arc-tree are intended for arbitrary curves, not for simple polylines. The Multi-Scale Line tree is intended for polylines, but it introduces a discrete number of detail levels and it is a multi-way tree, meaning that a node in the tree can have an arbitrary number of children. The BLG-tree is a binary tree for a variable-scale representation of polylines, based on the Douglas and Peucker (1973) line generalisation algorithm. Note that these line data structures cannot be used for spatial organisation (indexing, clustering) of multiple objects (as needed by variable-scale or multi-scale map representations), so they only solve part of the generalisation and storage problem.

One of the first multi-scale vector data structures designed to avoid redundancy was the reactive BSP-tree (van Oosterom 1989), which supports both spatial organisation (indexing) and multiple levels of details. Its main disadvantage, however, is that it is a static structure. The first dynamic vector data structure supporting spatial organisation of all map objects, as well as multiple scales, was the Reactive tree (van Oosterom 1992, 1994). The Reactive tree is an R-tree (Guttman 1984) extension with importance levels for objects: more important objects are stored higher in the tree structure, which makes more important objects more accessible. This is similar to the reactive BSP-tree, but the dynamic structure of the Reactive tree enables inserts and deletes—functions that the BSP-tree lacks. The BLG-tree and the Reactive tree are eminently capable of supporting variable-scale/multi-scale maps composed of individual polyline or polygon objects.

Generalised Area Partitioning: The BLG-tree and Reactive-tree structures are not well suited for area partitioning, since removal of a polygon results in a gap in the map and independent generalisation of the boundaries of two neighbour areas results in small slivers (overlaps or gaps). Overcoming this deficiency was the motivation behind the development of the GAP tree (van Oosterom 1993). The BLG-tree, Reactive-tree, and GAP-tree data structures can be used together, while each supports different aspects of related generalisation processes, such as selection and simplification, or for area partitioning (van Oosterom and Schenkelaars 1995).

Following the conceptualisation of the GAP tree, several improvements were published to resolve limitations of the original data structures (van Putten and van Oosterom 1998; Ai and van Oosterom 2002; Vermeij et al. 2003). The next section describes the background of the topological GAP tree, which combines the use of the BLG-tree and the Reactive tree and avoids the problems of the original GAP tree—namely redundant storage and slivers near the boundary of two neighbouring areas.

4.2.2 GAP Tree Background

The first tree data structure for generalised area partitioning (GAP tree) was proposed by van Oosterom (1993). The idea was based on first drawing the larger and more important polygons (area objects), so as to create a generalised representation. However, one can continue by refining the scene through the additional drawing of the smaller and less important polygons on top of the existing polygons (based on the Painters algorithm; Fig. 4.1). This principle has been applied to the Digital Land Mass System-Digital Feature Analysis Data (DLMS DFAD) data structure (DMA 1986), because it already had this type of polygonal organisation. When tested with the Reactive tree and the BLG-tree, it was possible to zoom in and zoom out, and obtain map representations with more or less detail of a smaller/larger region in constant time (Fig. 4.2, left).

Computing the GAP Tree: If one has a normal area partition (and not DLMS DFAD data) one first has to compute the proper structure. This is driven by two functions. First, the importance function [for example: *Importance(a)* = *Area(a)* * *WeightClass(a)*] is used to find the least important feature *a* based on its size and the relative importance of the class it belongs to. Then the neighbour *b* is selected based on the highest value of *Collapse(a,b)* = *Length(a,b)* * *CompatibleClass(a,b)*, with *Length(a,b)* being the length of the common boundary. Feature *a* is removed and feature *b* takes its space on the map. In the GAP tree this is represented by linking feature *a* as the child of parent *b* (and enlarging the original feature *b*). This process is repeated until only one feature is left covering the whole domain, forming the root of the GAP tree. Figure 4.1 gives a schematic representation of such a GAP tree.

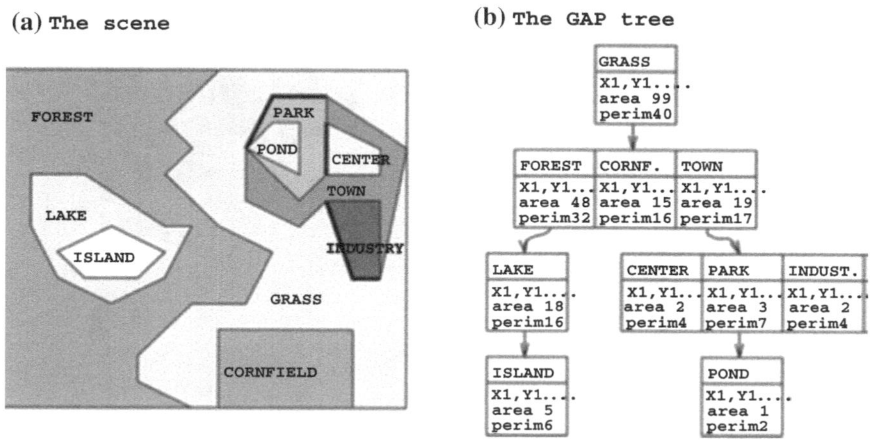

Fig. 4.1 The original GAP tree (van Oosterom 1993)

Work by van Smaalen (2003) focuses on finding neighbour patterns, which might in turn be used for setting up an initial compatibility matrix. Bregt and Bulens (1996) give area generalisation examples in the domain of soil maps, based on the same principles. Both van Smaalen (2003) and Bregt and Bulens (1996) use an adapted classification for the higher (merged) level of objects, instead of keeping the original classification at all levels of detail. For example, deciduous forest and coniferous forest objects are aggregated into a new object classified as "forest" or "garden," while "house" and "parking place" objects form the new object "lot". This could also be done in the GAP tree.

Implementations and Improvements of the GAP Tree: Though the GAP tree may be computed for a source data set, which has a planar partitioning topology, the GAP tree itself is not a topological structure. Each node in the GAP tree is a polygon, and this introduces some redundancy as parents and child may have some parts of their boundary in common. The first GAP-tree construction based on topologically structured input was implemented by van Putten and van Oosterom (1998) for two real world data sets: Top10vector (1:10,000) and GBKN (1:1,000; Fig. 4.2 right). It turned out that finding the proper importance and compatibility functions (which drive the GAP-tree construction) is far from trivial and depends on the purpose of the map. In addition, two improvements were presented in the 1998 paper (at the conceptual level): (1) adding parallel lines to "linear" area features, and (2) computing a GAP tree for a large seamless data set.

Ai and van Oosterom (2002) presented two other possible improvements to the GAP tree: One improvement was that the least important object should not only be assigned to one neighbour, but subdivided along its skeleton and the different parts assigned to different neighbours/parents (the result is not a tree but a directed acyclic graph: GAP-DAG). The second improvement concerned extending the neighbourhood analysis by considering non-direct (sharing a common edge)

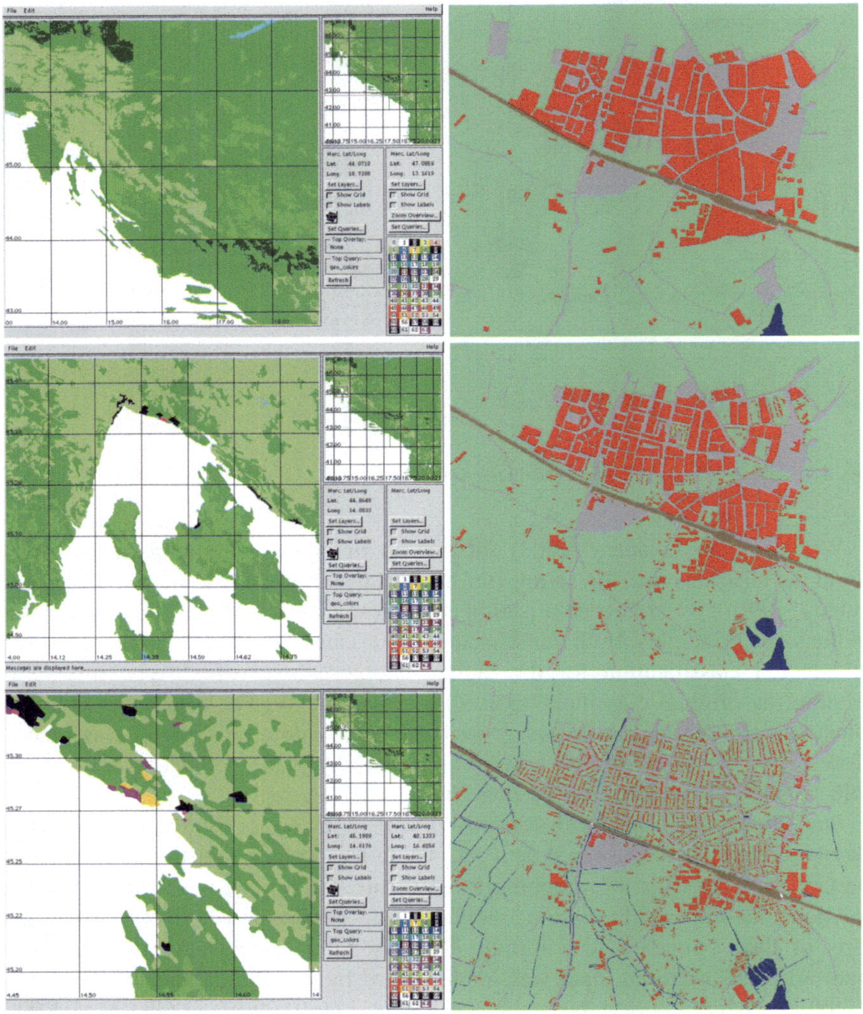

Fig. 4.2 *Left* GAP-tree principle applied to DLMS DFAD (adds detail when zooming in). *Right* GAP tree applied to large scale topographic data set (shown at the same scale)

neighbour areas as well. Both suggestions are based on an analysis using a Triangular Irregular Network (TIN) structure.

Topological Version of the GAP Tree: All improvements still result in a non-topological GAP structure, which means that it contains redundancy. Vermeij et al. (2003) presented a GAP-tree structure that avoids most redundancy by using a topological structure for the resulting GAP tree, not only for the input: thus the edges and the faces table both have attributes that specify the importance ranges in which a given instance is valid. The 2D geometry of the edges (and faces) is

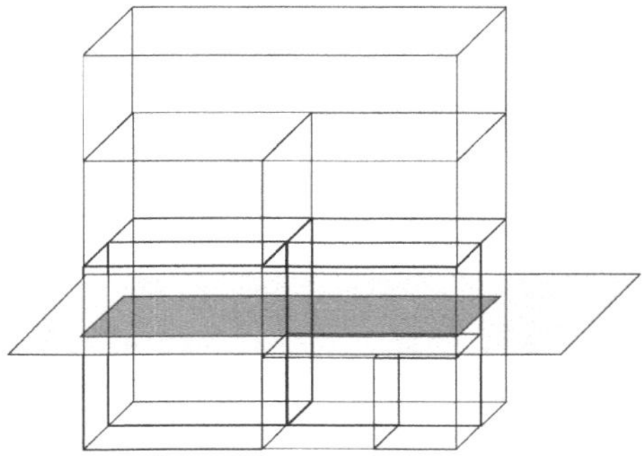

Fig. 4.3 Importance levels represented by the third dimension [at the most detailed level (*bottom*) there are several objects, while at the most coarse level (*top*) there is only one object]. The hatched plane represents a requested level of detail, and the intersection with the symbolic 3D volumes then gives the faces

extended by the importance value range (on the z-axis) for which it is valid (Fig. 4.3). One drawback of this approach is that it requires considerable geometric processing at the client side—clipping edges, forming rings, and linking outer and possible inner rings to a face. A second drawback is that there is some redundancy introduced via the edges at the different importance levels: i.e., the coordinates of detailed edges are again present in the edge at the higher aggregation level. Finally, van Oosterom (2005) introduced a data structure which also avoids the redundant storage of geometry at different levels of detail (in the edges). Figure 4.4 shows the generalisation process for the tGAP structure in which a simplified version of the edges is available, without explicitly storing them.

4.3 Space Scale Cube for Vario-Scale

This section introduces a truly smooth vario-scale structure for geographic information: a small step in the scale dimension leads to a small change in representation of geographic features that are represented on the map. Continuous generalisations of real world features can be derived from the structure that can be used for presenting a smooth zoom action to the user. Furthermore, mixed-scale visualisations can be derived with more and less generalised features shown together in which the objects are consistent with each other. Making such a transition area is one of the principal difficulties for 3D computer graphic solutions (e.g. using stitch strips based on triangles, such as used by Noguera et al. (2011).

The remainder of this section is structured as follows: Sect. 4.3.1 contains a discussion on how the classic tGAP structure can be represented by a 3D space-scale cube and how this can be adapted to store a more continuous generalisation. The third dimension is used to encode the tGAP representation, resulting in a

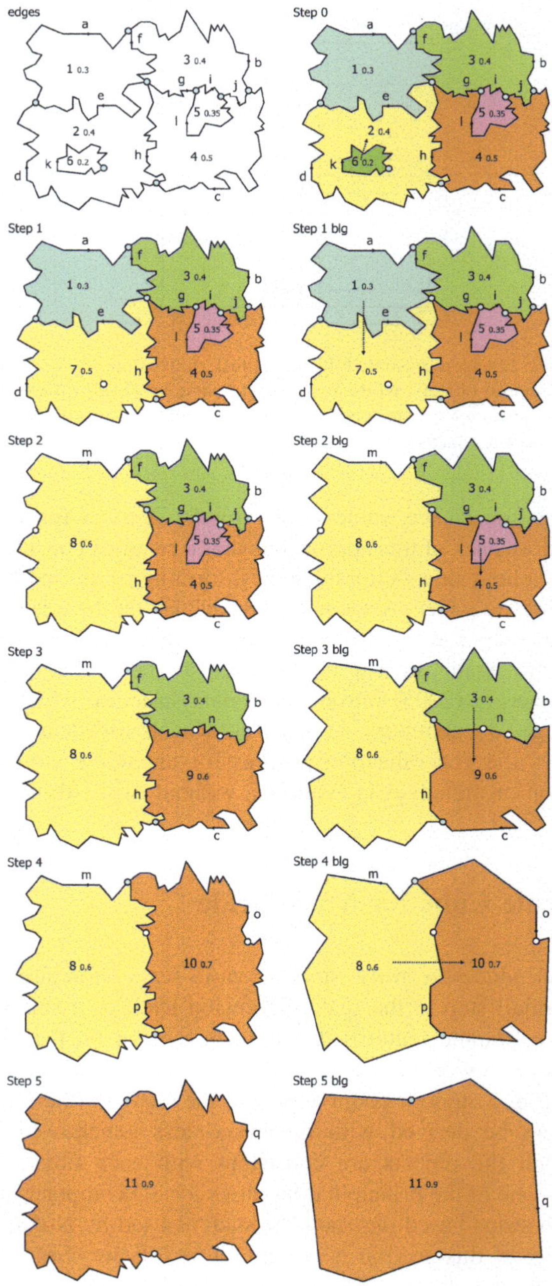

Fig. 4.4 Generalisation example in *five steps*, from detailed to course. The *left side* shows the effect of merging faces, while the *right side* shows the effect of also simplifying the boundaries via the BLG tree. Note that nodes are depicted in *green/blue* and removed nodes are shown for one next step only in *white*

volumetric partition of stacked prisms (polygons extruded according to their valid scale range). Maps are created by horizontal slices and continuous generalisation corresponds to gradually moving the slicing plan (and scaling). The classic encoding of the tGAP with stacked prisms, has horizontal faces causing sudden local shocks when moving up or down the slicing (zoom) plane. The truly smooth tGAP structure is obtained by removing the horizontal faces and replacing them with tilted faces. The support of additional generalisation operators in a SSC is described in Sect. 4.3.1. Section 4.3.2 illustrates how a mixed scale representation is obtained from the smooth tGAP structure by taking non-horizontal slices. The presented data structures and methods are valid for both 2D and 3D data.

4.3.1 The tGAP Structure Represented by the 3D Space-Scale Cube

The tGAP structure has been presented as a vario-scale structure (van Oosterom 2005). The tGAP structure can be seen as a result of the generalisation process and can be used efficiently to select a representation at any required level of detail (scale or height). Figure 4.5 shows four map fragments and the tGAP structure in which the following generalisation operations have been applied:

1. Collapse road object from area to line (split area of road and assign parts to the neighbours);
2. Remove forest area and merge free space into neighbouring farmland;
3. Simplify boundary between farmland and water area.

The tGAP structure is a DAG and not a tree structure, as the split causes the road object to have several parents (Fig. 4.5e). In our current implementation the simplify operation on the relevant boundaries is combined with the remove or collapse/split operators and is not a separate step. However, for the purpose of this chapter, it is clearer to illustrate these operators separately. For the tGAP structure, the scale has been depicted in the third dimension—the integrated space-scale cube (SSC) representation (Vermeij et al. 2003; Meijers and van Oosterom 2011). We termed this representation the space-scale cube in analogy with the space–time cube as first introduced by Hägerstrand (1970). Figure 4.6a shows this 3D representation for the example scene of Fig. 4.5. In the SSC the vario-scale 2D area objects are represented by 3D volumes (prisms), the vario-scale 1D line objects are represented by 2D vertical faces (for example the collapsed road), and the vario-scale 0D point object would be represented by a 1D vertical line. Note that in the case of the road area collapsed to a line, the vario-scale representation consists of a compound geometry with a 3D volume-part and 2D face-part attached.

There are many small steps from the most detailed to the most coarse representation—in the classic tGAP, $n - 1$ steps exist, if the base map contains n objects. Nevertheless, this could still be considered as many discrete generalisation actions approaching vario-scale, but not truly smooth vario-scale. Split and

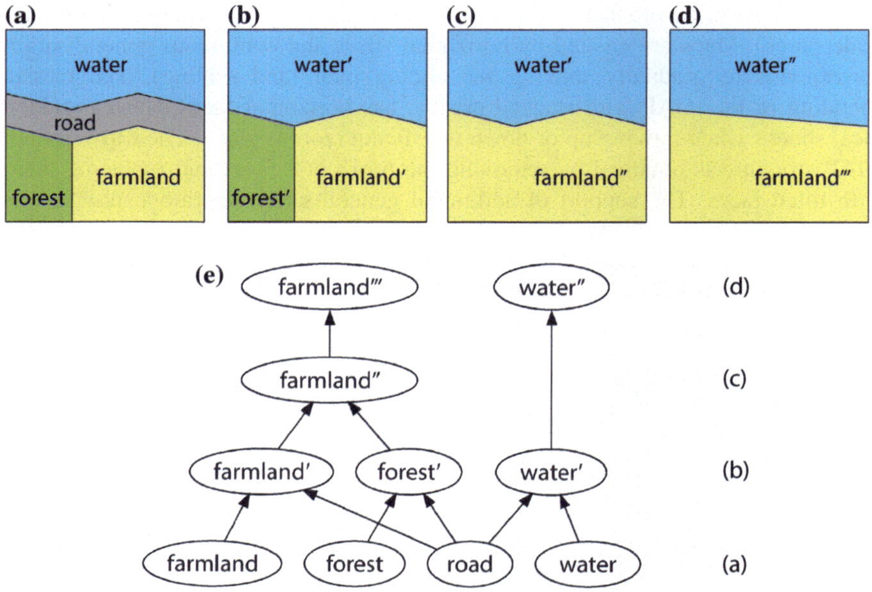

Fig. 4.5 The 4 map fragments and corresponding tGAP structure. **a** Original map. **b** Result of collapse. **c** Result of merge. **d** Result of simplify. **e** Corresponding tGAP structure

Fig. 4.6 The space-scale cube (*SSC*) representation in 3D: **a** SSC for the classic tGAP structure. **b** SSC for the smooth tGAP structure

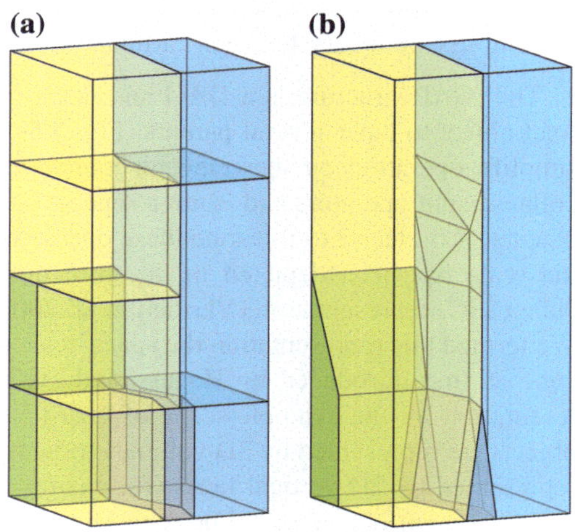

merge operations cause a sudden local ‘shock’: a small scale change results in a not so small geometry change where, for example, complete objects disappear; (Fig. 4.7a, b). In the space-scale cube this is represented by a horizontal face; a sudden end or start of a corresponding object. Furthermore, polygon boundaries

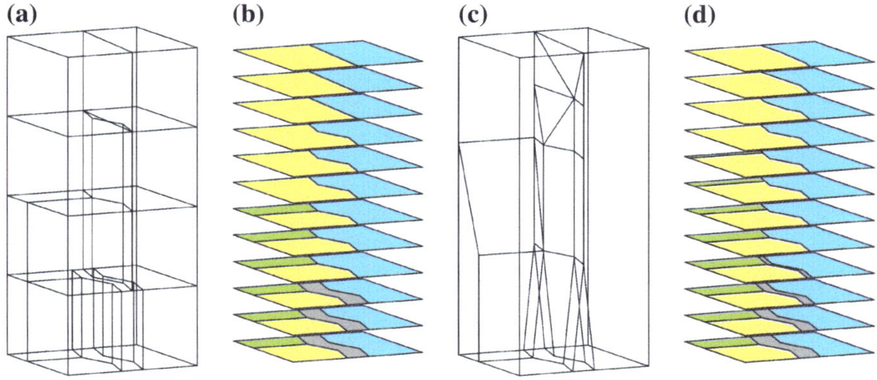

Fig. 4.7 Comparing the classic and smooth tGAP/SSC structures: **a** wireframe classic, **b** map slices classic, **c** wireframe smooth, and **d** map slices smooth. Note that nothing changes until a tGAP event has happened in classic structure (**a**–**b**)

define faces that are all vertical in the cube, i.e. the geometry does not change at all within the corresponding scale range, which result in prisms constituting a full partition of the space-scale cube. Replacing the horizontal and vertical faces with tilted faces as depicted in Fig. 4.6b, results in smooth transitions; (Fig. 4.7c, d).

4.3.2 Smooth Zoom and Progressive Transfer

In an online usage scenario where a 2D map is retrieved from the tGAP structures, the amount of vector information to be processed has an impact on the processing time for display on the client. Therefore, as a rule of thumb, we strive to show a fixed number of (area) objects on the map, independent from the level of detail the objects have, in such a way that optimal information density is achieved. This number is termed here the *optimal number* of map objects and will be used for retrieving data in such a way that the amount of data to be retrieved on average remains constant per viewport (independent from which level of detail is retrieved) and thereby the transfer and processing times stay within limits.

The optimal number can be realised because the generalisation procedures that create tGAP data lead to progressively less data in the hierarchy, i.e. less data are stored near the top of the SSC where the extent of area objects is considerably larger than at the bottom. A slice (cross section) in this cube leads to a 2D map. The extent of the viewport (i.e. the window through which the user is looking at the data) also implies that it is necessary to clip data out of such a slice. When a user is zoomed out, the viewport of a user will lead to a big geographic extent (the area to be clipped is large) and when a user is zoomed in, this extent will become considerably smaller. For a user that performs a panning action it is necessary to

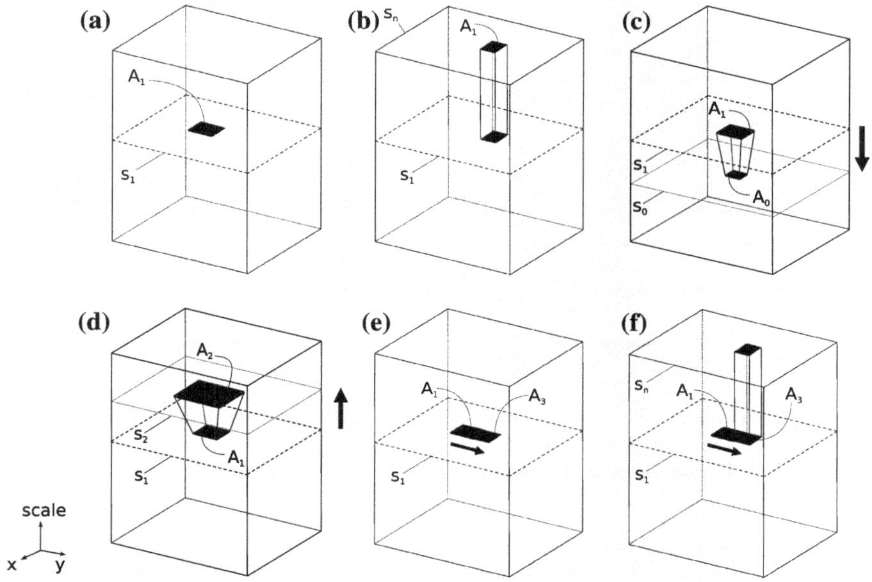

Fig. 4.8 Zooming and panning with vario-scale data explained via the SSC [after van Oosterom and Meijers (2011a)]: **a** Initial request (non-progressive). **b** Initial request (progressive). **c** Smooth zooming in (progressive). **d** Smooth zooming out (progressive). **e** Panning (non-progressive). **f** Panning (progressive)

move the extent of the clip within the horizontal slicing plane. Figure 4.8 illustrates this idea. A smooth tGAP based server can be arranged to respond to the following types of requests from a smooth tGAP-aware client (illustrated for 2D maps represented by a 3D space-scale cube):

1. A request to provide an initial map based on simple 2D spatial range overlap selection of the relevant 3D polyhedra representing the vario-scale 2D objects in the requested area $A1$ for the requested scale $s1$ as illustrated in Fig. 4.8a. Note that the number of selected objects may be relatively large, so it can take some time before a map covering a requested area $A1$ can be created by the client.
2. A request to provide an initial 2D map with progressive transfer of data (coarse to fine) is based on overlap with a 3D block orthogonal to the axes. The server sends the selected 3D polyhedra smallest scale first (Fig. 4.8b). The client can quickly start drawing an initial course representation, while still receiving additional detail.
3. A request to provide the 3D polyhedra for a progressive zoom-in as shown in Fig. 4.8c. Note the shrinking of the spatial selection range from an area $A1$ at scale $s1$ to an area $A0$ at scale $s0$ (a truncated pyramid up-side-down).

Alternatively it is possible to provide data for a simple zoom-in. In that case the client does not need to receive 'intermediate' 3D polyhedra (this alternative is not depicted in Fig. 4.8).

4. A request to provide the 3D polyhedra for a progressive zoom-out as shown in Fig. 4.8d. Note the growing of the spatial selection range from an area $A1$ at scale $s1$ to an area $A2$ at scale $s2$. In this case the 3D polyhedra are sorted based on largest scale value from the larger to the smaller scale without sending 'intermediate' 3D polyhedra (again, not depicted in Fig. 4.8).
5. A request to provide the 3D polyhedra for a simple pan operation from a first area $A1$ to an adjacent area $A3$ represented a the same scale $s1$ as shown in Fig. 4.8e. In that case the server immediately transmits the object data for the new map at the required level of detail.
6. A request to provide the 3D polyhedra for a progressive transfer pan as shown in Fig. 4.8f from a first area $A1$ to a second area $A3$. This case is comparable to the case 2 'initial 2D map with progressive transfer' for the new part of requested spatial range $A3$. In this case the server subsequently transmits more and more detailed data (gradually changing from scale sn to scale $s1$) for the requested spatial range $A3$, and the client can gradually increase the displayed level of detail.

Note that the client has to be smooth tGAP-aware, after receiving the polyhedra (in sorted order) it has to perform slicing before the actual display on the screen takes place. In case of 'smooth' zoom-operations, the slicing operations are repeated (at slightly different scale levels) before every display action. Note that an efficient implementation may exploit the fact that the slicing plane is moving monotonically in a certain direction (up or down) and may avoid repeating parts of the needed geometric computations. This is similar to plane-sweep algorithms as used in computational geometry.

Positioning the slice in the cube, together with taking the clip, should lead to an approximately constant number of objects to be visualised. To realise the position of the cross section means that the question to be answered is 'which height corresponds to the map scale on the client side?' In practice this will mean that a thin client will only have to report the current extent (plus its device characteristics) to a server and then it can be sure to receive the right amount of data for a specific level of detail as the server can translate this extent to a suitable height value to query the data structures. Note that this on average is the right amount of information, as there may be regions with more or with less dense content; e.g. rural versus urban areas. For a thick client, that has more capabilities and where it is possible to perform more advanced processing, it should be possible to first receive a coarse map, and then to incrementally receive additional details while a user waits longer, leading to a more detailed map (with additional map objects). The mapping of map scale to height in the SSC in this scenario is necessary to determine when to stop sending additional details. Independent from whether a thin or thick client is used, it is key to know what the height value is that is implied by the current map scale of the client (i.e. positioning the height of the slice).

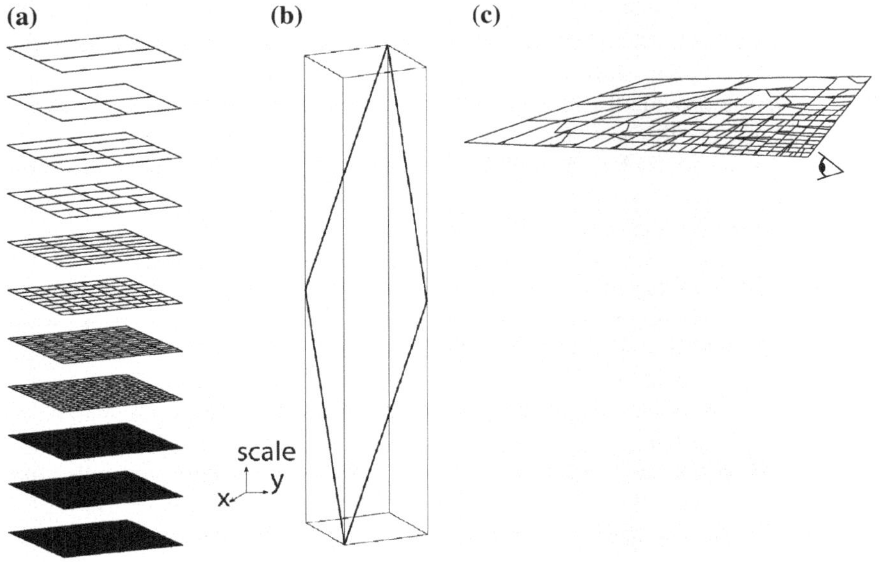

Fig. 4.9 Checker board data as input: each *rectangular* feature is smoothly merged to a neighbour. Subfigures show: **a** a stack of horizontal slices, **b** taking a non-horizontal slice leads to a 'mixed-scale' map and **c** one mixed scale slice (*non-horizontal plane*)

The filled tGAP data structure will be able to supply data for a specific scale range. This range is dependent on the optimal number of objects we want to show together with device characteristics.

4.3.3 Mixed Scale Representations

So far, only horizontal slices were discussed for creating 2D maps with homogenous scale, however nothing prevents us from taking non-horizontal slices. Figure 4.9 illustrates a mixed-scale map derived as a non-horizontal slice from the SSC. What does such a non-horizontal slice mean? More detail at the side where the slice is close to the bottom of the cube, less detail where the slice is closer to the top. Consider 3D visualisations, where close to the eye of the observer lots of detail is needed, while further away not so much detail. Such a slice leads to a mixed-scale map, as the map contains more generalised (small scale) features far away and less generalised (large scale) features close to the observer.

The mixed-scale representation can also be obtained by slicing surfaces that are non-planar; e.g. a bell-shaped surface that could be used to create a meaningful 'fish-eye' type of visualisations (Fig. 4.10). This is likely to be the desired

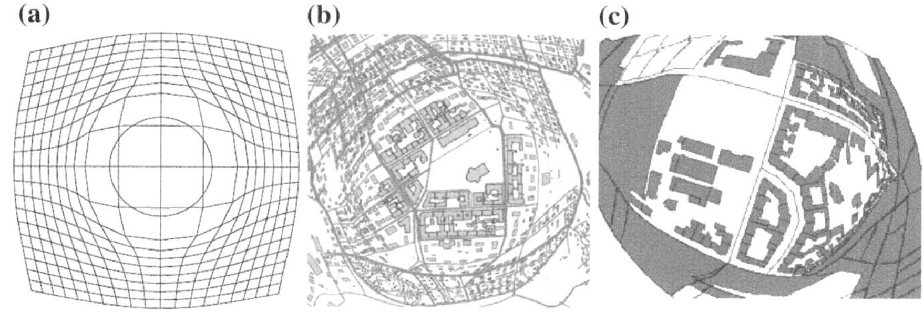

Fig. 4.10 A 'mixed-scale' map. Harrie et al. (2002) term this type of map a 'vario-scale' map, while we term this a 'mixed-scale' map. Furthermore it is clear that there is a need for extra generalisation closer to the borders of the map, which is not applied in (**b**), but is applied in (**c**). With our solution, this generalisation would be automatically applied by taking the corresponding slice (*bell-shaped, curved surface*) from the SSC. Illustrations (**a**) and (**b**) taken from Harrie et al. (2002) and (**c**) from Hampe et al. (2004)

outcome in most cases, but it might be true that a single area object in the original data set might result in multiple parts in the slice (but no overlaps or gaps will occur in the slice). What are other useful slicing surface shapes? A folded surface seems to be nonsensical as it could lead to two representations of the same object in one map/visualisation.

The artificial checker board 'map' is a kind of worst case example as the map reader would expect to see only rectangular shapes in the output. The combination of smooth transitions (tilted faces) which are intersected by the mixed scale diagonal slicing plane results in some non-orthogonal shapes. By contrast the (non-smooth) classic tGAP encoded in the SSC in Fig. 4.6a for generating the mixed scale output will only generate rectangular shapes (and in that sense looks more natural). However, this classic tGAP does not support very well, truly smooth zooming. More usability tests will be needed to evaluate the dynamic map types the users prefer with more realistic map data.

4.4 Case Study I: Dutch Large Scale Basic Topographic Data in Constraint tGAP[1]

The tGAP (topological Generalised Area Partitioning) structure is designed to store the results of the gradual generalisation process and to support smooth zoom as described in Sect. 4.2. In this section the cartographic quality of the tGAP is improved by using a known smaller scale high quality representation 'as target' when taking the

[1] Case Study I is a selection, by the chapter authors, from the publication (Dilo, van Oosterom, Hofman 2009).

generalisation decisions on the large scale input data when constructing the tGAP structure. With the additional smaller scale input data (the constraints), the iterative algorithm can be controlled to obtain higher quality (intermediate) maps.

The generalised dataset (smaller scale input) can be obtained via two routes: (1) from an external, independent source or (2) via a different generalisation algorithm. The first option is explored in this section where the idea was implemented with real topographic datasets from the Netherlands for the large- (1:1,000) and independent medium-scale (1:10,000) data. The second option is investigated in Sect. 4.5, where the small scale input is derived from the same base data (via an optimisation approach using mixed integer programming).

4.4.1 Topographic Datasets of Large and Medium Scale

Generalisation for the tGAP data structure is performed on an area partition; thus the large-scale dataset is required to be an area partition. Area objects of the smaller scale data set act as region constraints in the generalisation process, i.e. they restrict the aggregation of large-scale objects only inside the region constraints. The method proposed here consists of two stages: the first stage matches objects of the large-scale dataset to objects of the smaller scale dataset, which act as region constraints in the next stage; the second stage compiles additional information needed for the constrained tGAP and performs the generalisation.

In this study we use large- and medium-scale topographic data from the Netherlands, at 1:1,000 and 1:10,000 scales, respectively. Large-scale topographic data in the Netherlands are mostly produced by municipalities, while smaller scale maps are produced by the Netherlands' Kadaster (who also holds the mapping agency). IMGeo is the information model for large-scale data at the scale 1:1,000. Top10NL is the model for topographic data at the scale 1:10,000. The datasets in this study follow the two models, IMGeo and Top10NL. All the illustrations in this section were prepared with data from the region of Almere in the Netherlands.

The two models, IMGeo and Top10NL, both endorse this meaning of topography (relief is not included as we focus on the area partitioning). The main feature classes of IMGeo are Road, Railway, Water, Building, LayoutElement and Terrain. Class LayoutElement has 11 subclasses, each representing a different kind of topographic element, e.g. Bin, StreetFurniture and OtherBuilding. The class Terrain contains anything that is not buildings or infrastructure, for example, forest, grassland, fallow land, etc. Although the (area) objects of IMGeo should form an area partition, this was not the case for the test data. We processed the test data and created a partition. The overlaps were resolved by imposing an importance order on classes: layout elements have the highest importance, followed by roads and water, while terrain has the lowest importance. This order is translated to rules. For example, 'if a Terrain object overlaps with another object, subtract the overlap from the Terrain object'; see Hofman (2008) for a complete set of rules. The IMGeo test data allows categorisation mainly on classes. The class Terrain

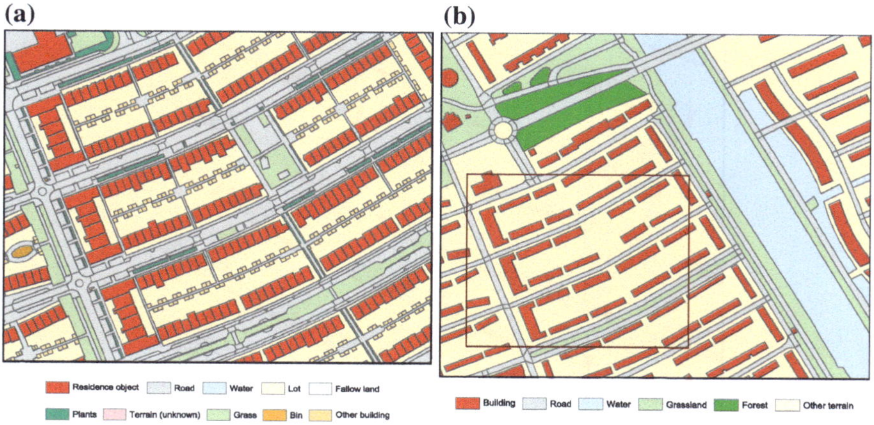

Fig. 4.11 **a** Large-scale topographic data from the city of Almere, the Netherlands. **b** Medium-scale topographic data from the city of Almere, the Netherlands. The *rectangle* in the *middle of the image* shows the extent of the map of (**a**)

can be further categorised by the LanduseType attribute values, and the class LayoutElement allows categorisation based on its subclasses. We created a new attribute class to store these categories: building; road; water; lot, fallow land, plants, other terrain, and grass, for the land-use values of Terrain; bin, and other buildings as subclasses of the LayoutElement class. Figure 4.11 shows the IMGeo data for part of the city of Almere, the Netherlands.

The Top10NL model contains all the IMGeo classes, which have similar attributes to their corresponding classes in IMGeo. We would expect the topographic model of a smaller scale to be less detailed than the model of a larger scale. This is not always the case for Top10NL compared to IMGeo, e.g., Top10NL differentiates between different kinds of forests, while IMGeo classifies them all as forest. The Top10NL objects were categorised based on the main classes. Additionally, objects of class Terrain were further categorised according to the LanduseType attribute values. A new attribute, region-class, was created for the Top10NL data, with the following values: building; road; water; grassland, forest, and other terrain from the land-use values of Terrain. Terrain objects had overlap with objects of other classes. There were also overlaps between road and water objects. We also processed the Top10NL data using the importance order of classes, and created a partition. Figure 4.11b shows Top10NL data from the city of Almere, covering a larger area than Fig. 4.11a. The rectangle in the middle of the map shows the extent of the map in Fig. 4.11a.

Some of the categories have the same semantics in both models, e.g. Road and Water. Some categories have similar semantics, e.g. Building in IMGeo is a residence unit, while in Top10NL it is a building block, Grass in IMGeo is similar to Grassland in Top10NL. There are categories of IMGeo that do not have a one-to-one match to the categories of Top10NL. For example, Other building from

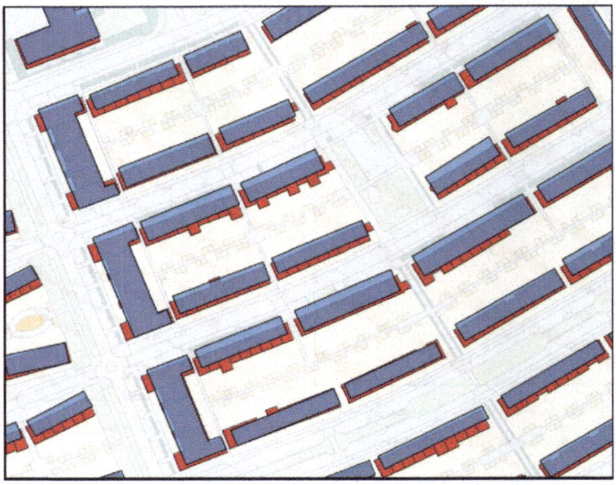

Fig. 4.12 Large-scale IMGeo building objects (in *red*) overlaid with medium-scale Top10NL buildings blocks (in *blue*, semi-transparent). The background is the large-scale dataset in *dimmed colours*. It should be noted that for other feature classes the matching is often also very challenging

IMGeo can be a Building or Other terrain in Top10NL, while Other terrain of Top10NL can be Lot or Fallow land in IMGeo. Semantic matching is considered in the object matching described in the next section.

4.4.2 Data Pre-Processing and Object Matching

The constrained tGAP has some assumptions, which put requirements on the data we want to generalise. The first assumption is that data has to form an area partition, which was not the case for IMGeo data of Almere and Rotterdam as overlapping areas did occur, which had to be corrected. The second assumption is that we know in advance to which region (constraint) each object of the original data belongs. We want to use Top10NL objects as region constraints for the IMGeo data, while the two are created independently. Figure 4.12 shows overlay of Top10NL and IMGeo data, illustrating the matching problem. We need to assign a Top10NL object to every IMGeo object before we can run the constrained tGAP code. After pre-processing, IMGeo and Top10NL, each have become a partition. To assign IMGeo objects to region constraints, i.e. Top10NL objects, several methods were investigated and implemented. Four possible methods to assign IMGeo objects to Top10NL regions were developed:

1. The *simple overlay* method: An intersection between the models where every IMGeo object is split at the borders of the overlapping Top10NL region. In the end result only Top10NL geometry will be visible.

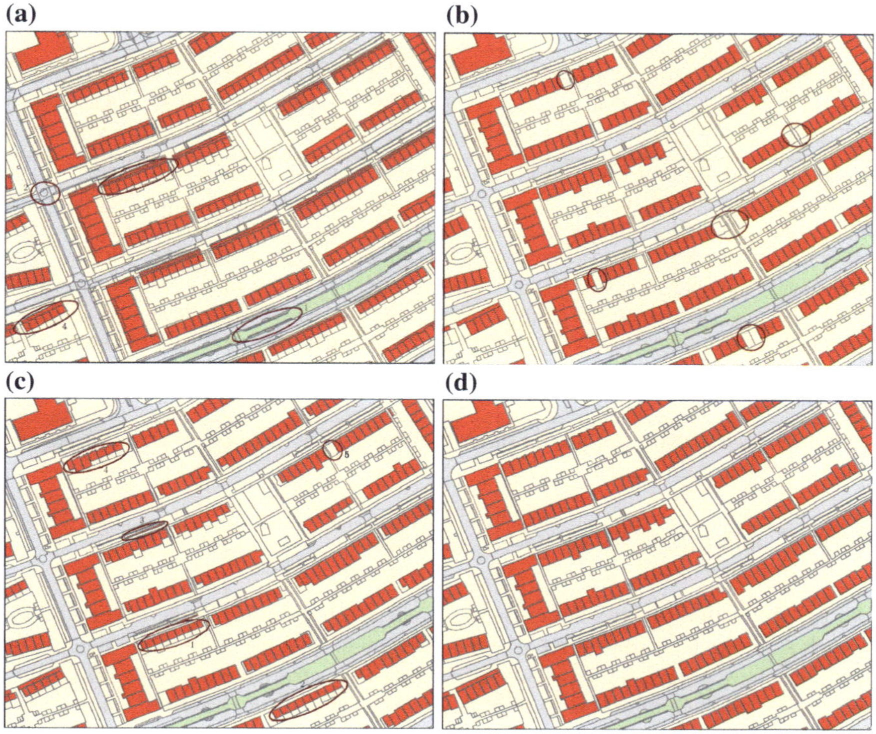

Fig. 4.13 The results of the four object matching methods visualised: **a** Simple overlay (note slivers). **b** Maximum Area method (note missing building in some blocks). **c** Split method (note that some blocks have been deformed by slit 'don't understand what you mean by slit'). **d** Building First method (least negative side effects). The IMGeo objects are coloured according to the region they belong to, which gives an indication of the constrained tGAP end result

2. The *maximum area* method: The Top10NL region which overlaps the IMGeo object the most is the shape to which the whole IMGeo object is assigned to. The IMGeo geometry is kept in this method.
3. The *35 %-split* method: If an IMGeo object belongs for more than 35 % to two Top10NL regions we consider this Top10NL geometry as enrichment of the structure; therefore the IMGeo object is split and new IMGeo objects are created. For all other IMGeo objects the maximum area method is applied.
4. The *building first* method: This method assigns IMGeo-buildings to a building region in the case where there is some overlap with a Top10NL building block, without considering the amount of overlap. The other IMGeo objects are processed as in the maximum area method.

As the last method proved to give the best matching results (based on our visual inspection of applying the various methods to our test data (Fig. 4.13), we decided to continue with matching results using this method.

Table 4.1 Class weights determined for the vario-scale IMGeo and compatibility values for the vario-scale IMGeo

From-class code →		1001	2001	3001	4001	4002	4003	4004	4005	5001
Weight		**13.0**	**1.20**	**1.30**	**9.00**	**1.00**	**0.93**	**0.90**	**0.88**	**1.00**
Class name	To-c↓									
Building	1001	1.00	0.50	0.00	0.90	0.90	0.50	0.50	0.00	0.99
Road	2001	0.00	0.99	0.00	0.00	0.00	0.00	0.00	0.00	0.00
Water	3001	0.00	0.00	1.00	0.00	0.00	0.00	0.00	0.00	0.00
Lot	4001	0.50	0.90	0.00	1.00	0.95	0.95	0.95	0.80	0.95
Fallow land	4002	0.90	0.90	0.00	0.50	1.00	0.90	0.90	0.50	0.95
Plants	4003	0.50	0.50	0.00	0.50	0.50	0.99	0.95	0.90	0.50
Terrain unk.	4004	0.90	0.50	0.00	0.50	0.90	0.00	1.00	0.99	0.95
Grass	4005	0.50	0.90	0.00	0.80	0.50	0.95	0.95	0.99	0.50
Other Building	5001	0.99	0.50	0.00	0.50	0.90	0.50	0.50	0.00	1.00

Motivation was readability as Table 4.1 shows both class weight (in bold) and class compatibility matrix

4.4.3 Constrained tGAP Generalisation

The generalisation process is performed in steps very similar to the normal tGAP. In each step, the least important object is merged to its most compatible neighbour, forming a new object. This pair wise merging is controlled by region constraints: *objects are allowed to merge only if they belong to the same smaller scale region.* Inside the constraints, the generalisation results are driven by the importance and compatibility values of objects. The importance value of an object *v* is calculated from the area size, and the weight of the object's category: *Imp*(*v*) = *area_size*(*v*) * *weight*(*class*(*v*)). The compatibility between two objects *u* and *v* is calculated from the length of the shared boundary, and the compatibility values between their categories: *Comp*(*u*,*v*) = *bnd_length*(*u*,*v*) * *compatib*(*class*(*u*), *class*(*v*)). The generalisation stops when all the objects are merged up to the region constraints. The initial values for weights and compatibilities were based on the work of van Putten and van Oosterom (1998). Next, these values were tuned in order to get better generalisation results. The tuning of values was by trial and error, based on visual inspection of the results, with the aim of a gradual transformation of the large-scale dataset toward the medium-scale dataset. The optimum values found for the weights (line in bold) and the compatibilities are given in Table 4.1. Buildings are important and should be retained through all the scales of the maps. We ensure this by granting a high weight value to the category Building. The IMGeo dataset of Almere considers parking places and sidewalks as roads, while the Top10NL dataset does not. This results in more road objects in the IMGeo data as compared to the Top10NL data. To achieve a gradual decrease of roads during the generalisation, we attach a relatively small weight to roads, as well as a very low compatibility of other categories to roads (for row 2001, most of the values are 0, Table 4.1) while allowing relatively high compatibilities of roads

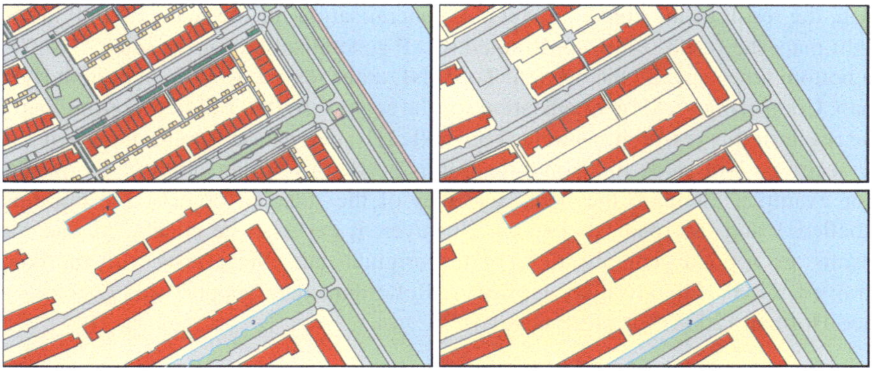

Fig. 4.14 Results constrained tGAP: (*top left*) original IMGeo 1:1,000, (*top right*) result scale 1:5,000, (*bottom left*) result scale 1:10,000, (*bottom right*) Top10NL 1:10,000

to other categories (see column 2001). We consider water to be compatible only with itself, and incompatible with other categories: compatibility of water to any other category equals 0, and also compatibility of any other category to water equals 0. The IMGeo categories, lot, fallow land, terrain (unknown) and other building have a similar semantics with the category terrain of Top10NL, thus we want the objects of these categories to aggregate together. We make Lot an attractive category by granting it a high weight value and high compatibilities of the other categories to lots, while allowing it to merge to the other categories, which is ensured by the medium–high compatibility values of lots to the other categories.

The tGAP structures are stored as Oracle tables, and the constrained tGAP generalisation is performed by code written in PL/SQL. Oracle tables store data for the constrained tGAP, their relations, as well as primary and foreign keys. Region constraints subdivide the whole domain into parts that can be processed independently and therefore in parallel. In addition, due to this the tGAP operations are local (i.e. look for least important feature within region constraint in contrast to looking for globally least important feature). This is relevant when considering updates: the original tGAP iteratively looks for the globally least important feature, and a small change in the largest may cause quite a different resulting structure. This is not the case for the constrained tGAP, where the effect of a change is limited to the area of the involved region constraint.

4.4.4 Constrained tGAP Generalisation: Conclusions

The maps of Fig. 4.14 illustrate results of the generalisation using weight and compatibility values as discussed in the previous section. The maps show the same part of Almere, for comparison: the original IMGeo data (scale 1:1,000) in top left map, the result of the constrained tGAP generalisation for the scale 1:5,000 in top right map, the result of the constrained tGAP generalisation for the scale 1:10,000 in bottom left map, and the original Top10NL data (scale 1:10,000) in bottom right map. Comparing tGAP generalisation for the scale 1:10,000 (middle-bottom map), the result of this research, with the Top10NL 1:10,000 dataset (bottom map), it can be seen that still some line simplification is needed to further improve the results (for example, compare the boundary lines of the selected objects: the building labelled 1 and the road labelled 2). However, it can also be concluded that the results are a good generalisations of the original data. More tests with the constrained tGAP generalisation applied to Rotterdam data support this conclusion [see Hofman (2008) for more details].

4.5 Case Study II: German Land Cover Data in Constraint tGAP[2]

Instead of using large scale topographic data the second case is using land cover data from ATKIS DLM50, which starts at the medium scale (1:50,000). There is no smaller scale available as a constraint. However, via an optimisation procedure a 'one-step/one-scale' smaller scale representation (1:250,000) is derived. This derived representation is then used as constraint for the vario-scale structure. The advantage of this derived smaller scale over an independent smaller scale map, is that there are rarely any mismatches between the larger scale objects.

In order to create a better tGAP-generalised dataset (without having a second medium or smaller scale dataset available as a constraint), we propose a method of two stages: First, we create a generalised representation from a detailed dataset, using an optimisation approach that satisfies certain cartographic rules (e.g. minimum size for objects that are shown on the map). Secondly, we define a sequence of basic tGAP merge and simplification operations that transforms the most detailed dataset gradually into the generalised dataset. The obtained sequence of gradual transformations is stored without geometrical redundancy in a structure that builds up on the previously developed tGAP (topological Generalised Area Partitioning) structure. This structure and the algorithm for intermediate levels of detail (LoD) have been implemented in an object-relational database and tested for land cover data from the official German topographic dataset ATKIS at a scale of 1:50,000 to the target scale 1:250,000. Results of these tests allow us to conclude that the data at the lowest LoD and at intermediate LoDs is well generalised. Applying specialised heuristics the applied optimisation method can cope with large datasets; the tGAP structure allows users to efficiently query and retrieve a dataset at a specified LoD.

[2] Case Study II is a selection, by the chapter authors, from the publication (Haunert, Dilo, van Oosterom 2009).

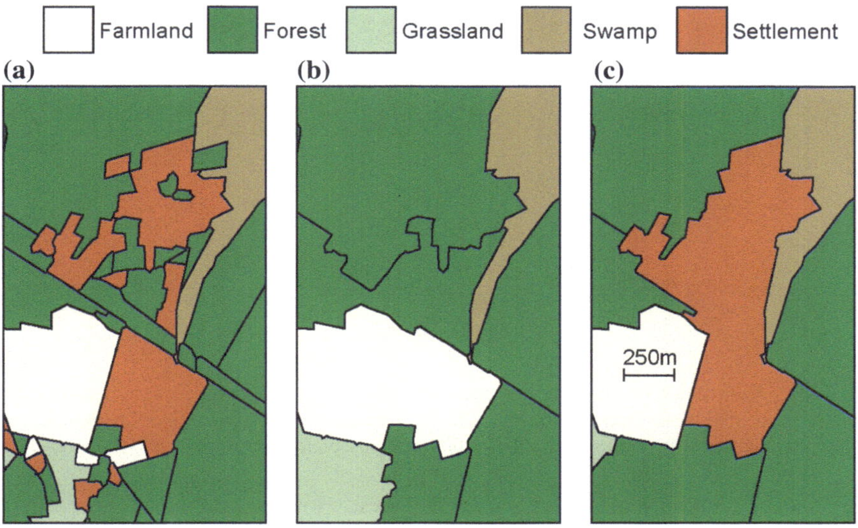

Fig. 4.15 Comparison of two aggregation methods (ATKIS): **a** input dataset DLM50, **b** iterative merging towards DLM250, and **c** optimisation towards DLM250

4.5.1 Create Constraint Dataset Via Optimisation

Figure 4.15 illustrates the generalisation challenge. The sample in Fig. 4.15a was taken from the German topographic database ATKIS DLM50, which contains the same amount of details as a topographic map at the 1:50,000 scale. In order to generalise these data, we need to satisfy size conditions defined for the lower level of detail (LoD) that we aim to achieve. In many countries, such conditions are defined in specifications of topographic databases, for example, in Germany (Arbeitsgemeinschaft der Vermessungsverwaltungen der Länder der Bundesrepublik Deutschland: ATKIS-Objektartenkatalog: http://www.atkis.de).

Obviously, size conditions can be satisfied by aggregation, that is, by combining the input areas into fewer composite regions. This can be done, for example, by iteratively selecting the least important area in the dataset and merging it with its most compatible neighbour until all areas have sufficient size. Different measures of compatibility and importance have been proposed. Figure 4.15b shows the result of the algorithm when applying size conditions defined for the ATKIS DLM250, which corresponds to scale 1:250,000. Though all areas in the output have sufficient size, the example reveals a shortcoming of the iterative algorithm: As the algorithm only takes the compatibility between adjacent areas into account, large parts of the dataset change their class. In particular, the group of non-adjacent settlement areas is lost. This is because each single settlement area is too small for the target scale and becomes merged with a

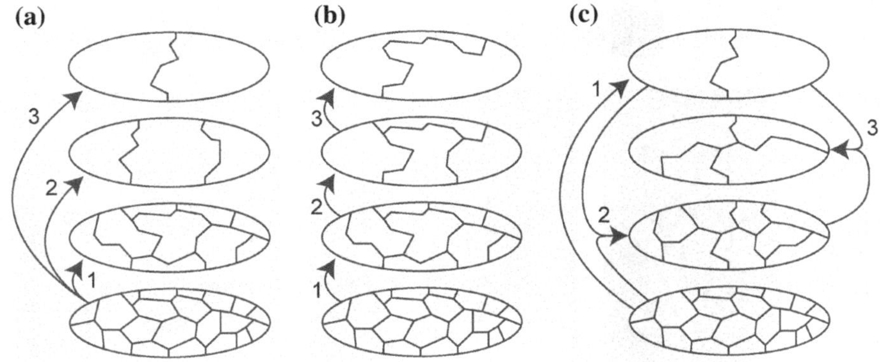

Fig. 4.16 Approaches to create a sequence of LoDs: **a** generalisation from a single source dataset (repeated optimisation), **b** successive generalisation (classic tGAP approach), and **c** intermediate representations fitting in the target representation (constraint tGAP approach)

neighbour of another class, for example, forest. However, together the group of settlements would reach the required size. In order to preserve such groups, we developed an optimisation method for aggregation that minimises the change of land cover classes while ensuring contiguous regions of sufficient size (Haunert and Wolff 2006). The result is shown in Fig. 4.15c: The small settlements are grouped into one large composite region. Generally, constraint- and optimisation-based approaches to map generalisation are considered highly flexible and capable of providing high-quality results.

4.5.2 Use Automatically Generalised (Optimised) Dataset to Create Constraint tGAP

Our basic assumption is that we are given an algorithm for the classical map generalisation problem, that is, for a given input dataset we can produce a dataset at any reduced LoD by appropriately setting the parameters of the algorithm. We can apply our optimisation approach for this task or any other generalisation methods that are available. Figure 4.16 illustrates three different ideas to generate a sequence of LoDs by applying such an algorithm.

In Fig. 4.16a the most detailed dataset is used as input for the algorithm to generate all levels of the sequence. Though each single dataset is well generalised, the obtained sequence of datasets does not conform to the idea of gradual refinement. For example, a line boundary that appears at a smaller LoD may not be present at a higher LoD. An alternative approach is to generate the sequence of LoDs in small steps—each step using the previously generated dataset as input for the generalisation algorithm (Fig. 4.16b). The iterative method that we previously used to set up the tGAP structure follows this approach. Though it leads to a sequence of relatively small changes between two consecutive LoDs, it entails the

Fig. 4.17 Two examples of optimised constraint tGAP. From *left* to *right* most detailed dataset, a series of intermediate representations, and finally arriving at a dataset generalised by optimisation

risk of getting unsatisfactory results at low LoDs (end result). In particular, this iterative approach does not allow us to optimise global quality measures, for example, to minimise changes of land cover classes between the highest LoD and the lowest LoD. Figure 4.16c shows a third approach, which we propose to overcome the disadvantages of both the other methods: We first create the dataset at the lowest LoD (smallest scale) and then define a sequence of intermediate representations (Fig. 4.16c). Using our optimisation method for the first stage, we can ensure a well-generalised dataset at the lowest LoD. In order to define the intermediate LoDs, we apply a modified version of the iterative algorithm that we earlier applied to set up the tGAP structure. Some results for two examples are visible in Fig. 4.17.

4.6 Case Study III: Corine and ATKIS Data in the Space Scale Cube

Section 4.3 introduced the third dimension to represent the scale of the tGAP structure in the so called Space-Scale Cube (SSC). While Sect. 4.3 only used artificial data sets to illustrate the 3D principle, Sects. 4.4 and 4.5 used real map data to create a constraint tGAP structure. In this section, we will first show an artificial example, followed by some examples with real data from respectively the Corine data set (1:100,000) and the ATKIS DLM data set (1:50,000), illustrating the concept of the SSC. Finally, the results of the 3D SSC will be shown, where gradual-change line simplification is applied. For all data sets first a tGAP structure is created which is then converted into a SSC.

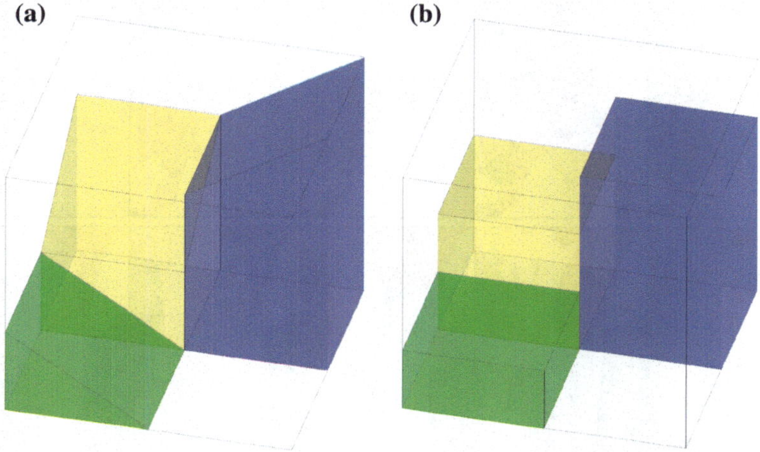

Fig. 4.18 Two tGAP to SSC construction alternatives: **a** prism based approach (transparent/ *white* object first taking over space from *green* neighbour, next from *yellow* neighbour and finally from *purple* neighbour and the whole area has become *white*), **b** the gradual transition approach

4.6.1 Artificial Data in SSC

The artificial data example in Fig. 4.18 illustrates the difference of a sudden-change ('discrete') merge operation versus having a smooth ('gradual') merge operation available. The prism based SSC is the result of sudden-change merge operation, resulting in horizontal faces closing the space-scale polyhedrons in the SSC. This leads to sudden changes in the derived 2D maps when the horizontal slicing place passes the horizontal top face of such a prism. Hence, the gradual change for a merge operation was proposed by removing the horizontal top faces of the prism and replacing them by tilted faces (Sect. 4.3.1).

When gradually moving the horizontal slicing plane, this gives a more continuous transformation of the map (a small change in scale leads to a small change of the map). By recursively repeating the same pattern (of Fig. 4.18), a larger artificial data set is created in the form of a checkerboard (Fig. 4.19).

Creating a perspective view by applying a tilted slicing plane we can see how it gradually changes from higher to lower detail in this mixed scale representation. The difference between the perspective views generated for the prism based SSC (sudden-change merge) and the gradual transition SSC is illustrated in Fig. 4.20. Note that the prism based (sudden-change) SSC might look better in this case, because the data are very artificial and the diagonal slicing plane is axis aligned and the rectangular pattern is better maintained. With real world data we expect that the smooth (gradual transition) SSC actually gives better results.

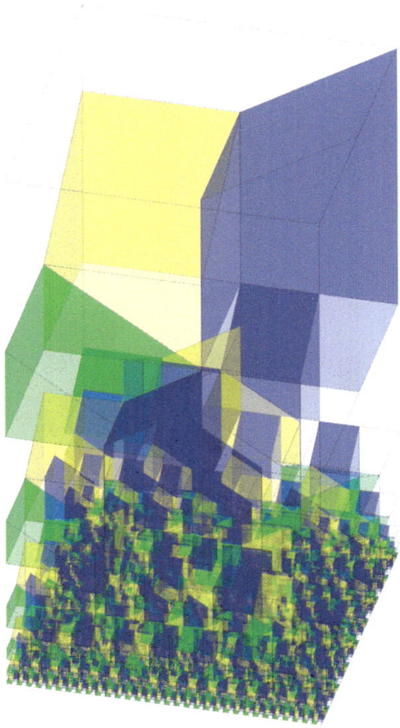

Fig. 4.19 The tGAP to gradual SSC construction for a larger artificial data set (recursive repetition of the same type of transitions)

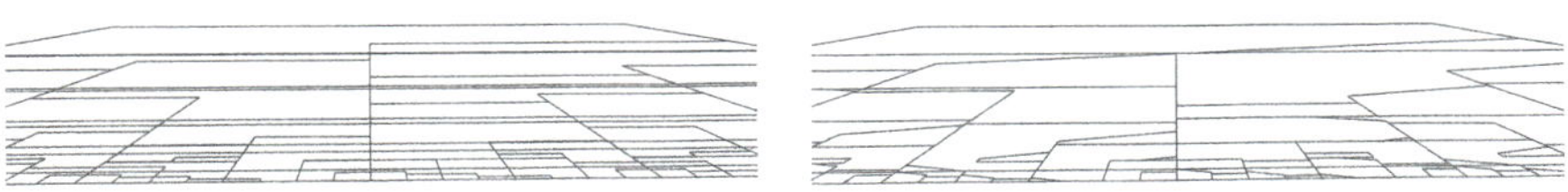

Fig. 4.20 Slicing the two tGAP to SSC construction alternatives (with slicing plane aligned with the *cube*). Note that slicing the more simple prism SSC looks better (no 'diagonals') compared to the more advanced gradual SSC. This is due to the artificial nature of starting with a regular grid

4.6.2 Corine Data in SSC

First, a tGAP was constructed based on polygons from the Corine dataset 2006. The Corine Land Cover (CLC) inventory is a Pan-European land use and land cover map. This dataset is intended to be used at a scale of 1:100,000. Once data are obtained for the tGAP structure, this data can be converted to an explicit 3D structure. Figure 4.21 shows the result for a small subset of the Corine data. For all edges in the tGAP structure vertical faces are constructed. This leads to a SSC that is filled with prism-shaped polyhedrons. From the resulting SSC slices we can see

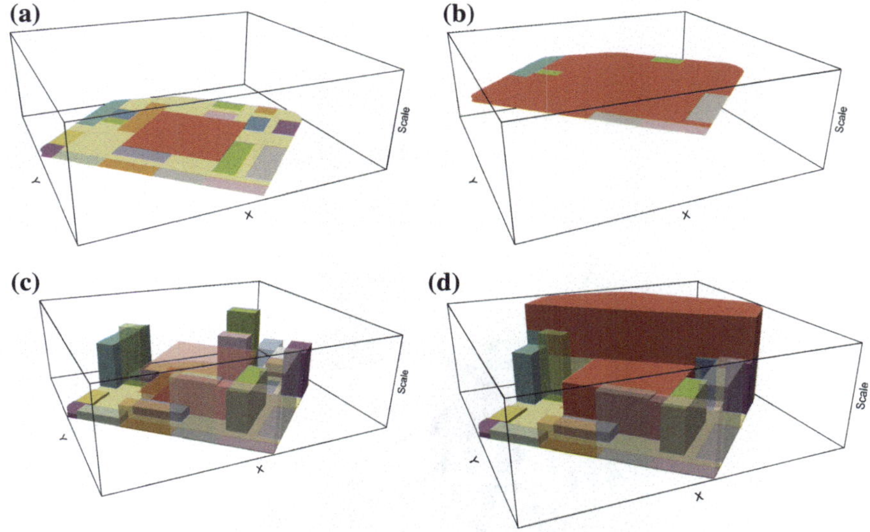

Fig. 4.21 Small Corine subset (30 objects, start scale 1:100,000), based on the SSC representation populated with the prism directly derived from the tGAP: **a** (*top left*) detailed slice low, **b** (*top right*) overview slice high, **c** (*bottom left*) objects in 3D SSC—more important objects are higher and **d** (*bottom right*) objects and tilted plane slice in 3D SCC—the red object are deemed to be the most important and are therefore retained until the end

more detail is available near the bottom (original data) and less (more generalised) data is available near the top of the cube.

4.6.3 ATKIS Data in SSC

Figure 4.22 shows the use of generalised ATKIS data in the SSC. Again the result is a prism based SSC, which was also constructed by extruding all the tGAP edges into vertical faces. Furthermore, Fig. 4.22b illustrates taking a non-horizontal slice, that leads to a map of mixed scale. Such a map is intended to be used in perspective view; Fig. 4.20 gives an impression of how this looks, using artificial data.

4.6.4 Smooth Line Simplification in the SSC

Up till now, the prism-based SSC data was presented (based on only merging neighbours and no line simplification). This section illustrates that the inclusion of line simplification in the gradual-change map generalisation leads to non-vertical faces in the SSC. Figure 4.23 shows two variants of a single space-scale

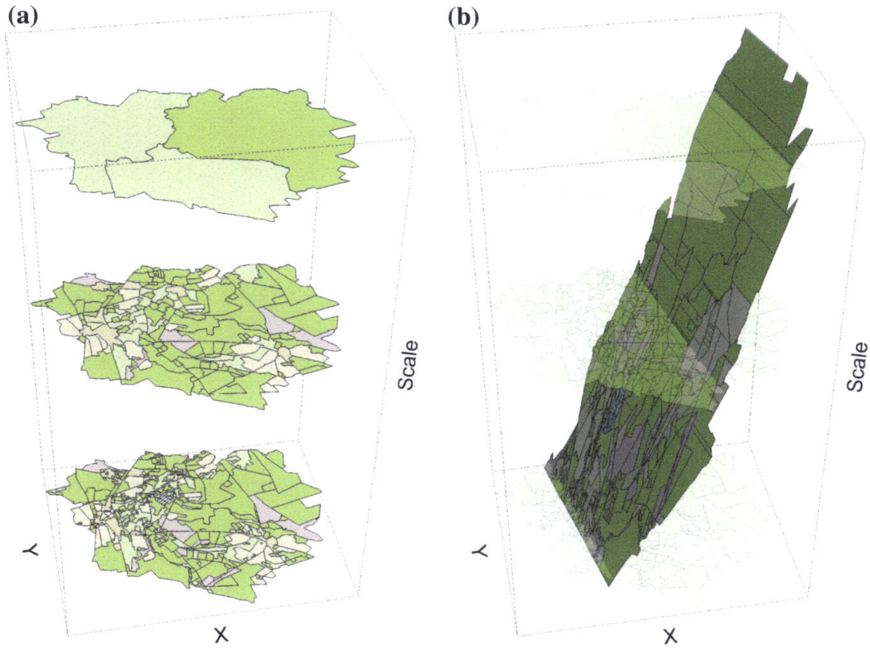

Fig. 4.22 Medium ATKIS subset (1,049 objects, start scale 1:50,000), based on the SSC representation populated with the prism directly derived from the tGAP: **a** detailed slices low, medium and overview slice high, **b** titled plane slice

polyhedron (selected from the complete SSC). The first polyhedron does not show any simplification to its boundaries (Fig. 4.23a). The boundaries are the same for the whole range of the map scales where this polyhedron is used (Fig. 4.23b–d). The second polyhedron shows the space-scale representation for the same object, when the boundary of the object is gradually simplified (Fig. 4.23d). The line simplification was performed on the edges of the tGAP structure, using the approach of Meijers (2011), in which a set of polylines is simplified simultaneously, but in which any changes in the topology are not allowed.

Construction of the first polyhedron, takes as input the polygon and outputs vertical polygons for every segment of the boundary (similar to how the illustrations of real world data in Sects. 4.6.2 and 4.6.3 were made). In the second case, the result of the line simplification algorithm is stored in a BLG tree in the edge table of the tGAP structure. This binary tree structure captures all the knowledge to be able to construct the 3D surface boundaries for every edge in the tGAP structure, making the space-scale cube an explicit 3D representation. If a slice is taken and gradually moved through this 3D model, the result is a smooth changing 2D representation of the map object. This idea is illustrated in Fig. 4.23f–h.

Whether true 3D data storage is to be applied for vario-scale data, depends hugely on the purpose. In an editing environment, a real 3D implementation of 2D

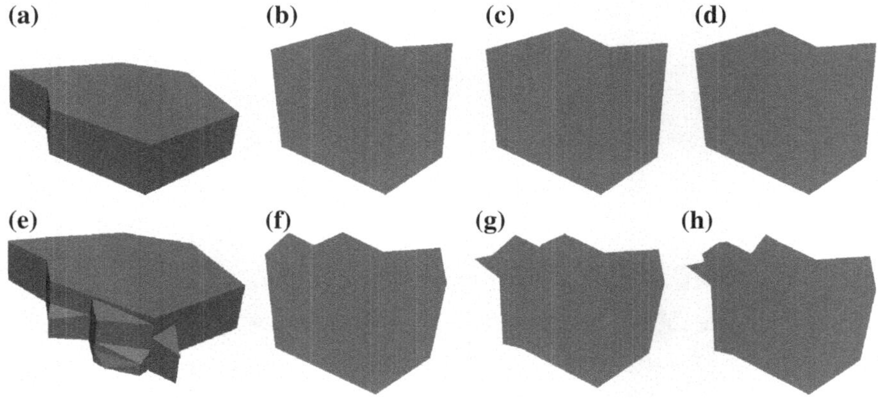

Fig. 4.23 Comparing a space-scale polyhedron (**a**) without and (**e**) with line simplification. The resulting slices (**b**), (**c**) and (**d**) through the polyhedron without line generalisation. All are the same, while the resulting slices (**f**), (**g**) and (**h**) through the polyhedron with line simplification have gradually reducing detail

space and 1D scale data can have benefits for interactive editing sessions. With the right tools available (for example, the editing environment does not allow the creation of intersecting surfaces), this might provide an intuitive way of editing vario-scale data. However, for real time use of the data for end users, where they are using the data as a backdrop map, it may be more efficient if the data is compactly encoded in the original tGAP data structures. 3D surfaces in the Space Scale Cube can be represented efficiently by storing the sequence of generalisation steps that were performed during line simplification in a (set of) BLG tree(s). Such a binary tree can subsequently be compactly serialised into a binary data format, which saves bandwidth during transmission from a client to server. It remains to be seen what is the best approach (explicit 3D cube using a general 3D data structure, or highly specialised 2D data structure, together with separate support for 1D scale). Conceptually however, the explicit 3D model makes it easier to reason about the desired effects of the map generalisation process from a geometric point of view.

4.7 Conclusions

Section 4.2 presented a geometrical non-redundant, variable-scale data structure. The previous versions of the GAP tree had some geometry redundancy, primarily between the polygons at a given scale and/or between the scales. The key to the solution presented in this section was applying a full topological data structure, though this is far more complicated than topological structures designed for the traditional single-scale data sets. The topological GAP tree is very well suited for a web environment–client requirements are relatively low (no geometric processing of the data at the client side) and progressive refinement of vector data is supported (allowing quick feedback to the user).

Section 4.3 has introduced a truly smooth vario-scale structure for geographic information: a delta in scale leads to a delta in the map continuously down to an infinitesimally small delta for all scales. The smoothness is accomplished by removing all horizontal faces of the classic tGAP structure. Algorithms and corresponding intuitive proofs of correctness were given on how to obtain data for the smooth tGAP structure by performing generalisation operations in such a way that the output gradually changes the boundaries between the features being generalised.

The smooth tGAP structure delivers vario-scale data and can be used for smooth zoom. The proposed new structure has very significant advantages over existing multi-scale/multi-representation solutions (in addition to being truly smooth vario-scale). The first advantage comes from tight integration of space and scale resulting in guaranteed consistency between scales (it is one integrated space-scale partition) and when using non-horizontal slices the resulting 2D maps will be a valid, mixed-scale planar partition. This is useful in 3D computer graphics. Secondly it is relatively easy to implement smooth zoom, and thirdly, it provides a compact (good for storage, transfer and CPU use) and object-oriented encoding (one higher dimensional object for a complete scale range).

In Sects. 4.4 and 4.5, we presented an approach to the integration of datasets of two different scales and generalisation between the scales based on the constrained tGAP structure. Objects of the medium-scale dataset are used for both approaches as constraints in the generalisation process. They restrict the aggregation of the large-scale objects only inside these constraints, resulting in a gradual transformation of the large-scale dataset into the medium-scale dataset. In Sect. 4.4, we used large- and (independent) medium-scale topographic data from the Netherlands. The first stage of the process is object matching between the two datasets. The output of the matching process is used in the second stage, the generalisation that is performed by the constrained tGAP. In Sect. 4.5, the constraints were derived from the large scale data set, by applying an optimisation method. The values that remain crucial for the quality of GAP-tree generalisation are the importance value of the selected feature classes (and importance function) and the compatibility values between two different feature classes (and compatibility function). More research is needed in this area to automatically obtain good generalisation results for real-world data. The constrained approach however results in a significant improvement of the generalisation quality (compared to unconstrained tGAP).

Section 4.6 has shown how a conceptual 3D model is straightforward for deriving smooth representations. The state-of-the-art implementation engineered for 2D + 1D scale separately. However, our initial implementations applied to real data have shown that the approach is indeed feasible.

Although the smooth tGAP structure encoded in the SSC as presented in Sect. 4.3 is a breakthrough vario-scale data structure supporting smooth zoom, there is still a myriad of open research questions: Efficient engineering implementation issues, cf. van Oosterom et al. (2002), Meijers et al. (2009), formalise the structure and prove mathematical proofs to be based on Meijers and van Oosterom (2011) and Thompson and van Oosterom (2012), tune the creation process (an option is to use the constrained tGAP; Haunert et al. 2009), test with larger real world data sets and appropriate graphical user interfaces, investigate the effects of the Collapse/Split operator, improve the cartographic generalisation quality (include more generalisation operators such as displacement, typification, and symbolisation; Cecconi and Galanda 2002), parallel generalisation (and not tGAP-style 'one by one sequencing' of the generalisation operations), cartographic styling of smooth tGAP selections (which are mainly a DLM; Stoter et al. 2010), dynamic map labeling Been et al. (2010), support for non-linear primitives (e.g. circular arcs, cylinder/sphere patches, non-uniform rational B-splines, NURBS; Pu and Zlatanova 2006), and make the structure dynamic 3D objects, scale and time lead to 5D hypercube; van Oosterom and Stoter 2010.

Acknowledgments This research is supported by the Dutch Technology Foundation STW (project numbers 11,300 and 11,185), which is part of the Netherlands Organisation for Scientific Research (NWO), and which is partly funded by the Ministry of Economic Affairs.

Many thanks to all students and colleague researchers involved in the development of the vario-scale tGAP concepts: Vincent Schenkelaars, Judith van Putten, Tinghua Ai, Maarten Vermeij, Arjen Hofman, Arta Dilo, Marian de Vries and Jan-Henrik Haunert. Thanks to Wilko Quak, Edward Verbree, and Rod Thompson for a critical review of parts of draft version of this text. Finally, authors would like to thank the editors of this book for the initiative and for, together with the anonymous reviewers, providing many constructive suggestions on the earlier versions of this chapter. Special thanks to William Mackaness for his meticulous English proof reading and spotting unclear sections. All comments were of great help, but all (remaining) errors are the sole fault of the authors.

References

Ai T, van Oosterom P (2002) GAP-tree extensions based on Skeletons. Paper presented at the advances in spatial data handling, 10th international symposium on spatial data handling, Ottawa, Canada, 9–12 July 2002

Ballard DH (1981) Strip trees: a hierarchical representation for curves. Commun Assoc Comput Mach 14(5):310–321. doi:10.1145/358645.358661

Been K, Nöllenburg M, Poon S-H, Wolff A (2010) Optimizing active ranges for consistent dynamic map labeling. Comput Geom 43(3):312–328

Bertolotto M, Egenhofer MJ (2001) Progressive transmission of vector map data over the World Wide Web. Geoinformatica 5(4):345–373. doi:10.1023/a:1012745819426

Bregt A, Bulens J (1996) Application oriented generalization of area objects. M Methods for the generalization of geo-databases. Netherlands Geodetic Commission, Delft, The Netherlands, pp 57–64

Buttenfield B (2002) Transmitting vector geospatial data across the internet. In: Egenhofer M, Mark D (eds) Geographic information science, vol 2478. Lecture notes in computer science. Springer Berlin Heidelberg, pp 51–64. doi:10.1007/3-540-45799-2_4

Cecconi A, Galanda M (2002) Adaptive zooming in web cartography. Comput Graph Forum 21(4):787–799. doi:10.1111/1467-8659.00636

DMA USDMA (1986) Defense Mapping Agency product specifications for digital feature analysis data (DFAD) Level 1-C and Level 3-C (DFAD): level 1 and level 2. DMA Aerospace Center, St. Louis, Mo

Douglas DH, Peucker TK (1973) Algorithms for the reduction of the number of points required to represent a line or its caricature. Cartographica: Int J Geograph Inf Geovisual 10(2):112–122. doi:10.3138/FM57-6770-U75U-7727

Gunther O (1988) Efficient structures for geometric data management, vol 337. Lecture notes in computer science. Springer, Berlin

Guttman A (1984) R-trees: a dynamic index structure for spatial searching. SIGMOD Rec 14(2):47–57. doi:10.1145/971697.602266

Hägerstrand T (1970) What about people in regional science? Pap Reg Sci 24(1):6–21

Hampe M, Sester M, Harrie L (2004) Multiple representation databases to support visualization on mobile devices. In: Proceedings of the 20th ISPRS congress, vol 35 of international archives of photogrammetry, remote sensing and spatial information sciences, Istanbul, Turkey, vol 35(6), pp 135–140

Harrie L, Sarjakoski T, Lehto L (2002) A variable-scale map for small-display cartography. Int Arch Photogrammetry Remote Sens Spat Inf Sci 34(4):237–242

Haunert J-H, Wolff A (2006) Generalization of land cover maps by mixed integer programming. Paper presented at the proceedings of the 14th annual ACM international symposium on advances in geographic information systems, Arlington, Virginia, USA

Haunert J-H, Dilo A, van Oosterom P (2009) Constrained set-up of the tGAP structure for progressive vector data transfer. Comput Geosci 35(11):2191–2203

Hildebrandt J, Owen M, Hollamby R (2000) CLUSTER RAPTOR: dynamic geospatial imagery visualisation using backend repositories. In: Proceedings of the 5th international command and control research and technology symposium (ICCRTS) 2000

Hofman A (2008) Developing a vario-scale IMGeo using the constrained tGAP structure. MSc thesis geomatics, Technical Delft University

Jones CB, Abraham IM (1987) Line generalisation in a global cartographic database. Cartographica: Int J Geograph Inf Geovisual 24(3):32–45. doi:10.3138/0666-1648-61L4-3164

Jones C, Abdelmoty A, Lonergan M, van der Poorten P, Zhou S (2000) Multi-scale spatial database design for online generalisation. In: 9th international symposium on spatial data handling, pp 34–44

Kreveld M (2001) Smooth generalization for continuous zooming. In: Proceedings of 20th international cartographic conference, Beijing (China), pp 2180–2185

Lazaridis I, Mehrotra S (2001) Progressive approximate aggregate queries with a multi-resolution tree structure. Paper presented at the proceedings of the 2001 ACM SIGMOD international conference on management of data, Santa Barbara, California, USA

Meijers M (2011) Simultaneous and topologically safe line simplification for a variable-scale planar partition. In: Geertman S, Reinhardt W, Toppen F (eds) Advancing geoinformation science for a changing world. Lecture notes in geoinformation and cartography. Springer Berlin Heidelberg, pp 337–358. doi:10.1007/978-3-642-19789-5_17

Meijers M, van Oosterom P (2011) The space-scale cube: an integrated model for 2D polygonal areas and scale. Paper presented at the 28th urban data management symposium, volume 38 of international archives of photogrammetry, remote sensing and spatial information sciences

Meijers M, van Oosterom P, Quak W (2009) A storage and transfer efficient data structure for variable scale vector data. In: Sester M, Bernard L, Paelke V (eds) Advances in GIscience.

Lecture notes in geoinformation and cartography. Springer Berlin Heidelberg, pp 345–367. doi:10.1007/978-3-642-00318-9_18
Noguera JM, Segura RJ, Ogáyar CJ, Joan-Arinyo R (2011) Navigating large terrains using commodity mobile devices. Comput Geosci 37(9):1218–1233. doi:10.1016/j.cageo.2010.08.007
Pu S, Zlatanova S (2006) Integration of GIS and CAD at DBMS level. In: Proceedings of UDMS 2006, pp 9.61–69.71
Rosenbaum R, Schumann H (2004) Remote raster image browsing based on fast content reduction for mobile environments. Paper presented at the proceedings of the seventh Eurographics conference on multimedia, Nanjing, China
Samet H (1984) The quadtree and related hierarchical data structures. ACM Comput Surv Arch 16(2):187–260. doi:10.1145/356924.356930
Sester M, Brenner C (2005) Continuous generalization for visualization on small mobile devices. In: Developments in spatial data handling. Springer, Berlin Heidelberg, pp 355–368. doi:10.1007/3-540-26772-7_27
Stoter J, Meijers M, van Oosterom P, Grünreich D, Kraak M-J (2010) Applying DLM and DCM concepts in a multi-scale data environment. In: Buttenfield BP, Brewer CA, Clarke KC, Finn MP, Usery EL (eds) Proceedings of GDI 2010: symposium on generalization and data integration 2010, pp 1–7
Thompson RM, van Oosterom P (2012) Modelling and validation of 3D cadastral objects. Urban and regional data management—UDMS annual 2011, pp 7–23
van Oosterom P (1986, 1989) A reactive data structure for geographic information systems. In: AutoCarto 9, Baltimore, Maryland, pp 665–674
van Oosterom P (1990) Reactive data structures for geographic information systems. PhD Theses, Department of Computer Science, Leiden University, The Netherlands
van Oosterom P (1992) A storage structure for a multi-scale database: the reactive-tree. Comput Environ Urban Syst 16(3):239–247. doi:10.1016/0198-9715(92)90036-Q
van Oosterom P (1993) The GAP-tree, an approach to "On-the-Fly" map generalization of an area partitioning. GIS and generalization, methodology and practice. Taylor & Francis, London
van Oosterom P (1994) Reactive data structures for geographic information systems. Oxford University Press, Inc.
van Oosterom P (2005) Variable-scale topological data structures suitable for progressive data transfer: The GAP-face tree and GAP-edge forest. Cartogr Geogr Inf Sci 32(4):331–346. doi:10.1559/152304005775194782
van Oosterom P, Meijers M (2011a) Method and system for generating maps in an n-dimensional space. Dutch patent application 2006630, filed 19 April 2011, published October 2012
van Oosterom P, Meijers M (2011b) Towards a true vario-scale structure supporting smooth-zoom. In: Proceedings of 14th ICA/ISPRS workshop on generalisation and multiple representation 2011, pp 1–19
van Oosterom P, Schenkelaars V (1995) The development of an interactive multi-scale GIS. Int J Geogr Inf Syst 9(5):489–507. doi:10.1080/02693799508902052
van Oosterom P, Stoter J (2010) 5D data modelling: full integration of 2D/3D space, time and scale dimensions. Paper presented at the proceedings of the 6th international conference on geographic information science 2010, Zurich, Switzerland
van Oosterom P, Stoter J, Quak W, Zlatanova S (2002) The balance between geometry and topology. In: Richardson D, van Oosterom P (eds) 10th international symposium on spatial data handling. Springer, Berlin, pp 121–135. doi:10.1007/978-3-642-56094-1_16
van Putten J, van Oosterom P (1998) New results with generalized area partitionings. In: Proceedings 8th international symposium on spatial data handling 1998, pp 485–495
van Smaalen JWM (1996) A hierarchic rule model for geographic information abstraction. In: Proceedings, SDH'96, Delft, The Netherlands, p 4b.31

van Smaalen JWM (2003) Automated aggregation of geographic objects—a new approach to the conceptual generalisation of geographic databases. PhD Theses, Wageningen University, The Netherlands

Vermeij M, van Oosterom P, Quak W, Tijssen (2003) T Storing and using scale-less topological data efficiently in a client-server DBMS environment. In: Proceedings of the 7th international conference on geocomputation, pp 1–11

Zhou X, Prasher S, Sun S, Xu K (2004) Multiresolution spatial databases: making web based spatial applications faster. In: Proceedings of the 6th Asia-Pacific web conference, APWeb. Springer, pp 36-47

Chapter 5
Integrating and Generalising Volunteered Geographic Information

Monika Sester, Jamal Jokar Arsanjani, Ralf Klammer, Dirk Burghardt and Jan-Henrik Haunert

Abstract The availability of spatial data on the web has greatly increased through the availability of user-generated community data and geosensor networks. The integration of such multi-source data is providing promising opportunities, as integrated information is richer than can be found in only one data source, but also poses new challenges due to the heterogeneity of the data, the differences in quality and in respect of tag-based semantic modelling. The chapter describes approaches for the integration of official and informal sources, and discusses the impact of integrating user-generated data on automated generalisation and visualisation.

M. Sester (✉)
Institute of Cartography and Geoinformatics, Leibniz Universität Hannover, Hannover, Germany
e-mail: monika.sester@ikg.uni-hannover.de

J. Jokar Arsanjani
GIScience research group, Institute of Geography, University of Heidelberg, Heidelberg, Germany
e-mail: jokar.arsanjani@geog.uni-heidelberg.de

R. Klammer · D. Burghardt
Institute of Cartography, Dresden University of Technology, Dresden, Germany
e-mail: Ralf.Klammer@tu-dresden.de

D. Burghardt
e-mail: Dirk.Burghardt@tu-dresden.de

J.-H. Haunert
Institute of Geoinformatics and Remote Sensing, University of Osnabrück, Osnabrück, Germany
e-mail: janhhaunert@uni-osnabrueck.de

D. Burghardt et al. (eds.), *Abstracting Geographic Information in a Data Rich World*, Lecture Notes in Geoinformation and Cartography, DOI: 10.1007/978-3-319-00203-3_5,

5.1 Introduction

Data acquired by laymen, so-called Volunteered Geographic Information (VGI), and data from geosensor networks has lead to an increased availability of spatial information. Whereas until recently, authoritative data sets were dominating, VGI extends and enriches these data in terms of thematic variation and also by the fact that it is more user-centric. The latter is especially true for VGI collected by social media.

There are several reasons for integrating data from different sources rather than just relying on a single one: The heterogeneity of data models, sensors and application areas lead to complementary (and redundant) data covering our environment. Integration results in a higher completeness concerning geometric, temporal and thematic coverage, and quality. Especially the availability of sensors such as GPS-enabled devices and smart phones, leads to a plethora of new, previously unknown, data with various spatial characteristics: geo-located Flickr images, Tweets with location tags, Wikipedia articles, GPS-traces, just to name a few. In general, VGI acquisition can be distinguished between participatory and opportunistic. The latter refers to data, which users acquire more or less unconsciously, mostly with a non-specific purpose. A prominent example is the exploitation of mobile phone data to determine traffic information such as traffic jams. Participatory data acquisition requires the conscious participation of the user, as it is undertaken e.g. in the OpenStreetMap project, where users have the common goal to build up a map of the world in a joint, cooperative effort.

The general goal is the integrated and consistent use and visualisation of all sorts of available (spatial) data. The specific challenges posed by user-generated content are among others, the huge mass of data, which have to be handled and turned into valuable information. Other challenges are:

- The data are "as is"; often it consists of "unimportant" information chunks, and is of heterogeneous quality.
- The data are highly redundant, as many users may observe the same phenomenon. The challenge is to exploit and remove the redundancy in order to present a consistent result.
- The data are highly temporarily volatile: temporal analysis is needed in order to decide about the relevance of a chunk of information.
- The data are of variable, sometimes unknown explicit semantics, thus the thematic content has to be inferred from contextual information.
- The data has no explicit scale, although it can be said that most of the data is of large scale (e.g. Points of Interest—POIs); thus, the scale of the data has to be determined in order to integrate and visualise it.
- Often the data are raw data: in order to exploit it, it has to be analyzed and enriched with contextual information or by automatic interpretation (e.g. images of Flickr).

The goal is to provide integrated information with clear semantics, seamless geometry and consistent topology. There should be no visual conflicts, which would disturb and confuse the user—and possibly lead to incorrect decisions. Furthermore, the information should be carefully 'thinned' in order to provide no visual clutter. Thus, generalisation has to be applied. Finally, the heterogeneous nature and quality of the data should be made transparent to the user.

5.2 The Potential and Characteristics of User-Generated Content

In the following sections, two major sources of novel geographic information data are briefly sketched, namely geosensor networks and VGI.

5.2.1 Overview of Characteristics of Geosensor Network Data

A geosensor network is a network of geo-sensors, capable of measuring, communicating and calculating (Stefanidis and Nittel 2004). With these characteristics, it is possible that a geosensor network is not just a collection of sensors, but is able to develop some intelligence which allows it to solve complex tasks. Geosensor networks composed of a huge mass of possibly miniaturised sensors are able to sense the environment at unprecedented resolution. Matt Duckham (2012) called this capability "close sensing".

There are many application areas for geosensor networks, especially in environmental modelling, where other sensors are not available or cannot measure with the required sensitivity and resolution. Prominent examples are the observation of rare animal species, observation and monitoring of natural phenomena, risks (such as landslides, earthquakes), catastrophes (oil spills, fire evacuation), or traffic monitoring and control, to name but a few.

The underlying idea is that the sensors not just measure and communicate the data to a central server, where all information is aggregated and analysed. Instead, the sensors can process and aggregate local information and perform some analysis directly within the network. This has great advantages: the local measurements can be verified with respect to outliers, thus only valid information has to be transmitted and centrally processed. There is a series of algorithms, which can be completely processed in the network, such as calculating mean temperature, delineating the boundary of an area or calculating its area (Duckham 2012). In this way, the decentralised sensor network can generate a global perception while only performing local operations. In essence, today's geosensor networks vary in the degree by which neighbourhood information is exploited before the information is communicated to the central server: the extremes are defined by "no local communication", which corresponds to traditional networks, where all data are centrally processed, to "only local communication", where the computation is done in the network itself, requiring—in the worst case—communication between all sensors. It can be expected that future sensor networks will exploit local information to a certain degree, especially, when the amount of data is prohibitively large (e.g. images or point clouds).

In recent years, the OGC has developed specifications for the discovery of and access to sensor data: the SOS (Sensor Observation Service) defined in the Sensor Web Enablement (OGC SWE 2.0) specification (Bröring et al. 2011). This leads to a syntactic interoperability. In order to go beyond that and also allow semantic interoperability, the meaning of the sensors and the measured values has to be made explicit as well (Henson et al. 2009), leading to a semantic sensor web.

In this way, geosensor networks are a new source of spatial information exploitable for different purposes. Even if structural and also semantic annotations are possible through open standards, still, however, variations in accuracy, semantic richness, and redundancy are possible. This places additional demands in terms of interpretation and poses an integration and visualisation problem.

5.2.2 Overview on Aspects and Characteristics of VGI

Through VGI a new class of spatial data is available, which poses new opportunities, but also new challenges for cartography and GIScience. The wealth and diversity of current VGI data are presented in Case study I. We now present a brief description of the essential characteristics of VGI data.

The characteristics are somewhat different to those of authoritative data, such as topographic data or navigation data sets provided by commercial players.

1. **Thematic content/usage spectrum**: Authoritative data sets usually are collected based on a given feature catalogue, which describes in detail the attributes and characteristics of the features contained in the data set and thus provides a kind of ontology for the data. In contrast, most VGI data do not rely on a standardised vocabulary; instead, mostly free text is used. In OSM the feature catalogue is developed via a consensus based discussion. It is worth noting however, that there is a broad spectrum of VGI data available (Case study I), which leads to an unprecedented diversity and semantic richness of spatial data.
2. **Availability, coverage, and homogeneity**: Whereas authoritative data are acquired with the goal of completely covering an area of interest—in the case of topographic maps it is the whole country—this is not the case with VGI data. This is also different from a sensor network, which is usually designed to yield an optimal configuration of observations for the phenomenon of interest. In VGI, the data acquisition depends on the availability of volunteers, therefore the density, depth and semantic richness of the data acquired cannot be

guaranteed. This results in a substantial inhomogeneity. Evaluations concerning the quality of VGI data have been conducted especially w.r.t. OSM data (Haklay 2010; Mondzech and Sester 2011), revealing the dependency of the coverage on different factors, such as country (European users are very active as opposed to US Americans), ease of access to basic information, and availability of freely available, alternative data.

3. **Timeliness of data**: Authoritative and commercial data sets are updated in fixed cycles, depending on the usage of the data. For example, in topographic maps in Germany, the road network is updated every 3 months, whereas the other features only every 3 years. Potentially, user-generated information is of high recency, as users are quick to correct data which is no longer consistent with reality (or else they notify of a bug in the data set[1]). In this way, the community constantly takes care to correct and improve the information. The high timeliness of the data can be exploited for time-critical information such as earthquake-related information which can be gained from Tweets, for example.
4. **Scale range**: VGI data are often large-scale information: this is true for POIs or Tweets. Also OSM data acquired with GPS sensors is of high resolution. In contrast, authoritative data sets are available in different scales—often acquired separately, in recent years also in a (semi-) automatic way via model generalisation. A more detailed discussion on the derivation of different scales in OSM data is given in Case study II.
5. **Quality, reliability, trust, liability**: Authoritative data providers have set up a detailed quality assurance scheme to make sure that their data conforms to given specifications. Similarly, geosensor networks consist of calibrated sensors yielding information complying with standards (Poser and Dransch 2010). In VGI data, this quality assurance process is performed by the users themselves, who notice errors in the data and are able to immediately correct them. Furthermore, there are mechanisms that guard against fraud by excluding users, who have intentionally included errors in the data sets. A critical issue remains one of liability over the data (Rak et al. 2012).
6. **Redundancy**: Some types of VGI do not explicitly seek to provide place information, but more place-related information (e.g. Twitter as opposed to OSM). This will—by nature—lead to redundant data (e.g. ratings of restaurants). Redundancy can also be produced by people who upload their GPS traces.
7. **User-centric information**: Some types of VGI provide highly personal, user-centric and even emotional information, e.g. Flickr images or Tweets. This opens the way to automatically extracting opinions or emotions from the data (Gartner 2012; Tauscher and Neumann 2012).
8. **Spatial reference and geometric representation**: Often VGI data is given in terms of point features with geographical coordinates or addresses, e.g. POIs or Flickr Images. In order to fit those features into the right spatial context, they

[1] www.maps4debugs.openstreetmap.de

have to be visualised on top of a base map, e.g. an authoritative map or with internet maps as background. In this process, the correct topological relationship between the features has to be assured.

These characteristics are relevant for the usage and especially for the visualisation of such data. Mechanisms are needed to enrich it with semantics and to embed it into a spatial context in order to make sense of the data. It is also necessary to filter out irrelevant and incorrect data and to automatically identify the quality of the data. This can be achieved by exploiting the usually high redundancy in the data.

The main challenges are therefore the handling of large amounts of data by providing adequate mechanisms for data generalisation, the extraction of implicit information, and the integration of the data with other, georeferenced, information.

5.3 Aspects of Data Integration

In general, data integration is needed to bring together complementary data sets referring to the same spatial extent, often of the same physical objects, acquired by different organisations, at different times, with different underlying data models and quality levels. The simplest way of data integration is a simple overlay. This is adequate, as long as the geometry of the involved data sets (perfectly) fits, and no geometric and topological errors are introduced through the integration process. Otherwise, matching approaches have to be used to identify corresponding objects and mutually adjust them. Integrating data has several benefits: first, a richer information base is generated, as information (e.g. attributes) from both data sets can be integrated; second, it can also lead to a more consistent data set with higher quality, as through integration possible errors can be detected and eliminated.

The problem of data integration consists of several subproblems: objects from different sources can be integrated, when they match at both a semantic and a geometric level. The first task is to identify the semantic correspondences between object classes. For this, either existing ontologies or object descriptions are used, or the semantic correspondence is inferred from the data themselves (e.g. Noy 2004; Volz 2005; Kieler et al. 2009). Then, features have to correspond with respect to geometric location and geometric properties (e.g. Yuan and Tao 1999), as well as topological characteristics (e.g. Walter and Fritsch 1999; Siriba et al. 2012).

After matching of individual object instances is achieved, different processes can follow from this:

- Information transfer: the attributes of the data sets can be mutually exchanged;
- Data fusion and harmonisation: an integrated geometry of objects can be calculated, possibly leading to a better quality of the averaged geometry. Furthermore, a general transformation of the whole data set can be conducted, and thus also other features, which are not present in the other data set can be

integrated based on their spatial context using rubber-sheeting and/or least-squares adjustment (Safra and Doytsher 2006; Warneke et al. 2011; Dalyot et al. 2012).

Whereas the typical matching cases treat objects of similar geometry (e.g. lines, polygons), in VGI integration often different geometric feature types have to be matched (e.g. matching points with lines, lines with areas, points with areas). This requires additional explicit specifications of constraints to be preserved in the matching process (e.g. the fact that a point has to be on a line, on the first 100 m of the line, or within a specified area; (also discussed in Chap. 3). Furthermore, individual information chunks might not make sense, only their aggregation with similar features may lead to a meaningful concept (e.g. many points at a certain location indicating the importance of a place). Thus, the heterogeneity of VGI data with respect to theme, scale and geometry, leads to the requirement that integration has to take matching, generalisation and interpretation issues into account. The specific challenges are as follows:

- Redundant information: e.g. many traces of travellers indicating traffic infrastructure. Here a statistical integration is required.
- Matching may only be possible at higher levels of aggregation, thus aggregation and/or interpretation of features is required.
- Geometric resolution of the data might differ; therefore, multiscale matching is required.
- Semantic, structural, geometric and topological characteristics have to be preserved while integrating data—with the special constraint, that these characteristics might not be known explicitly.

These aspects will be described in more detail in the following sections. Figure 5.1 shows different levels of correspondences between different representations of data: individual VGI data elements can be recorded multiple times; for example several lines (trajectories) may have been acquired which correspond to one individual river object. Similarly, it might be possible to extract information chunks containing "city centre" from different data sets such as POIs or Tweets. On an individual feature level, the data might not be accurate or reliable enough, however, when the features are confirmed by many other (redundant) measurements, this information can be aggregated and transformed to a new level.[2] In the second level of Fig. 5.1, the hull around the point features delineates the city centre, whereas the aggregated hull around the individual trajectories represents the areal river feature. At this level, the data can be matched to a large-scale topographic data set; it can also be further generalised, for example by using the collapse operation to derive an even smaller scale representation, which can be matched to a small-scale topographic map.

---

[2] Note that the notion of aggregation is used here to describe the grouping of similar instances of the same feature class to a new, combined representative object.

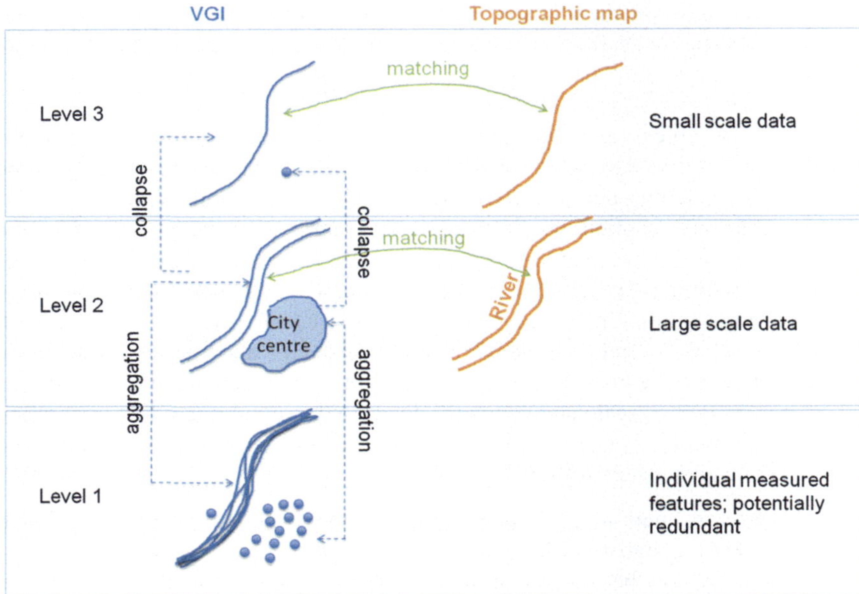

Fig. 5.1 Generalisation and matching of VGI and map data: levels correspond to different representations of the data; some of them can be linked to different scales

5.3.1 Aggregation of VGI

In recent years many researchers have analysed data collected by volunteers such as GPS-traces of hikers or car-drivers, which in principle can be considered as individual digitisations of roads or footpaths. The aggregation of GPS tracks mainly has to deal with the high degree of noise resulting from the low quality of the GPS measurements. This may make it difficult to separate nearby roads and thus reconstruct the underlying structure in the road geometry, e.g. the number of lanes, road width. In order to derive an integrated geometry from the collection of given tracks, aspects of reliability and trust (Sayda 2005) as well as geometric accuracy have to be taken into account (Zhang and Sester 2010).

In the case of road tracks, the goal is to reconstruct the centreline as well as the number of lanes from the noisy road data. Some approaches use histograms in profiles that run orthogonal to the hypothesised road. The mean of the intersection points of the profile with the traces delivers points of the centreline of the road. In order to separate different lanes, Schroedl et al. (2004) proposed a method that finds clusters in specific instances, corresponding to the typical width of lanes. Cao and Krumm (2009) use an approach based on a force model which optimises the displacement of individual tracks towards a modelled centre line. Chen and Krumm (2010) use a Gaussian mixture model (GMM) to model the distribution of

GPS traces across multiple lanes; prior information about lane width and corresponding uncertainty is introduced. Zhang et al. (2012) use a probabilistic relaxation for integrating crowd sourced data. Haunert and Budig (2012) extend an approach of Newson and Krumm (2009) using Markov map matching by also allowing missing road segments to be added (see also Case study III).

Davies et al. (2006) employ a raster-based approach—similar to occupancy grids used in robotics—in order to determine the geometry of roads. In their approach, they also include a temporal component by visualising the fading of roads that are not regularly frequented. In this way, abandoned roads can be identified. Thus they are able to describe the temporal change of objects.

Aggregation of VGI information has been proposed to delineate vernacular areas, i.e. regions that are well known to local people, but often are not represented on the map. They are often described by colloquial names, or may describe areas with fuzzy boundaries, such as the "Schwarzwald" or the "Midlands" (Jones et al. 2008). A popular approach is to use footprints of articles from search engines, Wikipedia articles or Flickr images to generate a collection of points. A kernel density estimation overlaying the points leads to a delineation of the fuzzy regions (Hollenstein and Purves 2010; Mackaness and Chaudhry 2013). In this way, Lüscher and Weibel (2013) successfully delineate city centres. Dahinden and Sester (2009) use the approach to determine the importance of geographic features in order to use it for cartographic visualisation and generalisation. The underlying idea is that the relative frequencies with which the geographic features are mentioned in public knowledge repositories gives an indication of their relative importance.

5.3.2 Integration of VGI and Topographic Data

VGI data are often associated with point locations, e.g. Flickr images, Tweets, POIs. In order to provide context for this kind of information, mostly background information in terms of topographic maps are used (as in Case study III). Thus, a challenge is to integrate VGI and the underlying reference data. To this end, ways of matching data are needed, which take known constraints and semantics into account. For example, Stigmar (2005) describes how topographic information (which was in a non-routable form) could be integrated with a routable data set. This was achieved using matching techniques, where the closest (and topologically best fitting) routes have been assigned to each other. In the final stage, only the routing instruction was transferred as an attribute—and not the geometry.

Another example is the semantic enrichment of 3D city models proposed by Smart et al. (2011). Geometrically exact ground plans from cadastres were augmented by information from Wikipedia to provide a semantically richer data set. The challenge in the approach was the matching of the VGI point information with polygons from the cadastre. To this end, the authors proposed a fuzzy matching approach taking the distance of the points to the corresponding polygons into account. Additional criteria, such as importance values derived on the one hand from the complexity of the ground plan (assuming that important or interesting objects such as a castle often have a complex ground plan) were also taken into account. The relative importance can also be indicated by the fact that several VGI sources relate to an object. In this way, implicit constraints can be stored. For example the fact that VGI annotations or POIs are mostly placed inside an areal object or at least in its vicinity is coded and resolved in a fuzzy matching approach.

Finally, Jaara et al. (2011) present a concept for integrating thematic data layers with background topographic information. They point out that constraints have to be preserved or even exaggerated. For example it is important to retain the position of a POI to the left or right of a road, or its location with respect to an inflection point of the road (see also Chap. 3).

5.3.3 Matching Multiscale Data

The integrated treatment of data from multiple scales has been investigated mostly in the context of topographic maps of different scales. One important research issue in the recent years was the creation and exploitation of MRDB—multiple resolution/representation databases. Different approaches have been proposed to generate the links between objects of different scales, e.g. by Devogele et al. (1996), Weibel and Dutton (2005), Hampe et al. (2004), Mustière and Devogele (2008), Shereen et al. (2009). The challenge is to take different geometric representations of the objects into account. For example, Kieler et al. (2009) implemented an algorithm that collapses rivers represented as polygons before matching them to the small-scale line objects in another data set. Similarly, different cardinalities in matching have to be taken into account, as typically several objects are aggregated to represent an object in the smaller scale, thus leading to an n:1 or n:m-matching situation.

5.4 The Visualisation and Generalisation of VGI

5.4.1 Visualisation of VGI

Visualisation of VGI data, possibly together with underlying base data, poses several requirements, which are substantially different from visualising conventional maps. The first issue considers the graphical design and adequate information selection, the second issue concerns the communication of the inhomogeneous quality of that information.

5.4.1.1 Graphical Design, Information Visualisation

In contrast to topographic maps, often no commonly accepted feature catalogue is available, nor an agreed signature catalogue or visualisation and generalisation rules, which describe how VGI should be graphically presented.

Another issue is the often heterogeneous information density of VGI, e.g. when visualising PoIs: the PoIs are often clustered in the city centre, whereas typically there are no or only a few features in surrounding regions. This requires an adequate information selection and placement. Different options have been proposed, for example by Slingsby et al. (2007). Other approaches try to use visualisations such as tag clouds to present the available information at a certain location (Paelke et al. 2012). These visualisations are also scale dependent and have to change as the user zooms in and out. Possible mechanisms for selection of adequate information and symbol sizes have been proposed by Bereuter et al. (2012) using a quadtree principle. Other approaches use focus maps to arrange (and aggregate) PoI-labels (Fink et al. 2012).

5.4.1.2 Visualisation of Data Quality

Trying to visualise the heterogeneity inherent in VGI data is a big challenge. A fundamental cartographic principle is to provide a homogeneous view of the real situation, which implies that a blank area in the map indicates the absence of information. This can of course not be guaranteed in VGI, as it can also indicate an area which just has not yet been mapped. Thus the challenge is firstly to identify where there are gaps in the data, i.e. areas, which have not yet been mapped. One option to detect these areas is to analyse the density of data (possibly in relation to the underlying topography) and thus gain a sense of areas that are unlikely to have been mapped. A second issue is to determine the level of detail of a certain piece of information. In other words, does a Wikipedia article or a photo refer to a small reference object or to a larger area? Here, references to gazetteers may help resolve this issue.

Once this meta-information about data quality is available, different techniques can be used to graphically present this information (MacEachren 1992). Senaratne and Gerharz (2011) describe an evaluation of different possibilities to visualise uncertainty or different qualities in the data. Kinkeldey and Schiewe (2012) use so-called noise annotation lines to create an overlay grid indicating the quality of the underlying data.

5.4.2 Data Generalisation and Adaptation of Scale

In order to integrate data, possibly different scales of base data and VGI data have to be taken into account. Different cases can occur: either the VGI data are more detailed or the base data are more detailed. In addition, the level of detail in the VGI data can vary from region to region, thus there is no homogeneous scale.

Broadly, generalisation is needed for several reasons (Jaara et al. 2011):

- In order to increase the consistency by changing the scale or the level of detail of layers;
- in order to decrease the level of detail in one or more layers, to adapt it to the context of use, the intended display scale and within the associated graphical constraints. Case study III presents an approach for the integration of VGI which has the potential of automatically adapting to the scale of the base map;
- generalisation is needed to visualise data at a smaller scale; this generalisation process has to take the characteristics of the individual objects, but also their relations into account by preserving or even enhancing them. Case study II describes the challenges of generalising VGI data—something that is quite different from authoritative data.

In order to create a consistent visualisation, we have the following options:

- creation of one homogeneous scale via generalisation;
- communication of difference in scales so that for the reader it is clear for which generalisation level the information is intended;
- creation of a 'focus lens', i.e. keeping both scales, but indicating, where which scale is used with the help of adequate graphical variables.

The main generalisation operations needed are as follows (the order indicates their importance and use in current approaches):

- **Selection and typification**: the most important operation, as it has the potential to significantly reduce the amount of data;
- **Aggregation/grouping**: this operation can be applied to generate a consistent object geometry by averaging several geometries; it can also be used to classify and aggregate objects of similar categories;
- **Dimensional change**: is an operation, which is applied in different ways the first is the grouping of several PoIs to form an area (e.g. the city centre). This can be considered as a change from point information to areal information. The second and more typical change is from higher to lower dimensions e.g. representing an areal market place as a point at a lower scale.
- **Displacement**: is needed in order to eliminate spatial conflicts between objects.
- **Enhancement**: can be applied to visually attract the attention to important, but small objects.

Case study II describes the use of generalisation operations in the case of OSM data.

5.5 Case Study I: VGI Platforms and Data Generalisation

Jamal Jokar Arsanjani

Undoubtedly, Web 2.0 technologies have provided a tremendous amount of user-generated content, which accounts for an interactive and alternative source of information from and for, the public. Among the contributed information, a large amount of information contains geolocated information or information about geographical objects. We are facing a new era in gathering information about objects and events disseminated via the web. Thanks to online mapping projects, individuals are able to digitise objects of interest from base maps via web mapping services (WMS) or alternatively collect information about geographical entities via GPS-enabled devices and then share them with the public. The term VGI was coined by Goodchild (2007). In the case where the individuals' contributions are taken without their awareness, this sort of information is called contributed geographic information (CGI). Up till now, citizens science projects, public participatory GIS projects (PPGIS: Carver 2001; Rambaldi et al. 2004), collaborative mapping projects (CMPs: Rouse et al. 2007), and crowdsourcing (Heipke 2010) have been the main sources of VGI and CGI. Although the new given possibilities by VGI, it does not offer the collected data via classical, consistent, and standardised data structure. Some examples such as georeferenced photographs (Flickr, Panoramio), videos, check-ins, and tweets can be mentioned here.

A large number of VGI-based services have been launched for covering a variety of disciplines—taking different forms of data from individuals and sharing them in different ways. It is important to understand the limits of VGI, and identify new research challenges. This Case study is intended to identify some representative VGI services and to consider their individual functionalities, fitness-for-use, and identify which types of information are provided by them. The accessible data from VGI will be considered from a cartographic perspective (e.g. generalisation and visualisation). The VGI services are categorised according to several classification systems based on their usage.

5.5.1 Characterisation and Categorisation of Popular VGI Services

Each individual VGI service collects a specific source of data and shares the collected information through a particular data type. For instance, OSM, Wikimapia, and Google Map Maker provide an opportunity to map the world into different feature classes, which are represented as point, line, polygon data types. Flickr and Panoramio provide geolocated photos along with their location and attribute tags. A set of VGI services is summarised below.

5.5.1.1 World Mapping Projects

OpenStreetMap is a collaborative mapping project (CMP), which seeks to create a free editable map of the world. It has attracted more than 1.3 million users so far. Users are able to digitise the geographical objects from WMS into points, polylines, and polygons in addition to importing GPS tracks recorded by smart devices and shapefiles from official sources (e.g. CORINE data in France and TIGER data in the USA). The features are structured as POIs, places, roads, railways, waterways, natural, land use, and buildings (http://wiki.openstreetmap.org/wiki/Map_Features). The compressed and uncompressed versions of OSM data can be downloaded via either OSM-API, or querying data through a plug-in in GIS softwares, or through Geofabrik and CloudMade websites (Ramm et al. 2010). OSM features have been further implemented in a variety of applications such as routing (www.Openrouteservice.org), 3D modelling (www.OSM-3d.org), disaster management (Yates and Paquette 2011), and land use mapping (Jokar Arsanjani et al. 2013b). OSM has been extended to include the mapping of the maritime—via *OpenSeaMap*.

Wikimapia is another popular CMP, which enables the marking of geographical objects and describing them in polygonal form. The project's motto is *Let's describe the whole world!* and has over 1.8 million users. Wikimapia API and (Motomapia.com) enable users to record the location, time stamp, category, tags, and descriptions of objects. *Google Map Maker* is also a CMP sponsored by Google to promote the quality of Google maps. Like OSM, users create and edit the objects into points, polylines, and polygons. It has two major differences compared with the above-mentioned services:

(a) it doesn't cover all regions in the world and only certain countries are covered (https://services.google.com/fb/forms/mapmakerdatadownload/);
(b) the contributed data are not downloadable.

5.5.1.2 Social Media Mapping

Social media mapping services allow users to share text-based messages, postings, photos, and videos captured by GPS-enabled devices (e.g. digital cameras and smart phones). *Flickr* helps people to share their photos with the public and has more than 51 million registered users and 6 billion uploaded photos, of which more than 200 million are geotagged. The coordinates are in point form (latitudes and longitudes). Via Flickr API, coordinates, time stamps, tags, and textual descriptions of the photos can be recorded. *Panoramio* is also a photo sharing platform which contains geolocated photos. Like Flickr, it provides tags assigned to photos and, the photos can be downloaded through its API.

Twitter is a platform which enables people to share their tweets, expressing ideas, reports, news, and events around the world. At present there are over 500 million registered users and over 300 million tweets are generated daily.

Geolocated tweets can be mapped as point features together with textual attributes. In addition to the coordinates and textual content of tweets, time stamps of the tweets and user profiles are available via the Twitter API. Twitter has frequently been leveraged in event detection e.g. earthquake, crowd behaviour, political campaigns, and disaster management.

Facebook is the most popular social network, which is dedicated to helping people connect and share their interests and information. Currently it has over one billion users. In addition to posts, geolocated photos and videos are available, which can be mapped as point features. The Facebook API provides the content, location, time stamp, user profile of the posts, photos and short videos. *Foursquare* is another popular social network that enables users to check-in at venues and share check-ins. Each check-in has coordinates, a time stamp, and attributes that can be accessed as point features via the Foursquare API. Possibilities and challenges of exploiting social networks are discussed by Roick and Heuser (2013).

YouTube is the most popular video-sharing website, which enables users to upload videos and edit videos by adding tags and descriptions. Similar to photos in Flickr, the videos themselves are uploaded to YouTube along with their metadata. Such metadata which contain time stamps, tags, coordinates, user names and textual descriptions are acquirable via the YouTube API as point features.

5.5.1.3 Environmental and Ecological Monitoring

Eye on Earth is a large and ambitious project supported by the European Environment Agency and dedicated to building a crowdsourced map for environmental monitoring. Volunteers make contributions by attributing their feelings on the quality of air, water, noise and nature. Observation sites marked on the base map of Eye on Earth are recorded as point features enabling various quality maps to be generated.

Geo-Wiki has been launched by the International Institute for Applied Systems Analysis (IIASA) in order to improve the accuracy of global land cover maps through volunteers (Fritz et al. 2012), who are asked to add their information on habitat and ecosystems. Volunteers validate the land cover maps by comparing global land cover maps with Google Earth together with their own knowledge. The data are recorded as point features and can be downloaded directly from the website.

eBird,Breeding Bird Surveys,Christmas Bird Count, and *Wildfinder* are free online databases of bird observations, recorded as point features and offering real-time data about bird distributions collected by citizens. Such data facilitates biological research (Sullivan et al. 2009; Scofield et al. 2012). The data are downloadable directly from the respective websites.

Tracks4Africa intends to map African ecosystem reliably from experienced and responsible eco-travellers who are based in the field. The main source is via GPS mapping of eco-destinations in rural and remote Africa of outdoor activities e.g.

hiking, mountaineering, river rafting, scuba diving, bird watching, paragliding, and green land travel through points of interest and thematic maps.

5.5.1.4 Weather Mapping

Weather mapping projects are platforms that enable users to provide real time descriptions of weather conditions as point records. The most practical weather mapping project is called 360.*org* which seeks to harvest weather information from weather stations, universities, and amateurs in order to broadcast weather predictions to the public. The website allows users to download the data directly.

5.5.1.5 Crisis and Disaster Mapping

Ushahidi is an application that collects reports on different crises from volunteers. Originally it was used to visualise incidents of violence and peace efforts throughout Kenya from 45,000 users, and now supports several events around the world, which can be downloaded from the Ushahidi website. An evident example of its effectiveness was proven in the Haiti earthquake (Zook et al. 2010; Yates and Paquette 2011). *Crowdmap* is also a platform used to collect crowd-sourced crisis information. *Did You Feel It?* is built to encourage people who actually experience earthquakes to estimate the effects and extent of damage of earthquakes. An intensity map is generated by combining the information provided by both official agencies and internet users. The geographic features are contributed as points and polygons representing the locations and intensities of affected regions (Atkinson and Wald 2007).

5.5.1.6 Crime Mapping and Tracking

WikiCrimes offers users interactive maps to anonymously report crimes and pinpoint their position in order to identify crime hotspots. Users are predominantly from Latin America. The location, the type and density of crime are accessible as point features. The data of WikiCrimes is downloadable from its website.

5.5.1.7 Outdoor Activity Mapping

Outdoor activity mapping projects allow citizens to share GPS tracks of their leisure trips and recreation spots such as running, walking, hiking, and biking. *Wikiloc* with more than 700,000 users is very popular for discovering and sharing the best trails for outdoor activities, which additionally includes PoIs, elevation profile, distances, accumulated altitude, and images. The information is recorded as polylines and points. Similar projects such as *EveryTrail, MapMyRun, Endomond,* and *Map My Tracks* have similar functionalities.

US Fish Finder allows users to add photos and fishing hot spots which are accessible to the public. The application provides a variety of information for people such as insight into fishing hotspots, plotting lakes, rivers and streams and directions to them, access and boat launches, uploading photos of specific locations, topographic maps, and tide information via point and polyline features.

5.5.1.8 Business Mapping

Yelp is a local directory service with social networking and user reviews which helps users to find local businesses such as restaurants, bars, hotels, and petrol stations. Similar to the venues in Foursquare, business venues in Yelp receive reviews, ratings and comments from users. Geolocated business venues are considered as geographic points. Via the Yelp API, the attributes such as business name, rating, review count, address, phone number, neighbourhood, category, latitude and longitude, and reviews are available. Similarly, *Where* is used by local businesses to reach local audiences by providing data of various features. It also enables check-ins on smart phones again considered as geographic points.

5.5.1.9 Transportation Mapping

TomTom Map Share enables users to avoid congested sections of roads. It allows users to report changes and to share them with others via their mobile devices or via an online Map Share reporter tool offered by TomTom. The reported information on road changes includes modifications of speed limits, new street names, blocked roads, new traffic directions and altered turn restrictions. The geolocated roads are considered as polylines. TomTom makes such information changes visible but not downloadable to the public. Similar examples include *TeleAtlas map insight* and *Navteq map reporters* (Coleman et al. 2009). *Here.com* also allows the public to edit Navteq and Nokia maps. *Waze* is a community-based traffic and navigation application which enables users to share real-time traffic and road information. The information includes the users' historical routings, real-time traffic jams, and petrol prices. *Trapster* provides the opportunity for people to report speed traps, red lights and speed cameras, accidents, and road hazards. It has a user base of nearly 18 million users. The geolocated traffic traps are mapped as geographic points.

Table 5.1 gives a summary of the dominant feature types acquired by the different VGI services. Predominantly the information is collected as point

Table 5.1 Dominant geometric features types acquired in different VGI services

Data source	Points	Polylines	Polygons
World mapping projects	X	X	X
Social media mapping	X	X	
Environmental mapping	X		X
Weather mapping	X		
Crisis and disaster mapping	X		X
Crime mapping	X		
Outdoor activity mapping	X	X	X
Business mapping	X		
Transportation mapping	X	X	

information. This characteristic is relevant for subsequent visualisation and generalisation procedures.

5.5.2 Data Quality Aspects

Accuracy indicates by how much the data on a map or in a geodatabase matches with the reference data/values, whilst *precision* notes the level of correctness of information in a database (van Oort 2006; Geravis et al. 2009). However, the concept of "data quality" is sometimes inappropriately interpreted as data precision, uncertainty, or error. Although geodata with high locational precision are often called high-quality data, the sense of quality is beyond just the concept of locational precision (van Oort 2006).

The quality of geodata should be internally and externally taken into account. Internal quality takes the data production standards and specifications into consideration by detecting the errors in the data. Some standard organisations such as ISO, ICA, FGDC, and CEN define internal quality based on five mutual aspects (a) attribute accuracy, (b) positional accuracy, (c) temporal accuracy, (d) logical consistency, and (e) completeness (Guptill and Morrison 1995). These data properties are usually given to the users through metadata files attached to datasets by the producers (Devillers et al. 2007).

External quality considers whether a dataset is suitable enough for a specific purpose—termed "Fitness for Use". In practice a dataset should meet or exceed some expectations from end users regardless each individual internal data quality aspect (Devillers and Jeansoulin 2006). Quantitative VGI data quality analyses have been applied mainly to OpenStreetMap e.g. Haklay (2010), Helbich et al. (2012), Jokar Arsanjani et al. (2013a) in comparison with proprietary datasets. However, Goodchild and Li (2012) propose using VGI as wisdom of crowd to evaluate the quality of VGI.

5.5.3 Challenges of Integration, Generalisation, and Visualisation of VGI

The possible challenges of integration, generalisation, and visualisation of VGI are:

(a) Integration: Data integration is intended to integrate data from different sources in order to provide a consistent database of objects. This concept should be considered differently in the case of VGI, because several users may separately report a single object to the same VGI platform with slightly different positional placement and therefore different location and attribute information. It might become even more confusing when different VGI sources exist and dissimilar versions of objects and their attributes are collected. The shared information might not necessarily match at some points and thus data quality analysis must be used to filter noisy, redundant, and inaccurate data (Yuan and Tao 1999). Applying majority filters enables us to resolve redundant attributes and data generalisation and matching resolves the positional redundancy of the objects (Fig. 5.2). Despite the difficulties of gathering the correct information from VGI, it certainly helps (a) to improve the positional accuracy of the objects or to record the latest positional shifts, and (b) to enrich the attributes and metadata of the datasets. Having more reports on the same objects might suggest new indicative concepts such as place popularity, landmarks, events, trajectories, users' interests and behaviours.

Fig. 5.2 Concurrent visualisation of some VGI data via Google earth

(b) Generalisation: The intended level of detail (LoD) can be an indicator for identifying the degree of generalisation required (Mackaness et al. 2007). Contributions to VGI come from diverse and heterogeneous sources (GPS tracks, online digitisation, shapefiles import) and therefore, each contribution has its own LoD. For instance, GPS tracks are captured from different devices at different precision and accuracy criteria and diverse signal receiving conditions. Contributions may have been digitised at dissimilar zoom levels. In other words, one side of a linear object could have dense nodes within a certain distance, which fits into a data scale 1:A, and the opposite side has widely separated nodes, which fits into data scale 1:B. This results in having features, which need generalisation techniques to make them homogeneous. Each shapefile has a certain scale and online digitisation is possible at certain levels of zoom. For example OSM has more than 10 (http://wiki.openstreetmap.org/wiki/Zoom_levels). The WMS contain different types of aerial and satellite images at different spatial and temporal resolutions. Therefore, the VGI data are heterogeneous in terms of source, producers' expertise, accuracy, and LoD (Girres and Touya 2010; Haklay 2010). This suggests that VGI data do not fall into certain scale ranges and accordingly cartographic generalisation might not be consistently applied. Since VGI might offer richer datasets than classical datasets in some cases, the aforementioned limitations must be tackled by developing new techniques such as identifying features with different accuracy, shape, and spatial relations of features in order to assign the most relevant scale and also switch between different scales and times tamps via multiple representation databases (MRDB: Balley et al. 2004; Hahmann and Burghardt 2010). This helps to harmonise the LoD of mapped features to facilitate the generalisation process.

(c) Visualisation: For VGI, the classical visualisation techniques need to be advanced in order to address these variabilities. Scale-dependent visualisation techniques enable us to represent objects at different scales and time stamps. We also need ways of visualising tweets that express events and trajectories as well as Flickr photos tags. Since most of VGI services provide their data through PoIs, the latent information must be extracted from the PoIs through point pattern analysis (Sengstock and Gertz 2012). Some VGI websites act as a pool in which data can be both collected and visualised. Google Earth is an example of this (Fig. 5.2).

5.5.4 Summary

VGI provides information from people on diverse objects, events, and activities. VGI is offering free, new, and never-collected information as well as new data types and structures that never existed before. VGI is a data pool with heterogeneous sources, heterogeneous data structures, heterogeneous producers, and heterogeneous quality. Therefore, it asks for new cartographical techniques on data integration, data fusion, data generalisation, and data visualisation. It is important to note that VGI is provided by people with varying expertise which raises some concerns on the credibility and reliability of VGI (Jokar Arsanjani et al. 2013a).

New techniques and algorithms on assessing its quality and fitness for use have been increasingly developed and continue to be improved. As more documents, practices, and researches on VGI are released and published, more advancements and developments of algorithms and approaches are expected.

5.6 Case Study II: Generalisation within the OpenStreetMap Project Compared to the Generalisation of Authoritative Data

Ralf Klammer and Dirk Burghardt

This Case study will provide a structured comparison between the application of generalisation strategies within the crowd controlled structure of the OpenStreetMap project to the officially controlled structure of NMAs. This will be done by illustrating the essentially applied generalisation processes of OpenStreetMap in contrast to the appropriate analogies of the conceptual model of digital landscape and cartographic models introduced by Grünreich (1992). Beginning with an outline of basic principles about both systems this Case study will compare the structure of OpenStreetMap and NMAs in the second section and illustrate the application of automated generalisation within the OpenStreetMap project in the third major section.

5.6.1 Basics Principles of Authoritative versus VGI Data

The basic concept of OpenStreetMap (OSM) is the supply of a freely available world wide spatial dataset by activating nonprofessional, volunteered, 'hobbyist-surveyors' to collect spatial information of their immediate environment using simple GPS devices or to convey spatial information of personally inaccessible regions using aerial images. The result is a continuously updated central database at global coverage. The quality and quantity of this free available dataset is mainly characterised by heterogeneous spatial density (e.g. *industrialised versus developing countries*), geometrical quality (e.g. *urban versus rural areas*) (Girres and Touya 2010; Neis et al. 2012) or feature type assignment (e.g. *local versus global tagging of administrative boundaries*[3]) (Codescu et al. 2011). Beyond that, OSM is

[3] http://wiki.openstreetmap.org/wiki/Tag:boundary%3Dadministrative#10_admin_level_values_for_specific_countries

a web based crowdsourcing project, with a huge variety of map applications[4] and utilised rendering software.[5] This diversity precludes the description of target data, maps and courses of action by a finite common generalisation strategy. Accordingly, this Case study will focus on the basic map visualisation of OSM data on openstreetmap.org as well as the most common usages of OSM datasets, described in the corresponding project documentations (e.g. *wikis*[6] *and tutorials*[7]).

The so called Grünreich-Model (Fig. 5.3), introduced in the early 1990s, is a fundamental conceptual model for map generalisation providing a basic overview on the applied generalisation strategy within official cartography. Its differentiation between object-, model- and cartographic generalisation facilitates a fundamental consideration of all generalisation procedures that arise from spatial data processing. It begins with the capturing of spatial data in the field and ends with their information visualisation on a map. This model is still generally accepted today (e.g. Stoter et al. 2010; Bobzien et al. 2007; Chap. 11) and is especially adapted to the applied data management and visualisation of the German authoritative topographic cartographic information system (ATKIS) (Grünreich 1992). Accordingly, it provides a good base for a structured comparison of the application of map generalisation to authoritative and crowd controlled data, as it is more generic than for example the conceptual models of and McMaster and Shea (1992) or Brassel and Weibel (1988).

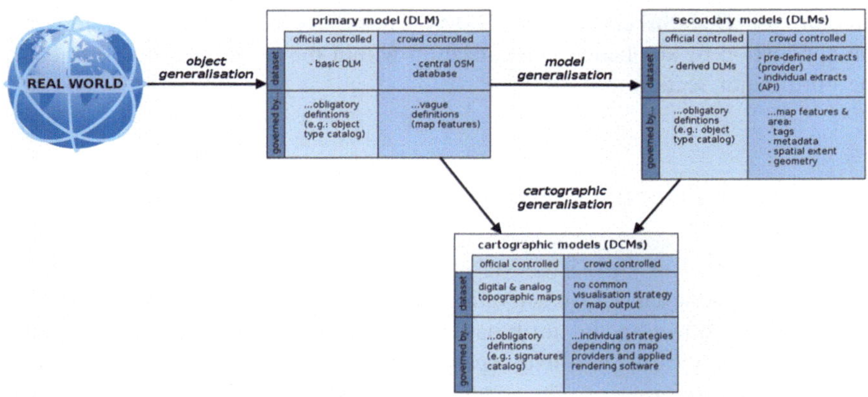

Fig. 5.3 The adapted Grünreich model in practice—NMAs versus OpenStreetMap

---

4 http://wiki.openstreetmap.org/wiki/Maps

5 http://wiki.openstreetmap.org/wiki/Renderers_feature_list

6 http://wiki.openstreetmap.org/wiki/Main_Page

7 http://www.osmfoundation.org/wiki/Main_Page

5.6.2 Comparison of Influencing Conditions on Generalisation

The basic conditions, that lead to the identification of structural differences in map generalisation defined by official and crowd structures, are contrasted in Fig. 5.3. NMAs traditionally pursue a central organised generalisation strategy with the primary aim of providing analog and digital topographic maps at different levels of detail, while efforts on supplying web-accessible maps are also aspired to (e.g. Sects. 11.6 and 11.7). At the other end of the spectrum, OSM is basically a decentralised crowdsourcing project with individual generalisation strategies, dependent on the data handler. Although a central unit exists in the form of the OpenStreetMap Foundation (OSMF), it does not control the project.[8] The primary map application of OSM data are interactive web maps commonly provided at up to 18 different scales, colloquially called zoom levels, mostly with a global coverage. NMAs provide maps of their country with up to 8 different scales (Chap. 11). The major difference between both is centered around how the data are processed. The basic tile-based map visualisation of OSM on openstreetmap.org is aimed at an immediate (Ramm 2012) visualisation[9] rendered completely automatically using data directly from the central OSM database.[10] The disadvantage of this performance driven rendering strategy is a rudimentary application of fundamental generalisation operators, resulting in restrictions to the graphical quality of maps (Touya 2012). The visualisation strategy of NMAs is generally geared to user oriented requirements (Chap. 2) which is why the whole set of generalisation operators is applied to produce map visualisations tailored to the user. As a consequence, the data processing is done, with exceptions outlined in Sect. 11.7, in a semi-automatic manner with interactive completion of the automated generalisation processes (Table 5.2).

5.6.3 Automated Generalisation in OpenStreetMap

To get a better sense of generalisation within the OSM project, this section will analyse OSM compared to the three basic operations—object, model and cartographic generalisation—of the Grünreich model.

Object generalisation is defined as the decision process by which spatial information has to be gathered, and which offers the highest spatial and semantic resolution (Fig. 5.3). This dataset is termed the primary digital landscape model (DLM). Object generalisation is realised in both the official and crowd controlled

---

[8] See Footnote 7

[9] http://munin.openstreetmap.org/openstreetmap/yevaud.openstreetmap/renderd_queue.html

[10] https://help.openstreetmap.org/questions/178/how-often-does-the-main-mapnik-map-get-updated

Table 5.2 Structural differences between official and crowd controlled map generalisation

	Officially controlled	Crowd controlled
Organisation	Central	Decentralised
Primary map application	Topographic maps (analog and digital)	Web maps (miscellaneous applications)
Scale range	~5–8 different scales	~17–20 different scales
Data coverage	National (regional)	Global
Primal guidance	Graphical quality (legibility)	High performance
Data processing	Partly automated with manually controlled interaction	Completely automated
Average update cycle	Yearly	By the minute
Modelling	MRDB-approaches (primary and derived secondary DLMs)	Rendering from a central database (primary DLM)
Applied operators	Complete set of generalisation operators	Selection, (Re-) classification

systems in a similar way. However, they are very different in detail. In officially controlled structures these decisions are determined by obligatory rules (e.g. *object type catalogs*) established within working groups, to which professionals must adhere when they gather data for the primary model. In OSM these rules are vaguely specified by OSM map features.[11] These are termed as "tags" and are used to assign features within the internal OSM data structure (Ramm et al. 2010). There is no obligation to respect any of them and each OSM member is generally able to define their own map features independently. Nonetheless, most members are geared to the commonly approved specifications in OSM practice and the introduction of new tags is often elaborately discussed within the OSM community via proposals.[12] The results of the object generalisation are in both cases a primary digital landscape model, while the primary model of OSM—the central OSM database—has a higher semantic resolution (e.g. *defibrillator*[13]). A second major difference is the type of person who collects the data: They vary from skilled and experienced surveyors, to unskilled volunteers with different levels of experience who control and increase the data quality collaboratively (Haklay et al. 2010) by relying on "The Wisdom of the Crowds" (Surowiecki 2004). This significantly influences the homogeneity of the primary DLM of OSM (Siebritz et al. 2012).

Model generalisation is defined as the derivation of secondary DLMs from the primary DLM, with a lower spatial and semantic resolution (Fig. 5.3). This is applied by different strategies in the NMA work practice but can be abstracted to a derivation of generalised databases adjusted to the production of small scale maps (Chap. 11). Accordingly, the model resolution is reduced by semantic (e.g. *feature*

[11] http://wiki.openstreetmap.org/wiki/Map_Features

[12] http://wiki.openstreetmap.org/wiki/Proposed_features

[13] http://wiki.openstreetmap.org/wiki/Tag:medical%3Daed

types) and geometric (e.g. *minimum size*) object selection. Applied generalisation operators aggregate areas, simplify geometries or build new object classes (classification). Many NMAs are already able to implement model generalisation as a fully automated process within their production workflow (e.g. Sect. 11.4). This automation effects the application of generalisation operators additionally, depending on the availability of appropriate algorithms and tools. In addition, storing and managing different forms of object representation implies the need for a specific MRDB concept (Kilpeläinen 1992). This is especially important for the update of secondary DLMs.

A comparable common course of action does not exist for OSM. Certain private initiatives and companies provide derived datasets, respectively secondary DLMs, with a lower spatial and semantic resolution than the central database. Two different types of derivation strategies exist. The first one is the derivation and supply of spatially and semantically filtered data extracts. The currently most common provider of such extracts is the company "Geofabrik",[14] who supplies daily[15] updated and completely automatically derived extracts from the central OSM database. This is especially advantageous for less experienced users of OSM data, as the whole OSM dataset is fragmented to administrative units (country specific datasets). The main disadvantage of these extracts is the lack of influence that a user has on the generalisation constraints as they are defined by the provider.

This limitation is overcome by the second derivation strategy, which provides the download of data extracts via application programming interface (API). The user can define individual data requests, adjust the degree of resolution reduction but needs programming skills and experience in using an API. Although the central database offers an integrated API,[16] the OverpassAPI[17] is currently the most stable and requested for deriving individual OSM data extracts. This implementation offers user defined reclassification of feature classes in addition to pure semantic filtering and uses a Quadtile[18] approach, based on Quadtrees (Samet 2005), for high performance execution of queries. Both strategies offer fully automated implementations of model generalisation with semantic selection—related to feature types or OSM metadata (e.g. *editor, time*)—and spatial selection—related to spatial extents or specific geometries (e.g. *country outline*). Consequently each extract can be termed as secondary OSM-DLM derived from the primary OSM-DLM (central OSM database). A specific MRDB concept for managing different forms of representations is not applied within the OSM project. However, objects can be linked by a unique identifier which is assigned to only one OSM feature and does not change during a features life-cycle. The uniqueness of features is a core concept of OSM as the identifier is used to build complex

[14] http://www.geofabrik.de/en/index.html

[15] http://blog.geofabrik.de/de/?p=75

[16] http://wiki.openstreetmap.org/wiki/API

[17] http://overpass-api.de/

[18] http://wiki.openstreetmap.org/wiki/QuadTiles

geometries from the specific OSM data format that only consists of nodes, ways and relations (Ramm et al. 2010).

Cartographic generalisation is characterised as the creation of digital cartographic models (DCM) by applying object symbolisation to the features of a specific DLM. Digital landscape models are just the preliminary stage of a final map visualisation where only the resolution of datasets is generalised without considering any legibility issues. Cartographic generalisation has, in contrast, a direct impact on the cartographic quality of a map as it includes the evaluation of feature representation and behaviour in a spatial context. This is done by contrasting object symbolisation with map specifications and user requirements, such as legibility (Chap. 2). This evaluation (Chap. 9) as well as appropriate geometric modifications can be applied generally to individual objects (independent generalisation) and sets of objects (contextual generalisation) implemented as manual, semi- or fully-automatic processes. An automatic computational recognition of graphical deficits needs accurately predefined constraints that characterise graphical object restrictions (e.g. *minimum sizes or distances*). These constraints would be evaluated by a map author visually within a manual evaluation.

Cartographic generalisation is driven by object symbolisation which is true for the official as well as for the crowd controlled structures. NMAs seek completely automatic workflows though the final evaluation and adjustment of cartographic quality is mostly done manually (Sect. 11.4). Completely automatic implementations have to accept a loss of graphical quality to provide a shorter refresh period (Sect. 11.3). Accordingly, the decision of implementing automatic or manual processing is currently a compromise between graphical quality and performance.

This also applies to the OSM project. Manual processes are almost completely ruled out in advance because of the huge amounts of data, in the specific case of the basic OSM map. Automatic evaluation or geometric modification is currently not aspired to in the general strategy or even implemented in the basic OSM map. The actual course of action for rendering a tile-based map, such as the basic OSM map, is as follows: Mapnik,[19] the main library, is applied to render all necessary map tiles, by obtaining OSM data from a regularly updated, locally managed version of the central database. The corresponding object symbolisation is defined within specific style files. Claims on the automatic application of generalisation operators have to be integrated to these definitions. Current implementations only contains fundamental generalisation operators—semantic selection and semantic reclassification—operators which are applied as model generalisation by NMAs. That means, the style definition contains a reference to the whole set of data (primary model), the corresponding data filter (e.g. [*highway*] = *'residential'*) and the declaration of the scale range for displaying the filtered features (e.g. *<MaxScaleDenominator>*25,000*</MaxScaleDenominator>*). The map author is responsible for the declaration of these definitions. The rendering tool (Mapnik) fulfills the data filtering. The generalisation process—semantic filtering and

[19] http://mapnik.org/

reclassification—is shifted to the rendering, and applied on-the-fly. Additionally, Mapnik has integrated methods for optimised label placement and simplification of linear geometries, but the OSM community has currently no common strategy on how to utilise these methods.

This course of action does not precisely fit with the definition of cartographic generalisation, as there is no evaluation and modification of graphical deficits that result from object symbolisation. Accordingly, it is only possible to include the described basic tile-based generalisation operations of OSM by weakening the initial definition of cartographic generalisation to a pragmatic definition. As a result, cartographic generalisation has to be characterised as the general application of generalisation operators for the visualisation of spatial information on a map.

5.7 Case Study III: Matching GPS Trajectories with Incomplete User-Generated Road Data

Jan-Henrik Haunert

5.7.1 Motivation

Map matching is the problem of finding a path in a given road data set that corresponds to a given trajectory. The trajectory is a sequence of positions with time stamps, which usually has been recorded with GPS. Several map matching algorithms have been developed that work well, even if the GPS measurements are noisy and the sampling rate of the GPS measurements is low (Newson and Krumm 2009; Quddus et al. 2007). The problem of incomplete or inaccurate road data, however, has seldom been addressed. Further research is needed, in particular to develop map matching algorithms that are suited for user-generated road data, which are often incomplete. In fact, user-generated data are often *more* detailed than commercial or official data and, in many regions, are suited for pedestrian or bicycle navigation. The quality of user-generated data varies, however, depending on the number of contributing users in a particular region, the time they spend on the project, and their skills.

Every user who contributes to a mapping project generalises, simply because he or she has to decide which objects to include in the map and how to represent them. Two users may decide differently, for example, on whether or not to include a small trail in the map. While the first user may consider the trail unimportant, the second user may recognise the trail as an important link in a network of hiking paths. For the success of a map matching algorithm, the inclusion of a short road segment in the data can be crucial. If the trajectory follows a road that is missing in the data, many map matching algorithms introduce long detours (Fig. 5.4). Similar

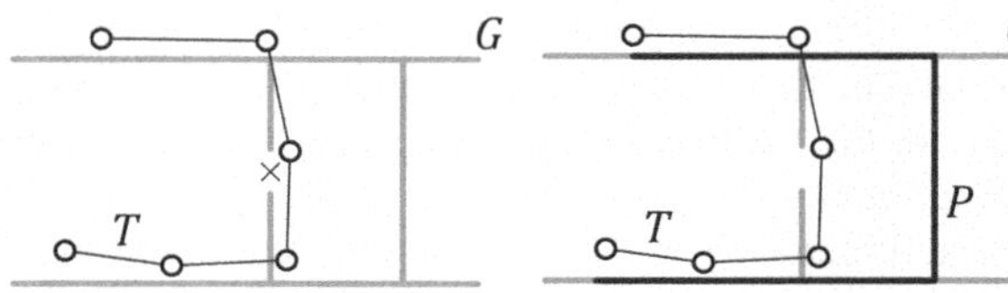

Fig. 5.4 Given a geometric graph ***G*** (the road network) and a polyline ***T*** (the GPS trajectory), the topological map matching algorithm of Alt et al. (2003), computes a path ***P*** in ***G*** that has the minimum Fréchet distance to ***T***. (Assuming that a dog runs along ***T*** and a man runs along ***P*** such that neither the dog nor the man are allowed to walk backwards, the Fréchet distance between ***T*** and ***P*** is the minimum length of a leash that allows the dog and the man to be kept connected.) Because the road network contains a small gap (*left*, marked with ×), the output path ***P*** (*right*) is much longer than ***T***

problems occur in the case of topological errors, which are often caused by users who add new paths without carefully integrating them into the present data.

In this Case study, we discuss experiments with a topological map matching algorithm that has been specifically designed to cope with incomplete road data (Haunert and Budig 2012). We review this algorithm and discuss experiments with user-generated road data from the OpenStreetMap (OSM) project and GPS trajectories recorded during four hikes.

5.7.2 Algorithm

Our map matching algorithm for incomplete road data (Haunert and Budig 2012) extends a basic map matching algorithm by Newson and Krumm (2009). For each point p_i of the GPS trajectory $p_1, \ldots, p_n$, the basic algorithm first selects a discrete set of k candidate matches $C_i = \{c_i^1, \ldots, c_i^k\}$, where k can be set by the user. Each candidate match is a point in a geometric graph G, which represents the road network. The output path is determined by selecting one candidate c_i^* of each set C_i and, for $i = 1, \ldots, n-1$, connecting c_i^* and c_{i+1}^* via a shortest path in G. Figure 5.5, left, illustrates this approach.

In order to select a sequence of candidates of maximum likelihood, Newson and Krumm (2009) apply a Hidden Markov Model (HMM). This requires that for each candidate match $c \in C_i$ an *observation probability density* is defined, that is, the probability density that GPS point p_i is observed if the actual position of the user is c. Furthermore, for each two candidates $c \in C_i$ and $d \in C_{i+1}$ a *transition probability* is defined, which models the probability that a user in c moves (within the time between two GPS measurements) to d. In the model by Newson and Krumm (2009), the probability of a transition from a candidate c to a candidate d is a function of the graph distance (the length of the shortest path) from c to d in G. Given the sets of candidate matches, the observation probabilities, and the

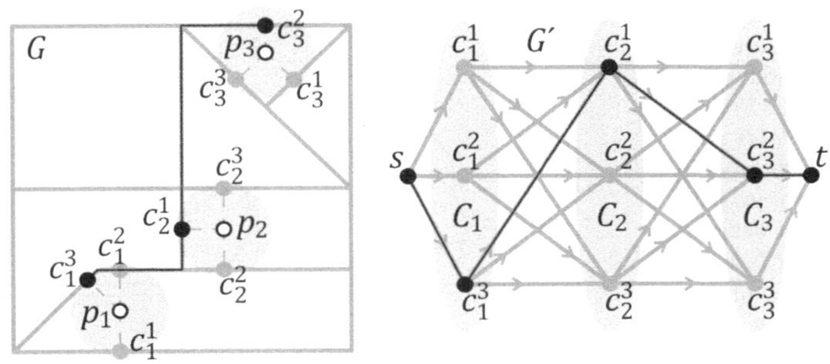

Fig. 5.5 *Left* The graph $\boldsymbol{G}$ representing the road network, three points forming a GPS trajectory $\langle \boldsymbol{P}_1, \boldsymbol{P}_2, \boldsymbol{P}_3 \rangle$ three candidate matches for each GPS point, and the output path (*black*) resulting from a selection of candidate matches. *Right* The directed graph $\boldsymbol{G}'$ for the problem instance in the *left* figure. Each $\boldsymbol{s}$-$\boldsymbol{t}$-path in $\boldsymbol{G}'$ corresponds to a solution to the map matching problem. The Viterbi algorithm selects the $\boldsymbol{s}$-$\boldsymbol{t}$-path that maximises the product of arc weights

transition probabilities, a weighted directed graph G' is defined, which is sketched in Fig. 5.5, right (G' contains a node for each candidate plus two dummy nodes s and t). Then, the Viterbi algorithm (Rabiner and Juang 1986) is applied to find the s-t-path in G' that maximises the product of arc weights. Given that the weights of the arcs of G' are appropriately defined (based on the observation probability densities and transition probabilities) this path can be interpreted as the sequence of candidate matches that explains the given GPS observations best.

Our map matching algorithm for incomplete road data (Haunert and Budig 2012) also uses the idea of a discrete set of candidate matches. The difference between our method and the method of Newson and Krumm is, however, that we include one additional candidate in the candidate set C_i of each GPS point p_i. We define this candidate as having the same position as p_i, thus it does not necessarily lie on an edge of the road data set. Therefore, we term such a point an *off-road candidate* and thereby distinguish it from *on-road candidates* that lie on edges of the road network. Our idea is to avoid long detours by allowing a GPS point to be matched with its off-road candidate. If we select the off-road candidate from a candidate set C_i and the off-road candidate from the candidate set C_{i+1}, we define that the output path contains the straight-line segment connecting both candidates. If, on the other hand, we select two on-road candidates for two consecutive GPS points, then our algorithm (just as the basic algorithm by Newson and Krumm) connects both candidates with a shortest path in G. In the case that for any two consecutive GPS points one on-road candidate and one off-road candidate are selected, we introduce a path connecting both candidates by concatenating a path in G with a straight-line segment. Among all possible concatenations, we choose the one of minimum total cost, where we define the cost of the path in G to equal its length and the cost of the straight-line segment to equal the product of its length and a constant factor. To avoid a situation in which the algorithm only selects off-road candidates, we define the transition probability from any node to an on-road

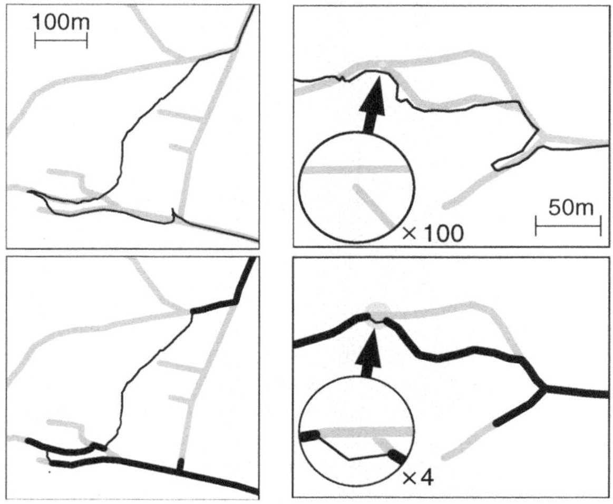

Fig. 5.6 *Left* A trajectory (*top*, *black*) with a road network (*gray*) and the output path (*bottom*, *black*) with on-road parts (*bold*) and off-road parts (*thin*) yielded by our algorithm. *Right* A sample from our data with a topological error (see close-up view in *circle*, *top*). The map matching algorithm selects the off-road candidate of one GPS point and thereby avoids a long detour

candidate to be higher (by a constant factor) than the transition probability to an off-road candidate.

5.7.3 Experimental Results

We implemented our algorithm in Java and tested it with OSM data and trajectories that we recorded with a GPS receiver of type Garmin Edge 710 during four hikes in the surroundings of Würzburg, Germany. In total, the GPS receiver yielded 7,799 GPS points over a distance of 65 km (on average, one GPS point every eight meters). In order to orient ourselves, we used a tourist map of scale 1:75,000 and the GPS receiver with a digital map based on the OSM data. In some cases, we took shortcuts along small trails, but we mainly used official hiking paths. We conducted experiments to find a good set of parameters, which we have used by default since. As a result of our test, we found 23 off-road sections in our output paths, that is, sections where we did not use edges of the road network. Figure 5.6 shows two examples with off-road sections.

To evaluate the robustness of our algorithm, we conducted additional tests in which we removed a randomly selected set of edges from the original road data. In the paths that we obtained by matching the original road data with two of our tracks (one in a rural and one in an urban area), we did not find significant errors and therefore used them as ground truth for our tests.

In each of our tests, we randomly selected and removed a certain percentage of the edges from our original road data set and matched one of the trajectories with the remaining roads. We term the percentage of the edges that we kept in the road network the *integrity* of the road network. Using this approach, we solved more

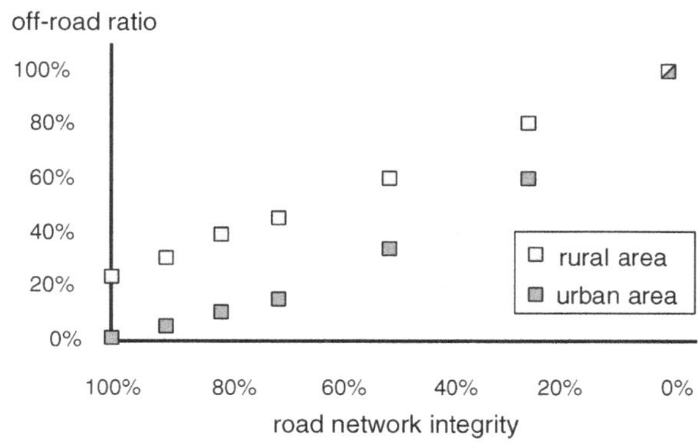

Fig. 5.7 Influence of the road network integrity on the off-road ratio

than 3,000 test instances with networks of different integrity. For each output path, we computed the *off-road ratio*, that is, the ratio between the length of all off-road parts of the path and its total length. Figure 5.7 summarises the results by showing averages over instances with the same trajectory and road networks of the same integrity. We observe that, as the integrity declines, generally more off-road edges are used in the output path, but we also observe differences between the results for the urban area and the rural area. In the urban area, the algorithm often used alternative road edges for edges we had removed from the road data. This is because the trajectory passed a public park with many mutually parallel paths. For example, the data set contained a bike track and a path for pedestrians next to each other. In contrast, for our track in the rural area, the algorithm seldom chose an alternative on-road route for a road we had removed from the road data. Instead, an off-road edge was introduced. In all cases, the total length of the output trajectory increased only marginally, for example, by a mere of 2 % for the urban data set of 80 % integrity.

5.7.4 Conclusions

Topological map matching algorithms often fail if applied to incomplete road data. In our experiments with OSM data and four GPS trajectories, we revealed 23 situations in which it was necessary to include off-road sections to the output paths, otherwise long detours would have been needed. Therefore, the problem cannot be neglected when dealing with user-generated data. With only a small modification of the existing map-matching algorithm by Newson and Krumm (2009), however, all four real-world instances on which we tested our algorithm were correctly solved. Even when we reduced the integrity of the road network drastically, our algorithm produced satisfactory results, but in urban areas sometimes matched the trajectory to a path that was parallel to the correct path. As an effect, when reducing the integrity of the road network to a certain percentage, the ratio between the length of all on-road sections and the total length of the path was reduced to a slightly lower percentage. Unlike most of the existing map matching algorithms, however, our algorithm does not require the output path to be fully contained in the road data and thus avoids long detours. Our experiment in which we removed a random set of road segments from the road data also suggests that the algorithm will cope with generalised data in which minor roads have been removed. In order to better assess the effect of map generalisation on the result of a map matching algorithm, however, further research is needed. For example, it would be interesting to investigate how line simplification affects the quality of the output path.

5.8 Conclusions

Visual presentation of the new (mass) data available gathered from human sensors and sensor networks poses a challenge for cartography. This new data source marks a fundamental change from mapping as an art performed by experts who produce maps as an end product for the general public—to the new situation where people both generate data and want to access it immediately. This implies that visualisation and map generalisation need to be simple to apply and fast. Also, mechanisms have to be available to clean the data due to frequent errors and redundancy. Furthermore, as most VGI is acquired as point data, the integration with known spatial data sets (topographic data) is needed, in order to provide the correct spatial context.

We conclude that the visualisation of VGI has great potential and provides great opportunities, but also presents us with some challenges. In order to exploit the benefits of the data, the following issues have to be tackled:

- (New) design principles to visualise VGI have to be developed, especially to visualise them in conjunction with background maps;
- Methods for automatically detecting and interpreting meta-information of VGI data are required. Meta information relate to semantics, scale, quality, timeliness, coverage, and consistency;
- Methods to visualise this meta-information have to be developed and made available;
- Generalisation methods for VGI have to be developed and made available. Especially methods for selection, aggregation and dimensional change, as they are the most relevant for frequently occurring point data;
- Data integration methods have to be developed, which take into account meta-data, known semantics and relations to background objects.

References

Alt H, Efrat A, Rote G, Wenk C (2003) Matching planar maps. In: Proceedings of the fourteenth annual ACM-SIAM symposium on discrete algorithms (SODA '03). Society for industrial and applied mathematics. Philadelphia, PA, USA, pp 589–598

Atkinson GM, Wald DJ (2007) "Did you feel it?" intensity data: a surprisingly good measure of earthquake ground motion. Seismol Res Lett 78(3):362–368

Balley S, Parent C, Spaccapietra S (2004) Modelling geographic data with multiple representations. Int J Geogr Inf Sci 18(4):327–352

Bereuter P, Weibel R and Burghardt D (2012) Content zooming and exploration for mobile maps. In: Proceedings 15th AGILE international conference on geographic information science, Avignon, France

Bobzien M, Petzold I, Burghardt D (2007) Automated derivation of a 1:300,000 topographic map from swiss dlm vector 200. In: Proceedings of the international cartographic congress, Moscow, Russia, International cartographic association

Brassel KE, Weibel R (1988) A review and conceptional framework of automated map generalization. Int J Geogr Inf Syst 2(3):224–229

Bröring A, Echterhoff J, Jirka S, Simonis I, Everding T, Stasch C, Liang S, Lemmens R (2011) New generation sensor web enablement. Sens Open Access J 11(3):2652–2699

Cao L, Krumm J (2009) From GPS traces to a routable road map. In: 17th ACM SIGSPATIAL International conference on advances in geographic information systems (ACM SIGSPATIAL GIS 2009), November 4-6, 2009. Seattle, WA, pp 3–12

Carver S (2001) Public participation using web-based GIS. Environ Plan B: Plan Des 28(6):803–804. doi:10.1068/b2806ed

Chen Y, Krumm J (2010) Probabilistic modeling of traffic lanes from GPS traces. In: 18th ACM SIGSPATIAL international conference on advances in geographic information systems (ACM SIGSPATIAL GIS 2010), November 2-5, 2010. San Jose, CA

Codescu M, Horsinka G, Kutz O, Mossakowski T, Rau R (2011) OSMonto—an ontology of OpenStreetMap tags. http://www.informatik.uni-bremen.de/~okutz/osmonto.pdf. Accessed 27 Feb 2013

Coleman DJ, Geogiadou Y, Labonte J (2009) Volunteered geographic information: the nature and motivation of producers. Int J Spat Data Infrastruct Res 4:332–358

Dahinden T, Sester M (2009) Categorization of linear objects for map generalization using geocoded articles of a knowledge repository. In: Proceedings of the international workshop presenting spatial information: granularity, relevance, and integration, held in conjunction with the conference on spatial information theory, COSIT

Davies JJ, Beresford AR, Hopper A (2006) Scalable, distributed, real-time map generation. IEEE Pervasive Comput 5(4):47–54

Dalyot S, Dahinden T, Schulze, MJ, Boljen, J, Sester, M (2012) Geometrical adjustment towards the alignment of vector databases. In: ISPRS annals of photogrammetry, remote sensing and spatial information sciences, vol. I–4, pp. 13–18

Devillers R, Jeansoulin R (2006) Fundamentals of spatial data quality. ISTE, London

Devillers R, Bédard Y, Jeansoulin R, Moulin B (2007) Towards spatial data quality information analysis tools for experts assessing the fitness for use of spatial data. IJGIS 21(3):261–282

Devogele T, Trevisan J, Raynal L (1996) Building a multiscale database with scale-transition relationships. In: Proceedings of the 7th international symposium on spatial data handling, advances in GIS research II. Delft, The Netherlands, pp 6.19–6.33

Duckham M (2012) Decentralized spatial computing: foundations of geosensor networks. Springer, Berlin

Fink M, Haunert JH, Schulz A, Spoerhase J, Wolff A (2012) Algorithms for labeling focus regions. Vis Comput Graph IEEE Trans 18(12):2583–2592

Fritz S, McCallum I, Schill C, Perger C, See L, Schepaschenko D, Van der Velde M, Kraxner F, Obersteiner M (2012) Geo-Wiki: an online platform for improving global land cover. Environ Model Softw 31:110–123
Gartner G (2012) openemotionmap.org—emotional response to space as an additional concept in cartography, international archives of the photogrammetry, remote sensing and spatial information sciences, vol XXXIX-B4
Gervais M, Bedard Y, Levesque M, Bernier E, Devillers R (2009) Data quality issues and geographic knowledge discovery, pp 99–116
Girres JF, Touya G (2010) Quality assessment of the French OpenStreetMap dataset. Trans GIS 14(4):435–459
Goodchild MF (2007) Citizens as sensors: the world of volunteered geography. GeoJournal 69:211–221
Goodchild MF, Li L (2012) Assuring the quality of volunteered geographic information. Spat Stat 1:110–120
Grünreich D (1992) ATKIS—a topographic information system as a basis for GIS and digital cartography in Germany. In: Vinken R (ed) From digital map series in geosciences to geo-information systems. Geologisches Jahrbuch, Federal institute of geosciences and resources, Hannover, vol A(122), pp 207–216
Guptill SC, Morrison J (1995) Elements of spatial data quality. Elsevier Science, New York
Hahmann S, Burghardt D (2010) Linked data—a multiple representation database at web scale? In: Proceedings of the 13th ICA workshop on generalisation and multiple representation, Zurich, Switzerland
Haklay M (2010) How good is volunteered geographical information? A comparative study of OpenStreetMap and ordnance survey datasets. Environ Plann B: Plann Des 37(4):682–703
Haklay M, Basiouka S, Antoniou V, Ather A (2010) How many volunteers does it take to map an area well? The validity of Linus' law to volunteered geographic information. Cartographic J 47(4):315–322
Hampe M, Sester M, Harrie L (2004) Multiple representation databases to support visualisation on mobile devices, international archives of photogrammetry and remote sensing, vol 35. Istanbul, Turkey
Haunert JH, Budig B (2012) An algorithm for map matching given incomplete road data. In: Proceedings of the 20th ACM SIGSPATIAL international conference on advances in geographic information systems (ACM-GIS'12), pp 510–513
Heipke C (2010) Crowdsourcing geospatial data. ISPRS J Photogrammetry Remote Sens 65(6):550–557. doi:10.1016/j.isprsjprs.2010.06.005
Helbich M, Amelunxen C, Neis P (2012) Comparative spatial analysis of positional accuracy of openstreetmap and proprietary geodata. Int. GI_Forum 2012. Salzburg, Austria
Henson C, Pschorr J, Sheth A, Thirunarayan K (2009) SemSOS semantic sensor observation service. International symposium on collaborative technologies and systems
Hollenstein L, Purves R (2010) Exploring place through user-generated content: using Flickr tags to describe city cores. J Spat Inf Sci 1(2010):21–48
Jaara K, Duchêne C, Ruas A (2011) Toward the generalisation of cartographic mashups: taking into account the dependency between the thematic data and the reference data throughout the process of automatic generalisation, 14th ICA workshop on generalisation and multiple representation, Paris, France, 30 June–1 July 2011
Jokar Arsanjani J, Barron C, Bakillah M, Helbich M (2013a) Assessing the quality of OpenStreetMap contributors together with their contributions. In: Proceedings of the 16th AGILE conference, Leuven, Belgium
Jokar Arsanjani J, Helbich M, Bakillah M, Hagenauer J, Zipf A (2013b) Towards mapping land use patterns from volunteered geographic information. Int J Geogr Inf Sci. doi:10.1080/13658816.2013.800871
Jones CB, Purves P, Clough D, Joho H (2008) Modelling vague places with knowledge from the web. Int J Geogr Inf Sci 22(10):1045–1065

Kieler B, Huang W, Haunert JH, Jiang J (2009) Matching river datasets of different scales. Advances in GIScience. In: Proceedings of 12th AGILE conference on GIScience, lecture notes in Geoinformation and cartography, Springer, Berlin, pp 135–154

Kilpeläinen T (1992) Multiple representations and knowledge-based generalization of topographical data. In: Waugh TC, Healey RG (eds) Advances in GIS research. In: Proceedings of the 6th international symposium on spatial data handling, IGU commission on GIS and association for geographic information, Edinburgh, GB, pp 882–893

Kinkeldey C, Schiewe J (2012) Visualisierung thematischer Unsicherheiten mit "noise annotation lines". Kartographische Nachrichten 62(5):241–249

Lüscher P, Weibel R (2013) Exploiting empirical knowledge for automatic delineation of city centres from large-scale topographic databases. Comput Environ Urban Syst 37:18–34. doi:10.1016/j.compenvurbsys.2012.07.001

MacEachren AM (1992) Visualizing uncertain information. In: cartographic perspectives, vol 13, p 10

Mackaness W, Ruas A, Sarjakoski T (2007) Generalisation of geographic information: cartographic modelling and applications. Oxford, Elsevier

Mackaness WA, Chaudhry O (2013) Assessing the veracity of TF-IDF for extracting place semantics from Flickr tags transactions in GIS (in press)

McMaster RB, Shea KS (1992) Generalization in digital cartography. Association of American Geographers, Washington

Mondzech J, Sester M (2011) Quality analysis of OpenStreetMap data based on application needs. Cartographica 46(2):115–125

Mustière S, Devogele T (2008) Matching networks with different levels of detail. GeoInformatica 12(4):435–453

Neis P, Zielstra D, Zipf A (2012) The street network evolution of crowdsourced maps: OpenStreetMap in Germany 2007–2011. Future Internet 4(1):1–21

Newson P and Krumm J (2009) Hidden markov map matching through noise and sparseness. In: Proceedings of the 17th ACM SIGSPATIAL international conference on advances in geographic information systems (GIS '09), pp 336–343

Noy N (2004) Semantic integration: a survey of ontology based approaches. SIGMOD Rec 33:65–70

Paelke V, Dahinden T, Eggert D, Mondzech J (2012) Location based context awareness through tag-cloud visualizations. Adv Geo-Spatial Inf Sci (CRC Press)

Poser K, Dransch D (2010) Volunteered geographic information for disaster management with application to rapid flood damage estimation. Geomatica 64(1):89–98

Quddus MA, Ochieng WY, Noland RB (2007) Current map-matching algorithms for transport applications: state-of-the art and future research directions. Transp Res Part C: Emerg Technol 15(5):312–328

Rabiner LR, Juang BH (1986) An introduction to hidden Markov models. IEEE ASSP Mag 3(1):4–15

Rak A, Coleman D, Nichols S (2012) Legal liability concerns surrounding volunteered geographic information applicable to Canada. In: Rajabifard, A, Coleman D (eds) Spatially enabling government, industry and citizens, USA: GSDI Association Press, pp 125–143. http://www.gsdi.org/gsdiconf/gsdi13/papers/256.pdf

Rambaldi G, Mccall M, Weiner D, Mbile P, Kyem P (2004) Participatory GIS. http://www.iapad.org/participatory_gis.htm

Ramm F (2012) Am Puls der Zeit—Minütliche updates bei OpenStreetMap. Paper presented at the FOSSGIS 2012, Dessau

Ramm F, Topf J, Chilton S (eds) (2010) OpenStreetMap: using and enhancing the free map of the world. UIT Cambridge, Cambridge

Roick O, Heuser S (2013) Location based social networks—definition, current state of the art and research agenda. Trans GIS. doi:10.1111/tgis.12032

Rouse LJ, Bergeron SJ, Harris TM (2007) Participating in the geospatial web: collaborative mapping, social networks and participatory GIS. In: Scharl A, Tochtermann K (eds) The

Geospatial Web. Springer, London, pp 153–158. Retrieved from http://dx.doi.org/10.1007/978-1-84628-827-2_14
Safra E, Doytsher Y (2006) Integration of multiple geo-spatial datasets. ASPRS 2006 annual conference, Reno, Nevada
Samet H (eds) (2005) Foundations of multidimensional and metric data structures. The Morgan Kaufmann series in computer graphics and geometric modeling. Morgan Kaufmann Publishers Inc. San Francisco, CA, USA
Sayda F (2005) Involving LBS users in data acquisition and update. In: Proceedings of the AGILE, Portugal
Schroedl S, Wagstaff K, Rogers S, Langley P, Wilson C (2004) Mining GPS traces for map refinement. Data Min Knowl Disc 9(1):59–87
Scofield RP, Christie D, Sagar PM, Sullivan B (2012) eBird and avifaunal monitoring by the ornithological society of New Zealand. NZ J Ecol 36:279–286
Senaratne H, Gerharz L (2011) An assessment and categorisation of quantitative uncertainty visualisation methods. The 14th AGILE international conference on geographic information science, Utrecht, Netherlands, 18–21 April 2011
Sengstock C, Gertz M (2012) Latent geographic feature extraction from social media. In: Proceedings of the 20th ACM SIGSPATIAL international conference on advances in geographic information systems, Redondo Beach
Sheeren D, Mustière S, Zucker JD (2009) A data-mining approach for assessing consistency between multiple representations in spatial databases. Int J Geogr Inf Sci 23(8):961–992
Siebritz L, Sithole G, Zlatanova S (2012) Assessment of the homogeneity of volunteered geographic information in South Africa. In: international archives of the photogrammetry, remote sensing and spatial information sciences, vol 39-B4. XXII ISPRS Congress, Melbourne, Australia
Siriba DN, Dalyot S, Sester M (2012) Geometric quality enhancement of legacy graphical cadastral datasets through thin plate splines transformation. Surv Revi 44(325):91–101
Slingsby A, Dykes J, Wood J, Clarke K (2007) Mashup cartography: cartographic issues of using Google earth for tag maps. ICA commission on maps and the internet, ICA, Warsaw, Poland, pp 79–93
Smart PD, Quinn JA, Jones CB (2011) City model enrichment. ISPRS J Photogram Remote Sens 6(2):223–234
Stefanidis A, Nittel S (2004) Geosensor networks. CRC Press, Boca Raton
Stigmar H (2005) Matching route data and topographic data in a real-time environment. In: Hauska H, Tveite H (eds) ScanGIS'2005—proceedings of the 10th scandinavian research conference on geographical information sciences, Stockholm, Sweden, pp 89–107
Stoter JE, Meijers M, van Oosterom P, Grünreich D, Kraak MJ (2010) Applying DLM and DCM concepts in a multi—scale data environment. Paper presented at GDI 2010: a symposium on generalization and data integration, Boulder, USA, 20–22 June 2010
Sullivan BL, Wood CL, Iliff MJ, Bonney RE, Fink D, Kelling S (2009) eBird: a citizen-based bird observation network in the biological sciences. Biol Conserv 142(10):2282–2292
Surowiecki J (ed) (2004) The wisdom of crowds: why the many are smarter than the few and how collective wisdom shapes business, economies. Societies and Nations Little, Doubleday
Tauscher S, Neumann K (2012) Combining web map services and opinion mining to generate sentiment maps of touristic locations, symposium on service oriented mapping, Wien, JOBSTMedia Präsentation Management Verlag, pp 277–286, 11/2012
Touya G (2012) What is the level of detail of OpenStreetMap? Role of volunteered geographic information in advancing science: quality and credibility. In: conjunction with GIScience 2012, 18 Sept 2012
van Oort P (2006) Spatial data quality: from description to application. PhD Thesis, Wageningen University, The Netherlands
Volz S (2005) Data-driven Matching of Geospatial Schemas. In: Cohn AG, Mark DM (eds) Spatial information theory. Proceedings of the international conference on spatial information theory (COSIT '05), Ellicottville, NY. Lecture Notes in Computer Science 3693, pp. 115–132

Walter V, Fritsch D (1999) Matching spatial data sets: a statistical approach. Int J Geogr Inf Sci 13(5): 445–473

Warneke H, Schäfers M, Lipeck U, Bobrich J (2011) Matching-based map generalization by transferring geometric representations. In: Proceedings Geoinformatik 2011, Münster, pp 71–77

Weibel R, Dutton G (2005) Generalizing spatial data and dealing with multiple representations. In: Longley P, Goodchild MF, Maguire DJ, Rhind DW (eds) Geographical information systems: principles, techniques, management and applications, 2nd edn (Abridged Edition). Wiley

Yates D, Paquette S (2011) Emergency knowledge management and social media technologies: a case study of the 2010 Haitian earthquake. Int J Inf Manage 31(1):6–13

Yuan S, Tao C (1999) Development of conflation components. In: Proceedings of the international conference on geoinformatics and socioinformatics. Ann Arbor, MI, pp 1–12. http://www.umich.edu/yiinet/chinadata/geoim99/Proceedings/yuan_shuxin.pdf

Zhang Y, Yang B, Luan X (2012) Automated matching crowdsourcing road networks using probabilistic relaxation. In: ISPRS annals of photogrammetry, remote sensing and spatial information sciences, vol 1–4

Zook M, Graham M, Shelton T, Gorman S (2010) Volunteered geographic information and crowdsourcing disaster relief: a case study of the haitian earthquake. World Med Health Policy 2:7–33. doi:10.2202/1948-4682.1069

Chapter 6
Generalisation Operators

Lawrence V. Stanislawski, Barbara P. Buttenfield, Pia Bereuter, Sandro Savino and Cynthia A. Brewer

Abstract This chapter summarises cartographic generalisation operators used to generalise geospatial data. It includes a review of recent approaches that have been tested or implemented to generalise networks, points, or groups. Emphasis is placed on recent advances that permit additional flexibility to tailor generalisation processing in particular geographic contexts, and to permit more advanced types of reasoning about spatial conflicts, preservation of specific feature characteristics, and local variations in geometry, content and enriched attribution. Rather than an exhaustive review of generalisation operators, the chapter devotes more attention to operators associated with network generalisation, which illustrates well the logic behind map generalisation developments. Three case studies demonstrate the application of operators to road thinning, to river network and braid pruning, and to hierarchical point elimination. The chapter closes with some summary comments and future directions.

L. V. Stanislawski (✉)
Center of Excellence for Geospatial Information Science,
United States Geological Survey, Rolla MO, USA
e-mail: lstan@usgs.gov

B. P. Buttenfield
Department of Geography, University of Colorado, Boulder CO, USA
e-mail: babs@colorado.edu

P. Bereuter
Department of Geography, University of Zurich, Zurich, Switzerland
e-mail: pia.bereuter@geo.uzh.ch

S. Savino
Department of Information Engineering, University of Padova, Padova, Italy
e-mail: sandro.savino@dei.unipd.it

C. A. Brewer
Department of Geography, Pennsylvania State University, University Park PA, USA
e-mail: cbrewer@psu.edu

D. Burghardt et al. (eds.), *Abstracting Geographic Information in a Data Rich World*,
Lecture Notes in Geoinformation and Cartography, DOI: 10.1007/978-3-319-00203-3_6,

6.1 Introduction

While there is a rich history of developments in manual generalisation discussed by numerous authors (for example McMaster 1983, 1989; McMaster and Shea 1992; Kilpeläinen 1997; Sarjakoski 2007), automated strategies for generalisation have been demonstrated to reduce manual workloads, and to minimise errors and inconsistencies. The challenge for automatic generalisation was, and continues to be, the development of rules and constraints which control the intensity of generalisation, the sequencing of operators, and conflict detection and resolution. Recent advances in computer science have introduced capabilities for artificial intelligence, amplified intelligence, pattern recognition, and automated spatial reasoning into generalisation software tools. Previous research mainly focused on orchestrating the logical sequence of operators, but models for some cases are still missing (e.g., for a broad range of geographic features, across a wide range of scales, or tailored to landscape differences). Practical application of these advanced techniques requires faster processing speeds as well as data enrichment, which enables context-sensitive generalisation that can be tailored to specific geographies (such as urban or rural areas, or dry and humid landscapes).

Regnauld and McMaster (2007) give a thorough overview of frameworks for generalisation operators. For the purposes of discussion throughout this chapter, a generalisation operator is defined as a generic descriptor for the type of spatial or attribute modification to be achieved on some set of geospatial data (Regnauld and McMaster 2007; Roth et al. 2011). An algorithm refers to a specific method by which one or more operators are implemented, and several algorithms usually exist for the same operator.

There are a myriad of ways in which generalisation may be implemented. Generalisation operators or algorithms may affect data at a micro or macro level, or any level in between. For instance, an operator may affect data at a micro (or atomic) level in a uniform way, such as resampling a digital terrain model, or reducing the number of vertices in linear features by simplification. A different operator may affect features at a subatomic level in a non-uniform way, such as simplification of linear features divided into parts based on the level of coalescence of each part. At the other end of the spectrum, 'super-operators' exist that may affect meso- or macro-level objects in uniform or non-uniform ways with regard to a local context (Ruas and Duchêne 2007). Examples include enrichment or thinning of road or hydrographic networks.

An exhaustive review of generalisation operators is not possible in the space of this chapter. The chapter devotes more attention to operators associated with network generalisation than to other operators. Network generalisation does illustrate the logic behind map generalisation developments. In the past, less has been written about network operations than other operators because the network graph data structure is complex to build and manipulate. However, conceptual and technological advances have enabled more network capabilities in recent years and additional research has been undertaken. Three case studies demonstrate the application of operators to road thinning, to river network and braid pruning, and to hierarchical point elimination. The chapter closes with some summary comments and future directions.

6.2 Generalisation Operators: Chronology of Typologies

A chronology of typologies developed for generalisation operators mirrors advances in processing power, in establishment of linked multi-representation databases, and in improved methods for automatic reasoning about spatial conflicts and geographic context. The shift from early typologies to subsequently published frameworks reflects a still accepted view that generalisation encompasses more than invocation of a list of isolated operators.

Early typologies emphasised paper map production. Ratajski's (1967) model distinguished between quantitative operations that reduced feature details for scale-changing and qualitative operations that transformed symbol designs. Ratajski's model was guided by the concept of a *generalisation point* identifying a critical scale at which representation methods begin to fail, mandating a change in geometry or cartographic symbols. Robinson and Sale's (1969) typology included four elements that were re-sequenced by Morrison (1974) into classification, simplification, symbolisation and induction, in order to focus specifically on abstraction and implementation of cartographic symbols. Nickerson and Freeman (1986) developed a typology similarly suited to map production, with feature deletion, simplification, merging, and reclassification as a first stage of processing, followed by symbol scaling and placement, scale reduction, and label placement. Throughout these developments, the challenge for automation is grounded in the need for 'intelligent' decisions driving the choice of algorithms and parameters. 'Intelligence' in this context relates to rules or constraints which are based upon reasoning on the spatial context.

Sarjakoski (2007, p. 19) reported that the earliest proposals for automatic reasoning and rule formation appeared in Britain, Germany, and the United States, with an operational prototype (GENEX) developed at Hannover. Buttenfield and Mark (1991) proposed an expert system for map design which included an inference engine to drive generalisation. Their typology for generalisation included operators to simplify geometric detail, classify attribute detail, and enhance detail with "… the purposeful and controlled introduction of information to augment or emphasize structures already present in the data" (Buttenfield and Mark 1991, p. 137). Their system was never implemented. Weibel (1991) argued that the lack of progress in operationalising an expert system was due in part to an incomplete understanding about generalisation operators and their interactions. He proposed amplifying the human acuity for "holistic reasoning, visual perception, [and] design" (Weibel 1991, p. 177) with machine acuity for handling repetitive or tedious tasks.

Various proposals for generalisation based upon constraints, rule formation, and amplified intelligence happened concurrently with adoption of object-oriented programming methods, and continued advances in processing speeds and data storage. Beard (1991) identified graphic, structural, application, and procedural constraints on generalisation operators, which respectively preserve display legibility, protect spatial relations, constrain generalisation according to map purpose, and ensure correct operator sequencing. The AGENT project (Ruas and Duchêne 2007; Ruas 1999) encapsulated strategies to eliminate spatial conflicts introduced by generalisation operators. Constraints guided the selection and trial of strategies within which the objects could self-modify. The AGENT project significantly advanced formalisation of automatic reasoning, and initiated examination about the impacts of varying geographical context upon generalisation.

Finally, Roth et al. (2011) propose a detailed typology of operators for multi-scale mapping that includes map design modifications on symbols and labels with modifications of feature geometry and attributes. Operators are categorised based on their impact on map content, geometry, symbol, and label. The typology has greater emphasis on map design than earlier approaches with the expectation that additional operators may be included as advancements are made in multi-scale and web-based mapping.

6.2.1 Vector-Based Operators

Figure 6.1 places the operators into three categories based on the intended function of the operator. Vector data may be pre-processed through enrichment or reclassification operators to enable subsequent operators that affect the amount of detail (visual quantity) or aesthetics (visual quality) of features retained for cartographic display. Arrows suggest the functional sequence of operators (or operations) typically implemented in workflows, with larger arrows representing more common approaches. Operators that reduce the quantity of content, or clutter on a visual display usually occur before retained features are massaged through operators that affect the visual clarity intended for the display, ergo visual quality. Less often, visual quality operators may require additional removal of content to achieve desired results, e.g. constraints for displacement or alignment may not be satisfied with

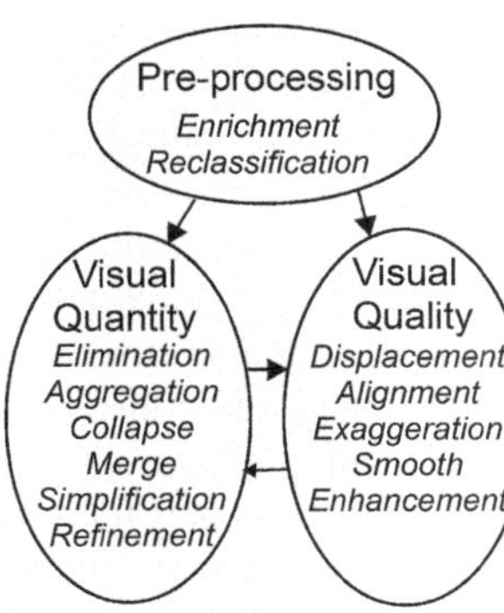

Fig. 6.1 Functional classification of primary vector generalisation operators. *Arrows* represent typical sequences of operators used in generalisation workflows

existing content mandating further feature elimination. This set of vector operators is presented here to provide a quick and simple overview. It follows Ratajski's (1967) concepts for map production, but with the addition of pre-processing functions. It closely resembles Morrison's (1974) typology, and the Roth et al. (2011) typology without the symbol and label operators except enhancement.

6.2.1.1 Pre-Processing Operators

Enrichment adds feature attribution to describe aspects of geometry, existing attributes, and/or local geography that are not explicitly stored in the original database. For instance, prominence of hydrographic network features may be derived by adding upstream length or drainage area, stream order, or hydrologic index (Verdin and Verdin 1999; Ai et al. 2006; Stanislawski 2009; Savino et al. 2011a; Wu et al. 2012). Information gathered from "structure recognition" or "structure analysis" (Steiniger and Weibel 2007) also provides enrichment information.

The *Reclassify* operator groups features based on existing attribution, including enriched data. Reclassification is a common form of pre-processing that facilitates the subsequent application of generalisation operations.

6.2.1.2 Operators Affecting Quantity of Visual Information

Elimination refers to removal of one or more features without replacement. It is generally used to reduce feature content for display at a reduced scale. It has also been referred to as *selection* (McMaster and Shea 1992; Jiang and Claramunt 2004; Gülgen and Gökgöz 2008; Touya 2010), *class selection* (Foerester et al. 2007), *extraction* (Wu et al. 2012), *thinning* (Punt and Watkins 2010; Briat et al. 2011; Stanislawski et al. 2012a; Brewer et al. 2013a, b), or *pruning* (Stanislawski 2009; Stanislawski and Savino 2011).

An *Aggregation* operation replaces many related features with a representative feature of increased dimension, such as replacing many point features with a single polygon feature. To clarify, point features may be considered zero dimensional, lines one dimensional, and polygons two dimensional. In slight contrast McMaster and Shea (1992) exclusively define aggregation as replacement of multiple points with a polygon.

Merge replaces more than one feature with a representation of equal dimension, such as creating a single envelope around multiple proximal polygons. In other typologies merge operates exclusively on linear features, while *amalgamate* operates exclusively on polygons. Some authors refer to aggregation only when combining isolated features, regardless of dimension, and refer to merging when combining connected or adjacent features. Here, as in Roth et al. (2011), we use change in dimension, or lack of, by an operator for a more concise distinction between operators where possible.

Collapse replaces a feature with a representation of lower dimension, such as replacing a polygon with a point or a line. *Refinement* is used to reduce multiple features or sets of features to a more simple representation of fewer features. For instance, a network of streams and canals near the shoreline of a body of water may be converted to polygonal delta feature through refinement. McMaster and Shea (1992) consider refinement as a process to reduce clutter in a display, after the primary elimination process. When numerous features are replaced by fewer features (or symbols) of the same type, it is referred to as *typification* (Regnauld and McMaster 2007; Foerster et al. 2007; Roth et al. 2011).

Simplification reduces the number of points used to represent a line or polygon boundary. Algorithms that *filter* vertices from lines fit into this operator class.

6.2.1.3 Operators Affecting Quality of Visual Information

Displacement adjusts the location of a feature to avoid coalescence with nearby features while maintaining topological integrity with each feature. In some instances, displacement is affine, while other typologies may augment positional shifting with shape adjustment, such as for building generalisation.

Alignment rotates or adjusts the orientation of a feature to maintain or emphasize its relation to other, proximal or adjacent features. This operation may also be referred to as *rotation* or *squaring*, and is sometimes subsumed under displacement.

Smoothing removes small variations in the geometry of a feature to improve its appearance. Smoothing can insert additional coordinates to protect against abrupt changes in directionality of shape, or modify original coordinates as in the case of low- or high-pass filtering.

Exaggeration refers to amplification of a specific part of a feature to maintain the clarity or a particular aspect of the feature. *Caricature* or *enlargement* operators may be included in exaggeration.

Enhancement includes graphic embellishments around or within a symbolised feature to maintain or emphasise spatial relationships.

6.2.2 Raster Operators

We might argue that the compression of raster images is a form of map generalisation, but overall vector based approaches have come to dominate. However McMaster and Monmonier (1989) did distinguish four types of raster generalisation, and Jenny et al. (2011) applied multi-scale Laplacian pyramids to generalise terrain and used curvature coefficients, similar to Leonowicz et al. (2010), to preserve or accentuate edges or other important relief features. Numerical categorisation techniques (McMaster and Monmonier 1989) reduce content of raster data by image segmentation, cell-value slicing, classification, zonal statistics, and channel and ridge extraction. Applications of numerical categorisation include formation of morphologically similar terrain partitions (Chaudry and Mackaness 2008a), and construction of natural drainage density partitions for hydrologic generalisation (Stanislawski et al. 2012b).

Weibel (1992) focused specifically on terrain and identified three raster strategies including global filtering, iterative filtering, and a heuristic approach based upon structure lines. Schylberg (1993) introduced a set of raster ("area feature") operators that process contiguous pixels from the same category. Raster techniques see important application in altering image resolution through either resampling or aggregation. As an alternative to such global operators, neighbourhood operators can be applied in order to generalise portions of a raster based on a user-specified search radius, or a moving window (Fotheringham et al. 2000).

6.3 Operators in Commercial Software

The Euro SDR project (Stoter et al. 2010) was a major coordinated research effort by several European National Mapping agencies, academia, and software vendors to evaluate four state-of-the-art (at the time of testing) commercial generalisation software systems. The project identified strengths and weaknesses of the systems, which stimulated vendor enhancements; suggesting that further customisations are needed, particularly to detect and handle differing geographical contexts. Here we provide a brief description of five commercial geographic information system (GIS) software systems that offer tools or systems for cartographic generalisation. Table 6.1 provides a summary of vector operators that are available in each of these systems. Most of the systems were described in Regnauld and McMaster (2007) and evaluated in the Euro SDR project (Stoter et al. 2010).

Axpand™ ng is a product of Axes Systems in Switzerland. The system employs about 40 different algorithms in combinations or 'operators'. It is a constraint-based system founded on a multi-representation (MRDB) data model (Sarjakoski 2007; Bobzien et al. 2008) and includes automatic generalisation and incremental updating through selective re-generalisation of updates to the source data. Axpand ng is process-based and operates through workflow processing. Operators and constraints are invoked from within a workflow, which can contain sub-workflows for specialised generalisation of 'zones', or regions within the data with constraints specific to that region. This newer axpand ng system was not tested during EuroSDR research. Section 11.4 demonstrates axpand ng.

Esri's ArcGIS® software (version 10.1) has a cartography toolbox that includes tools for cartographic generalisation. It furnishes a geoprocessing environment with numerous functions that can be sequenced into algorithms and tools through the Python scripting language. The environment enables automated partitioning for processing large datasets.

GIPS is Intergraph® Corporation's Geospatial Intelligence Production Solution set of software products that is built upon the GeoMedia® family. GIPS provides a

Table 6.1 A summary of vector generalisation operators available in commercial GIS software systems: Axes axpandTM ng, Esri ArcGIS®, Intergraph GeoMedia® GIPS, Leibniz University of Hanover's CPT, and 1Spatial's Radius Clarity

Operators/algorithms	Axes expand ng	ArcGIS 10.1	Geomedia GIPS	Hanover CPT	1Spatial clarity
<u>Line simplification</u>					
Nth point	✔		✔		
Douglas	✔	✔	✔		✔
Lang			✔		
Reuman-witkam			✔		
House algorithms	✔	✔	✔	✔	✔
<u>Line smoothing</u>					
Brophy			✔		
Averaging	✔		✔		
McConalogue interpolation					✔
Bezier interpolation	✔	✔			
Akima interpolation					✔
Other cubic splines	✔	✔			✔
House algorithms	✔	✔		✔	
<u>Exaggeration/Bend caricature</u>					
Accordion	✔				✔
Min break	✔				✔
Max break	✔				✔
Bend removal	✔				✔
Exaggeration	✔			✔	✔
<u>All-in-one line generalisation</u>					
Plaster	✔				✔
Generalise-by-parts	✔				✔
<u>Line merging</u>					
Blend line	✔	✔	✔		
<u>Area simplification</u>					
Irregular shape	✔	✔			
Orthogonal shape	✔	✔	✔	✔	✔
Turn to rectangle	✔	✔	✔	✔	✔
<u>Area enhancement</u>					
Area extend	✔			✔	
Squaring	✔	✔	✔		✔
<u>Merge/Aggregation</u>					
Merging	✔	✔	✔		✔
Irregular amalgamation	✔	✔		✔	✔
Orthogonal amalgamation	✔	✔	✔	✔	
Point aggregation	✔	✔	✔		✔
<u>Collapse</u>					
Area > Point	✔	✔	✔		✔
Line > Point	✔	✔			✔

(continued)

Table 6.1 (continued)

Operators/algorithms	Axes expand ng	ArcGIS 10.1	Geomedia GIPS	Hanover CPT	1Spatial clarity
Points > Point	✔	✔			✔
Area > Line	✔	✔	✔		✔
Area > Edge	✔				
2 Lines > Line	✔	✔			✔
Refinement/Typification					
Points	✔			✔	✔
Network simplification					
Street network	✔	✔	✔		✔
River network	✔		✔		✔
Neighbourhood detection					
Hierarchical Blend line	✔		✔		
Exaggeration/Area enlargement					
Scaling	✔		✔	✔	✔
Enlarge to rectangle	✔		✔		✔
Enlarge bottleneck	✔				✔
Conflict detection	✔	✔			✔
Clustering	✔	✔			✔
Displacement					
Vertex	✔	✔		✔	✔
Holistic	✔	✔		✔	✔

rich set of data capture, review, and editing as well as product finishing capabilities for geospatial data. The GIPS Feature Cartographer product provides tools for using generalisation functionality in a production mapping environment.

Change, Push and Typify (CPT) provides parameterised batch tools which simplify, aggregate, displace, deform (i.e. object specific geometric operations, such as exaggerate), and typify (refine) objects. The system was developed by the Institute of Cartography and Geoinformatics, Leibniz University of Hanover. Change and Typify are designed for generalisation of building features, whereas Push performs holistic displacement of any feature type.

1Spatial's Radius Clarity (now rebranded as 1Generalise) is a constraint-based, object-oriented system, with a multi-agent framework designed to find optimum generalisation solutions in complex situations, especially where contextual relationships between geographic features are important. The operators iterate towards a solution which maximises the satisfaction of a set of constraints. A framework for operator sequencing is also available, as well as batch processing. The system requires configuration of numerous parameters and constraints.

6.4 Recent Advances in Operator Development

Since publication of the 2007 ICA book (Mackaness et al. 2007), much research and development has focused on advancing methods to automate generalisation over large areas, such as a country, or on large databases. Consequently work has focused on identifying and implementing rule-based constraints to control generalisation, enrichment, tailoring sequences of operations for contextual classes, and metrically assessing data and generalisation results to validate and refine processes.

6.4.1 Enrichment

Most developments in generalisation frameworks extend the premise that "place matters", developing generalisation strategies which vary depending on local spatial context or geographic conditions. These contexts are made explicit through the process of 'enrichment'—an essential pre-processing stage.

The most common reason for enriching data with regard to generalisation is to assign relative prominence estimates to data for use in subsequent elimination operations. In addition, data enrichment formalises spatial relations and adds data characteristics explicitly to the set of attributes (Bobzien et al. 2008). Detected and defined patterns in spatial data or local geography can guide generalisation to retain feature characteristics based on context, spatial distribution, and geographic conditions (Buttenfield et al. 2011; Touya et al. 2010). Steiniger and Weibel (2007) enrich cartographic data with vertical and horizontal relations to reduce subsequent processing. Neun et al. (2008) consider enrichment a labour-intensive yet mandatory operation for web-based generalisation services.

Enrichment can be applied to all forms of vector data (i.e., lines, networks, polygons, points) for generalisation purposes. Zhang et al. (2010) utilise enrichment to preserve shape and alignment in generalised building polygons. Steiniger et al. (2008) use supervised classification by way of discriminant analysis of geometric shape of buildings to assign urban structure characteristics for building generalisation. Use of geometric algorithms or statistical measures have been investigated for defining building patterns (Christophe and Ruas 2002; Zhang et al. 2010) or detecting island structures (Steiniger et al. 2006).

6.4.1.1 Partitioning

Partitioning is typically performed in order to subdivide data into manageable units. Partitioning data into manageable units is regularly performed for parallel processing of raster data (Wallace et al. 2010). With ArcGIS 10.1, Esri introduced partitioning to enable processing of large datasets for several algorithms, including road network thinning (Briat et al. 2011; Esri 2012).

In addition, partitioning may be used as an enrichment operator to form context-based geographic spaces that allow proper application of site-specific generalisation operations (Ruas 1995; Bobzien et al. 2008; Chaudhry and Mackaness 2008a; Stanislawski and Buttenfield 2011; Touya 2010; Touya et al. 2010). Partitioning for contextual classification may be completed on individual or multiple data themes, and the partitioning method depends on the generalisation strategy. For generalisation purposes, a partition is a logically-derived subset of a larger dataset which is to be separately processed with one or more generalisation operators or algorithms. Among others, partitions may be referred to as clusters, areas, regions, or groups.

For example, vector and raster algorithms were used to assign line-density partitions to hydrography and road network features to help retain density variations during generalisation (Stanislawski 2009; Stanislawski and Buttenfield 2011; Stanislawski et al. 2012a; Brewer et al. 2012).

Chaudhry and Mackaness (2008b) clustered buildings around road nodes to estimate settlement boundaries and develop partonomic relations for a multiple representation database and for generalisation. Werder et al. (2010) proposed an unsupervised classification to cluster buildings into settlement areas based on geometric and topological characteristics of the building outlines and road network data. Bildirici et al. (2011) combined buffers around building point features separated by road lines to build point clusters for typification algorithms.

Raposo et al. (2013) used a scale-dependent regular tessellation to subdivide topographic summit points for enrichment and thinning operations. Bereuter and Weibel (2013) apply a quadtree tessellation to subdivide point data for real-time generalisation (Sect. 6.7).

6.4.2 Transformations of Groups of Objects

It is often necessary to simultaneously process groups of objects (points, lines, polygons) through generalisation operators in order to retain relative geometric characteristics among the features through scale changes. This section describes some recent research or new approaches involving transformation of groups of objects.

Bildirici et al. (2011) proposed typification operators that use length and angle measures between the points (buildings) in each cluster to retain the geometric structure of each cluster through scale reduction. Also, an incremental approach for automatically displacing building points away from road lines and other points was proposed for generalisation (Aslan et al. 2012).

Working with point objects, Raposo et al. (2013) propose an automated way to separately thin each partition of points by summit prominence, which is enriched on the points through integration with elevation contours. Partitions are derived through a scale-dependent regular tessellation. The objective is to automatically select the most prominent set of summit points for legibly labeling summit points through multi-scale displays. Bereuter and Weibel (2013) partition point data using quadtrees and present operators for selection, simplification, aggregation, and displacement for on-the-fly generalisation (Sect. 6.7).

van Oosterom (2005) and van Oosterom and Meijers (2011) present a truly vario-scale structure for smooth zooming through a polygonal thematic map (Chap. 4).

6.4.3 Generalisation of Networks

Networks have a variety of geographical uses, which include mapping transportation and hydrography, routing materials and services, and hydrologic modelling. Increasingly, networks are being used as a reference framework for relating ancillary data or for managing supply chains (Long et al. 2012; Simley and Doumbouya 2012; Yager et al. 2012). Vector geospatial data for networks may be stored in a directed or undirected graph structure, which enable traversal and other network analysis functions. Because the spatial pattern and connectivity of a network affects interpretation and modelling, networks are typically generalised as a macro object through super-operators or processes, which combine several operators on a group of features.

Several methods to generalise hydrography or road networks have been developed, and research continues to refine these methods. Development of processes or algorithms for building a hierarchy of 'strokes' or paths of best continuation (Thomson and Richardson 1999; Thomson and Brooks 2007; Chaudhry and Mackaness 2005) is still a focus for network thinning (a vector elimination operator, Sect. 6.2.1) strategies. Prominence estimates help define and rank strokes in the network.

For river networks, some factors contributing to overall prominence are stream order, stream name, longest path, drainage area, straightness, and upstream branches (Ai et al. 2006; Touya 2007; Stanislawski 2009; Savino et al. 2011b; Gutman 2012). In the United States, each network feature in the National Hydrography Dataset (NHD) is assigned a unique permanent 'reach code' address that defines a continuous segment of surface water in the network, which often spans more than one feature. Instead of generating strokes to thin the NHD network, reach codes are assigned prominence estimates based on enriched upstream drainage area estimates. Local density is also assigned to NHD network features through a line density partitioning algorithm. A stratified pruning process then thins each partition to a prescribed target density, which ensures that natural density variations reflecting local geographic conditions are maintained (Stanislawski 2009; Buttenfield et al. 2011; Stanislawski and Savino 2011).

Recent research on network generalisation has focused on refining thinning strategies to account for local contextual variations related to local geography. Network thinning operations have been enabled or refined through enrichment of local line density (Savino et al. 2011a; Stanislawski et al. 2012a; Benz and Weibel 2013), pattern (Heinzle et al. 2005, 2007; Touya 2007; Savino et al. 2011b), road type (Balboa and Lopez 2008; Savino et al. 2010; Stanislawski et al. 2012a), and road network or block structure (Jiang and Claramunt 2004; Touya 2010; Gülgen and Gökgöz 2011). With regard to road networks, Zhou and Li (2012) evaluated several stroke-building strategies. Later, Li and Zhou (2012) combined stroke and mesh-density strategies (Chen et al. 2009) to thin road networks. A mesh is a region enclosed by network roads, and each mesh is assigned a density. Mesh-density thinning progressively removes edges of the highest density mesh until a minimum density is achieved. Benz and Weibel (2013) further researched combined stroke- and mesh-based road thinning, adding refinements to maintain density patterns of settlement areas. Stanislawski et al. (2012a), Brewer et al. (2013a, b) tested road network thinning stratified by density between rural and urban areas. Results indicate the process will support thinning of road and road labels for multi-scale display (Sect. 6.5).

Other generalisation research on road networks has focused on techniques or algorithms to remove excess detail (e.g., roundabouts, divided highway) at the appropriate level of detail, which can enhance network thinning and road labelling through scales (Brewer et al. 2013a, b; Weiss and Weibel 2013).

6.5 Case Study I: Generalisation of Road Networks

Lawrence V. Stanislawski and Cynthia A. Brewer

This section describes development of a workflow to implement a super-operator that thins (eliminates) road network data to multiple levels of detail for cartographic display and data delivery. A goal of the United States Geological Survey (USGS) is to implement automated generalisation to enable multi-scale display and delivery of transportation and other geospatial data available through the USGS *National Map* (Sugarbaker and Carswell 2011). *The National Map* transportation data displayed on 1:24,000 (24k) US Topo maps (Sect. 11.6) are currently compiled by TomTom® North America and the U.S. Forest Service and stored in the Best Practices (BP) Data Model using the Esri geodatabase format (USGS 2006). The goal is to thin the BP road network data, using commercially available tools, to levels of detail appropriate for scales ranging from about 1:20,000 (20k) to 1:1,000,000 (1M). This research is being conducted by the USGS Center of Excellence for Geospatial Information Science (CEGIS) in collaboration with University of Colorado-Boulder and Pennsylvania State University.

6.5.1 Context and Objective

Esri ArcGIS® version 10 and later includes the Thin Road Network tool that applies simulated annealing to automate road network thinning (Punt and Watkins 2010; Briat et al. 2011). Connectivity, general morphology of road patterns, and integrity of navigable routes are maintained in the thinned network. The thinning algorithm considers relative importance, significance, and density of input features. Importance is determined through the hierarchical road class (e.g., Interstate, State Route, local road, etc.) assigned as an attribute on the input data. Feature significance or functional relevance (Thomson and Richardson 1995) is determined as a feature's participation in long routes across the extent of the network. Features that are part of long network routes are deemed more significant than those required only for local travel. Density is computed similar to any density metric, as a ratio of the length of a street segment to its associated area.

The tool can be used to thin a topologically clean road network that has a hierarchy field. Therefore, use of the tool usually requires pre-processing to populate the hierarchy and importance fields and enhance the integrity of the input network features. The tool does not actually thin the network, but rather enriches the network with a binary "invisibility" field populated with one (invisible) or zero (visible) determined by the thinning algorithm and input parameters. The level of thinning from a single run of the tool is controlled by the importance field and "minimum length" parameters. The minimum length value is a tolerance that roughly corresponds to the shortest road segment that is visually sensible to include in the thinned network at the final scale (Esri 2011). The network can be thinned based on results stored in one or more invisibility fields populated by one or more runs of the tool.

The relation between minimum length and resulting network density is not consistent, and depends on conditions of the input road network, which vary with local geography. In addition, CEGIS researchers have noted that using the Thin Road Network tool with a uniform minimum length tolerance to thin an area of roads with substantial line density variations tends to homogenise density variations at smaller scales more than expected by the Radical Law (Töpfer and Pillewizer 1966). Consequently, a stratified network thinning strategy, similar to the methods applied to the NHD (Sect. 6.4.3), can be used to better maintain density variations, although this method requires multiple runs of the tool on the same dataset (Stanislawski et al. 2012a; Brewer et al. 2012).

6.5.2 Methods

To facilitate processing with the Thin Road Network tool, data partitions (Fig. 6.2, black lines) of up to 50,000 features were created for the BP road layer for the United States with Create Cartographic Partitions (Esri 2012). The cartographic

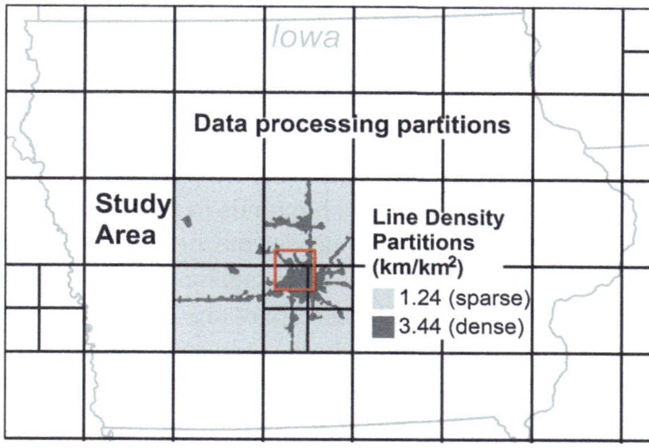

Fig. 6.2 Line-density partitions (*light and dark grey shaded areas*) generated for TomTom/USFS road features within the study area of seven data partitions (*bold black lines*) covering the Des Moines, Iowa area. *Redsquare* near middle of study area shows location of Fig. 6.3 examples

partitioning tool uses a quadtree-based approach to generate partitions, which seek to minimise the effect of partition shape on processing algorithms. A study area of seven data partitions (Fig. 6.2, light and dark grey shaded areas) covering the Des Moines, Iowa region were used to select a subset of BP road features for testing. This study area covers about 21,600 km^2 and includes over 104,000 BP road features.

The set of road features in the study area were prepared for the Thin Road Network tool to ensure that proper network topology and attribution exists. Pre-processing of feature geometry included projecting data from geographic coordinates to the North American Albers Equal Area Conic projection, removal of coincident (overlapping duplicate) features where necessary, conversion of multi-part to single-part features, and ensuring that features are split at all intersections. Feature attribution was pre-processed to transfer road names to retained features when removing coincident features, to populate the road class where needed based on feature names and types, and to assign importance values based on road class values. Custom tools were developed with Python and Esri geoprocessor functions to automate the pre-processing tasks.

Two line density partitions were generated for the test data using the raster-based partitioning process described by Stanislawski and Buttenfield (2011) with a density class break of 1.50 km/km^2 and a minimum polygon area tolerance of 45 km^2 (Fig. 6.2). Density partitioning assigns a density class to each road feature. The two partitions generated have overall road densities of 1.24 and 3.44 km/km^2 for the sparse and dense partitions, respectively.

Subsequently, the Thin Road Network tool was run in a ladder fashion on the test data, thinning with a series of minimum lengths incrementing from 500 m to 30 km. The resulting visibility settings were combined to produce a single 35-level attribute used to remove the least important roads in each partition through scale.

6.5.3 Results and Discussion

Table 6.2 lists rounded scales calculated for each thinning level for the dense and sparse partitions, as well as for the study area as a whole. Target scales are calculated by inverting the modified Radical Law (Buttenfield et al. 2011) to compare total length of visible roads to the original length of all roads at the source scale of 24k (listed as 24 across the first row of Table 6.2). Looking across each row, scales become smaller from sparse to whole to dense areas (left to right). This trend shows that the denser areas are being thinned more aggressively than the sparse areas causing the whole study to become more homogeneous in density at a single thinning level. This pattern across rows lessens as scale decreases, and at a laddered minimum length of 23 km the target scale for both partitions is fairly similar. We are not yet able to direct the Thin Road Network tool to produce a specific density of features, but this inventory of lengths helps understand the results across dense and sparse areas through scale, with the goal of producing general guidance on use of this thinning algorithm.

Looking at Table 6.2 to find shared approximate scales offers direction on how to combine thinning levels to retain a relative sense of dense and sparse character in the road network that is suitable for a particular scale. Examples of similar scales in different rows in the Table 6.2 are highlighted by hue. For example, aiming at 100k, orange highlights in the table show 10 km thinning is suitable for approximately 106k scale mapping overall, but thinning with 11 and 4 km best approaches 100k scale for the sparse and dense partitions, respectively. Pink highlights show a similar pattern for about 400k, with less contrast in the thinning levels combined for the two density partitions.

Table 6.2 also shows that at longer minimum lengths (greater than 23 km) the thinning scales invert between sparse and dense partitions. It has been demonstrated in later research that this scale inversion can be eliminated by merging divided highways in the road network prior to the running the Thin Road Network Tool (Brewer et al. 2013b). Furthermore, elongated patterns in the line density partitions that follow divided highway (Fig. 6.2) may be eliminated by merging divided highways prior to generating the line density partitions.

Figure 6.3 demonstrates the example thinnings for 100k (b and c) and 400k (d), with all maps drawn at 500k for comparison (Fig. 6.3a showing all roads has a target scale of 24k). Figure 6.3c has more representative densities in the urban areas to the lower left of the sample area (in the purple, dense partition shown in Fig. 6.3a) with all roads adequately thinned for small-scale display. Short dangling roads near partition boundaries were trimmed.

6.5.4 Next Steps

Using similar methods to thin other BP road data, Stanislawski et al. (2012a) noted that target 100k thinning densities estimated through the Radical Law are

Table 6.2 Levels of thinning for the study area shown in Fig. 6.2. For one level of thinning (row), each column lists suitable scales calculated using the total length of road segments by density partition (e.g., a scale of 34 represents 1:34,000). Suitable scales are computed through the Radical Law based on length (Töpfer and Pillewizer 1966; Brewer et al. 2013a)

Thin Road Network (minimum length)	scale sparse partition	scale whole study area	scale dense partition	
all roads	24	24	24	
500m	26	28	34	
1000m	28	32	47	
1500m	29	35	57	
2000m	30	36	66	
2500m	30	38	74	
3000m	31	39	82	
3500m	31	40	88	
4000m	32	42	96	
5000m	34	45	111	
6000m	43	56	128	
7000m	54	69	143	
8000m	64	81	167	
9000m	71	90	185	
10km	85	106	208	
11km	100	122	225	
12km	119	142	240	
13km	135	160	259	
14km	155	185	311	
15km	178	210	338	
16km	208	241	366	
17km	243	275	387	
18km	271	303	409	
19km	307	335	424	
20km	352	376	445	
21km	391	409	459	
22km	440	453	490	
23km	507	504	498	inverts
24km	567	549	509	
25km	623	593	527	
26km	834	734	552	
27km	928	792	562	
28km	1,045	859	570	
29km	1,143	912	573	
30km	1,205	976	630	

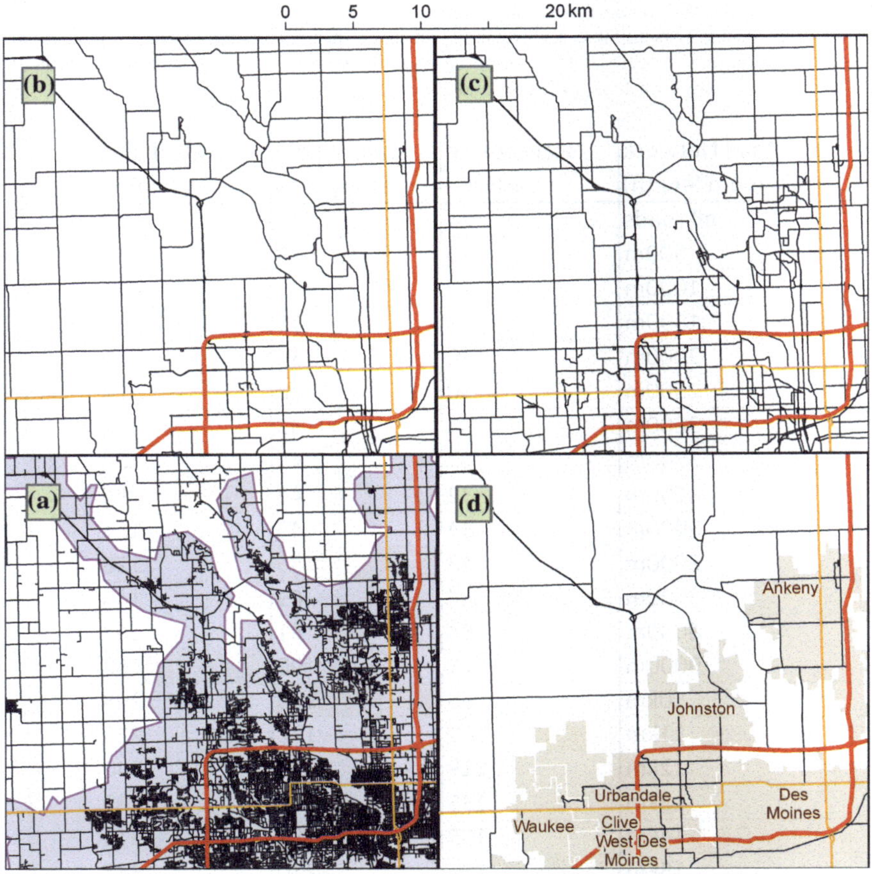

Fig. 6.3 Results of using Esri Thin Road Network tool on **a** original road data from USGS Best Practices data model, with line density partition in *purple*. Original roads thinned with minimum lengths of **b** 11 km, **c** 4 and 11 km in dense and sparse partitions, respectively, with trimming of short features at partition boundaries, and **d** 18 and 21 km in dense and sparse partition, respectively, with trimming of short features (incorporated places in beige, named for location reference). The four maps are each displayed at 500k

substantially less than benchmark densities from 100k USGS road data. USGS standards indicate that the vast majority of 24k road features are retained for 100k mapping (USGS 1985). The primary purpose of the map will govern the density and prioritisation of feature content. Scale-dependent densities governed by the Radical Law or other scaling rules (Jiang et al. 2013) may be adequate for interactive review of data through The National Map Viewer, but greater dominance of transportation features, as in historical USGS Topographic maps, may still be required for large and medium scale US Topo products. These decisions are yet to be determined, but enriching BP data with thinning levels as described in this section provides the flexibility to thin the data as needed for each purpose.

This approach is being tested on a variety of geographical locations across the United States to develop general thinning recommendations. Upon compiling results for other locations, an optimal implementation strategy will be designed to automate the process on the entire BP transportation theme of *The National Map*, including railroads. The design should minimise processing time (i.e. the number of runs of the tool), the number of density partitions, and the number levels of detail (LoD) datasets that must be generated and stored. Each LoD will be used for mapping a range of scales, which must be determined, and appropriate invisibility fields for each density must be joined to the attributes of each LoD. Among other details, research must identify appropriate simplification operators for each LoD and verify the LoD best suited for merging divided highways.

In other work on the same test data, Brewer et al. (2013a, b) identified a set of LoD scale ranges adequate for the target display scales of 20k to 1M and made use of thinning levels to prioritise road labels. Their system for road labelling will be further tested. Additional work with transportation data will examine how these thinning levels combine with other base information, such as hydrography, terrain, and populated place labels for complete mapping. Appropriate ranking of themes and features for displacement operators and integration (snapping) must be identified and available algorithms must be tested.

6.6 Case Study II: River Network Pruning by Enrichment and Density Analysis

Sandro Savino

This section illustrates an algorithm for the generalisation of river networks developed at the University of Padua, Italy. The network pruning algorithm enriches the network features with relevant information to assess the importance of each part of the network. Local and global density analyses are also used to further improve the selection process. The algorithm can generalise both natural and artificial streams, although an approach involving typification is probably best suited for generalising artificial streams that show regular patterns (e.g. irrigation systems) (Savino et al. 2011b). The algorithm was originally developed to generalise 1:5,000 scale data to 1:25,000 scale, but by tuning its parameters it was successfully adapted to data at different scales (Stanislawski and Savino 2011).

The workflow can be divided into four distinct operations: data enrichment, network pruning, braided sections generalisation, and density tuning. Where not otherwise stated, it is assumed that the network is composed only of linear elements. In cases where hydrographic drainage features include areal elements, their linear counterpart should be generated (e.g., by calculating the middle axis) in order to be processed (e.g., Regnauld and Mackaness 2006). In the following sections, a *feature* refers to a single linear element of the hydrographic network that may include one or more vertices between its endpoints. The term *river* or *stream* refers to an object spanning more than one feature, and the point where two or more features connect is called a *node*.

6.6.1 Data Enrichment

During enrichment, important data are extracted from the input hydrographic data and stored in a data structure that is used to drive the selection process. Extraction of important data components is based on input data attributes and channel morphological analysis.

The first step is to understand the relationship of each feature with the neighbouring ones. Given a feature R, all the features connected to it are classified either as 'fathers', 'children' or 'siblings': a father is a feature ending into R, a child is a feature beginning from R, and a sibling is a feature sharing a father or a child with R. The classification is based on the flow direction of each feature; in case this information is not explicitly available, it is computed by analyzing the values of the Z coordinate of each feature, when available.

Given a feature R, its two vertices with highest and lowest Z value pR_{zmax} and pR_{zmin} and their values R_{zmax} and R_{zmin}, and a feature C and its maximum and minimum Z values C_{zmax} and C_{zmin}:

- C is a father of R if $C_{zmax} > R_{zmax}$ and C is connected to R at pR_{zmax}
- C is a child of R if $C_{zmin} < R_{zmin}$ and C is connected to R at pR_{zmin}
- C is a sibling of R if $C_{zmax} > R_{zmin}$ and C is connected to R at pR_{zmin} or if $C_{zmin} < R_{zmax}$ and C is connected to R on pR_{zmax}

The "equal" case is handled as a special case in the algorithm (Savino et al. 2011a), though not discussed here for reasons of brevity.

We also observe that:

- if R has no fathers, it is a source,
- if R has no children, it is a sink,
- the flow direction of R is from pR_{zmax} to pR_{zmin}.

If the flow direction is known, the procedure above can be applied without comparing Z values, assuming for each feature R flowing from its vertex R_a to R_b, $pR_{zmax} = R_a$ and $pR_{zmin} = R_b$.

Once the relationship of each feature with its neighbours is known, the algorithm computes and stores the following measures for each feature R:

- the Strahler order S_r (Strahler 1952)
- the total distance to the furthest source uphill L_r
- the total number of branching points uphill B_r (where multiple features converge)

These values are computed as follows, where A and B are fathers of R:

$S_r = S_a + 1$ if $S_a = S_b$,
otherwise $S_r = \max(S_a, S_b)$
$L_r = \max(L_a, L_b) + \text{length}(R)$
$B_r = B_a + B_b + 1$

The process starts from a randomly chosen source and continues downhill on one of the children of R (each source has values $S_r = 1$ and $B_r = 0$); if R has no children the algorithm picks another source and starts the process from there; if one of the fathers of R has not been visited yet, R is not processed and the algorithm picks another source to process. The whole procedure ends when all the edges have been processed.

As a last step, the algorithm uses the information gathered to detect the rivers in the network. A river is a sequence of connected features, starting at a source and ending either in a sink or into another river. The river is conceptually similar to the idea of a "stroke" (Thomson and Richardson 1999; Thomson and Brooks 2002) and it is the basic unit on which the pruning process is applied.

The process is bottom-up and starts from one of the sinks: for each feature R the algorithm decides which father of R is the best continuation of the river; the choice is performed by giving a score to each father F of R based on the enriched data.

The score for F will be increased if:

- F has the highest value L_F
- F has the highest value B_F
- F and R have a high collinear (straight) alignment

and, if the data contains this information

- F has the same name of R,
- F belongs to the same hydrographic class of R,
- F has the largest width

On the other hand the score for F will be decreased if:

- F has a different name of R
- F has a lower Strahler order than R

The process ends when each feature has been associated to one river (there might be rivers spanning only one feature); each river inherits the S, L, B values of its most downhill feature (this is similar to Horton 1945).

6.6.2 Network Pruning

Pruning selects rivers deemed important enough for the generalised network, and removes the less important ones. The importance of a river is relative to the target scale and is modeled based on user-defined thresholds for S, L and B. When

Table 6.3 Parameters used to generalise data at different scales

	1:25,000	1:50,000	1:100,000
Minimum river length (m)	250	600	1600
Buffer size for density pruning (m)	120	400	1200
P_{max} percentage of buffer overlap (%)	50	50	50
S strahler order	3	3	3
L length to furthest source (m)	1000	1500	3200
B number of branches uphill	4	8	16

deleting a river, all the features composing the river or converging with it (i.e. its father features) are removed from the network.

Selection happens in two steps, the first deletes all the rivers shorter than a minimum length threshold (ideally, too short to be represented at the target scale), while the second prunes the network in the areas where it is too dense, improving the legibility of the output. The first selection deletes all the rivers having L smaller than a minimum value. Removal of full rivers, instead of single features, maintains the connectivity of the network.

The second selection requires assignment of local density to each river, which is performed by drawing a buffer around each river and calculating the percentage of its area that overlaps the buffers drawn around neighbouring rivers. The higher this percentage P, the closer the river is to other rivers. The algorithm sorts the rivers by decreasing values of P and, starting from the river with highest P, analyzes one river at a time, removing only the less important. The importance of a river is assessed comparing its S, L and B values with the user defined thresholds: the river is removed only if all the three values are below the thresholds; upon removal, the values of P of the neighbouring rivers are updated and the list sorted accordingly. The process continues until the highest value of P is below a threshold P_{max} or every river having P bigger than the threshold is deemed too important to be removed. Table 6.3 shows the threshold values used to generalise at different scales; these parameters have been found empirically; the process and results are shown in Fig. 6.4.

6.6.3 Braided Sections Generalisation

Where a river flows in a flat region its stream may split, forming many branches flowing downhill, which may merge and split again. This is referred to as a braided section of a river. Braided sections are characterised by the presence of braid bars, the islands sitting among the braids.

To deal with braids, the algorithm changes focus to generalise the islands instead of the river network (see also Touya 2007). The algorithm targets the islands whose size falls below a user defined threshold; these are either amalgamated with nearby islands or enlarged if isolated. Generalising a braided section,

Fig. 6.4 Different steps in the selection process: (*top left*) input network, (*top right*) detected rivers, (*bottom left*) buffers used to compute density, and (*bottom right*) pruned network

the algorithm mainly pursues two objectives: (1) create compact shapes to avoid narrow parts in the generalised islands that could lead to legibility problems, and (2) preserve the streamlined pattern of the network, that is, avoid hard bends in the generalised network.

Braided sections are detected by finding sets of one or more islands where the network lines form polygons. The size of each island is estimated as the area of the associated polygon less the area within the user-defined buffer around the network lines that represents the width of the river (or the actual area of the river polygons where available).

The algorithm processes (Fig. 6.5) each island, starting from the smallest one, looking for candidates for amalgamation among its neighbours. Islands with no neighbours are either deleted or enlarged, depending on their size; enlargement is performed by applying a scaling filter to the geometry that displaces the vertices based upon their closeness to the center of the polygon: vertices closer to the center are displaced further, thus producing shapes that are more compact than results from a buffer operation; the same filter is applied to the braids surrounding the enlarged island.

The algorithm evaluates each amalgamation candidate with respect to the compactness of the resulting amalgamated geometry (compact shapes are

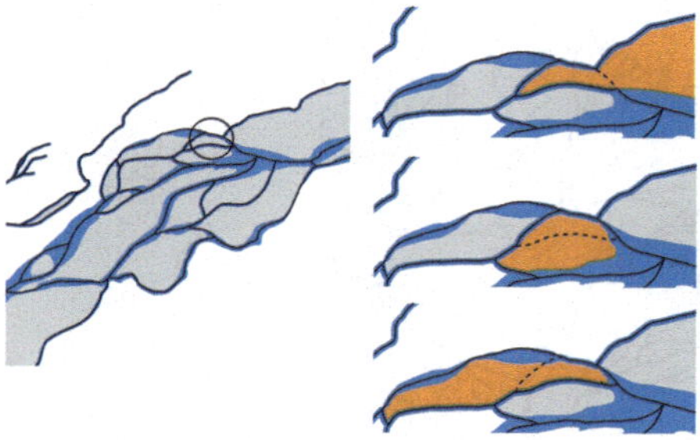

Fig. 6.5 Example amalgamation process of island polygons: (*left*) river polygons and islands of a braided section of the network; (*centre*) selection of the best amalgamation candidate (*orange*) for the island *circled at left*: middle solution is chosen because it produces an island that is more compact than that the top or bottom solution; (*right*) generalised braided section and the former network (*dashed lines*), with *arrows* pointing to two generalised isolated islands, one deleted and one enlarged

favoured, Fig. 6.5 middle center), the size of the braid between the island and the neighbour (narrow braids are favoured), and the angles in the resulting river network (hard bends are not allowed, Fig. 6.5 top center); the best candidate is selected through a scoring mechanism and upon amalgamation, the network and the list of islands are updated. The process ends when all the small islands have been generalised.

6.6.4 Selection Tuning by Density Analysis

In large datasets that span areas with different morphological traits, generalising the hydrographic network with fixed thresholds can produce an output too homogeneous, where local characteristics are lost (Stanislawski and Savino 2011). To handle relative density variations among different areas of the network, the last step of the generalisation process employs a technique that refines the selection process by comparing the local density in both the input and the generalised data.

Density is computed using a regular grid dividing the dataset in cells; the density is defined as the sum of the lengths of the geometries contained in each cell divided by the area of the cell; cells on the boundary of the dataset are not considered to avoid biasing the density calculation. Cells completely covered by braided sections are excluded from the process because braided sections are handled separately as previously described.

The algorithm computes the average cell density, D_{avg}, on the input and generalised datasets and then, for each cell, the difference as a percentage, between the cell density D_c, and the average D_{avg}: this difference, dD_c, can have a positive or negative value and marks the local variation of the density. By comparing the dD_c value of corresponding cells in the input and in the generalised dataset, it is possible to detect whether local variations have been lost: if dD_c is bigger in the

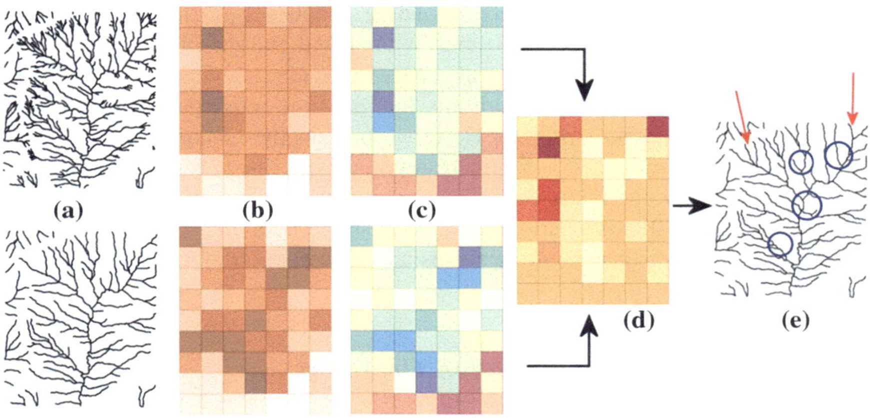

Fig. 6.6 The density tuning process: **a** the networks (*top* input data, *below* generalised data); **b** the density maps (bigger values have *darker colour*); **c** the density difference maps (the darker the colour, the bigger the difference); **d** the density difference comparison map (*dark colours* for over-generalised areas, *light colours* for under-generalised areas) and **e** the tuned network (removed rivers are *circled* while *arrows point* to added rivers)

generalised data, data are locally under generalised, if it is smaller, data are locally over generalised.

The algorithm compensates the density variation by removing or adding rivers in the cell until the target density difference is met. To perform this operation the scoring technique described before is used to choose the most important river to add, among those previously deleted, or to choose the least important river to delete. The algorithm keeps adding (or removing) rivers until the cell density is within a threshold around the desired value. The whole process is illustrated in Fig. 6.6.

6.7 Case Study III: Algorithms for On-the-Fly Generalisation of Point Data Using Quadtrees

Pia Bereuter

On-the-fly (or real-time) generalisation and adaptation to user interaction and content are essential for the dynamic use of web and mobile mapping. Typical applications, such as mashups or location-based services (LBS), usually encompass a thematic foreground layer predominantly in the form of points of interest (POIs) or large point collections (e.g. animal observations or twitter counts), against a spatial reference formed by background data such as a topographic map. Background data are typically rendered by a pre-generalised tile service to ensure seamless map interaction. On the other hand, the content of the foreground data, depending on user requests, requires dynamic adaption and therefore calls for cartographic generalisation in real-time (Weibel and Burghardt 2008). Bereuter and Weibel (2013) and Bereuter et al. (2012) provide a review of the relevant literature and propose several algorithms for real-time point data generalisation based on a quadtree data structure. Here, we present the generalisation of point groups as a Case study, focusing on results and ignoring the technical details. Below a short overview on the generalisation operators is provided, followed by a quantitative and qualitative cartographic analysis on data and conflict reduction, data enhancement, displacement measures and preservation of spatial patterns.

6.7.1 Overview of Generalisation Operators

The algorithms described by Bereuter and Weibel (2013) and Bereuter et al. (2012) provide implementations of the major generalisation operators that can be applied to point data (Sect. 6.2.1). Several algorithms are available for each of the generalisation operators summarised in Table 6.4. The basic idea of the quadtree-based generalisation approach is to apply generalisation operations to quadtree nodes according to the target level of detail (LOD), mapped to the depth of the quadtree and the selected point symbol size. Target LOD translates to the width of the quadnode side, which denotes the smallest required distance to resolve spatial conflicts. The LOD in this Case study is mapped to the zoom level of the background web map tile services (WMTS) and the zoom levels are named accordingly. Bereuter and Weibel (2013) also present a performance analysis of the algorithms.

Table 6.4 Point generalisation operators based on the quadtree (Bereuter and Weibel 2013)

Operator (see Sect. 6.2.1)	Description of implemented algorithms
Pre-processing operators affecting quantity of visual information	
Selection, elimination	Based solely on feature attributes, applying various selection functions per quadnode, such as rank, frequency or feature category distribution
Simplification	Returns one point feature per quadnode, governed by geometric criteria such as centrality, or weighted centrality
Aggregation	Reduces the number of points per quadnode by grouping together semantically similar or spatially close points, replacing the original points by a new placeholder feature, such as midpoint or based on clustering or collocation criteria
Typification (Refinement)	Replaces numerous points by fewer points of the same type within a quadnode
Operator affecting quality of visual information	
Displacement	Locally reconfigures as many point symbols as geometrically possible per zoom level to resolve spatial conflicts by moving points apart from each other using the quadtree for neighbour search

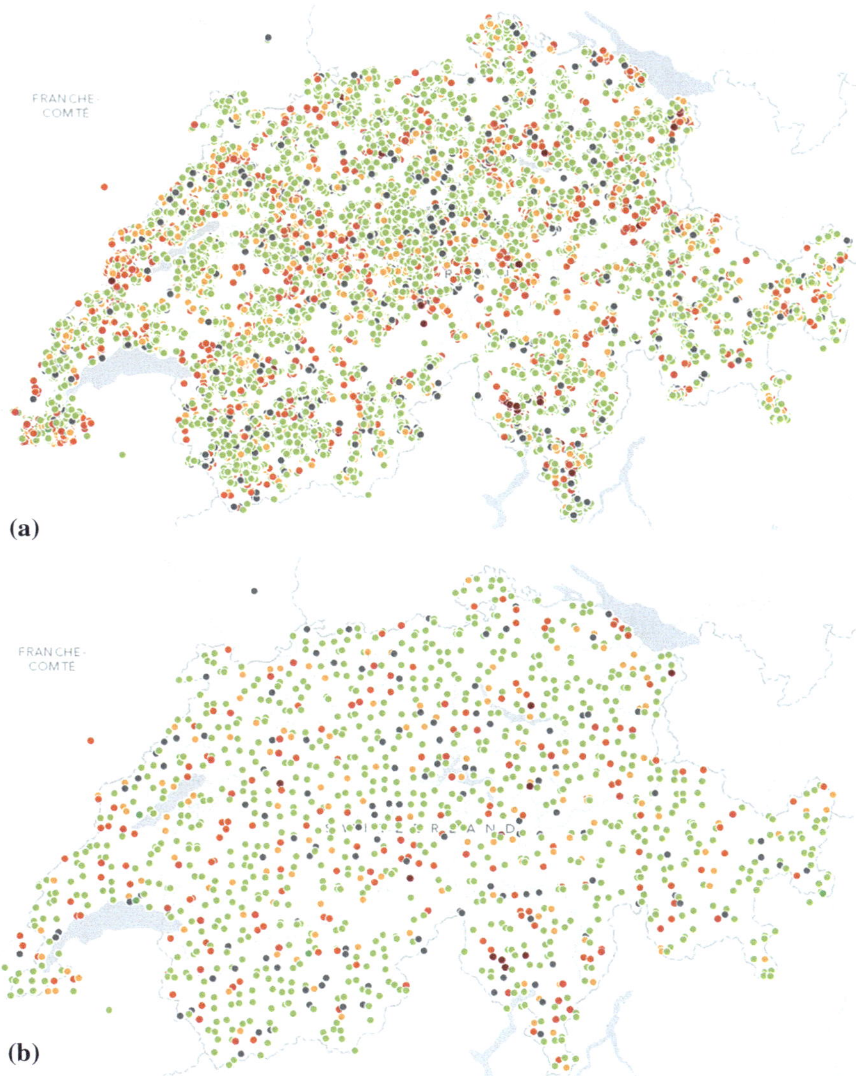

Fig. 6.7 Lichens observation in Switzerland at zoom level 8, color coded from least endangered (*green*) to most endangered (*red*) and no information (*grey*). (*Data* Stofer et al. 2012) *top* **a** raw data, *bottom* **b** centrality-based simplification (1,288 data points)

The following Case study presents several aspects of the introduced generalisation algorithms, implemented in a prototype generalisation platform based on Java and Processing (www.processing.org). The point collection used in this Case study originates from SwissLichens (Stofer et al. 2012), which is a database maintaining past and present population distributions of more than 500 different lichen species at over 86,000 locations within Switzerland.

An overview of the complete dataset is given in Fig. 6.7a, for the area of Switzerland. The map shows all observations of lichens in Switzerland and their red list status. Figure 6.7b shows a generalised view of the same data with the centrality-based simplification applied (Bereuter and Weibel 2013).

The following sections describes a quantitative analysis of the results obtained using the described generalisation algorithms. The analysis focuses on the main cartographic requirements for map generalisation by showing the data and conflict reduction rate, conservation of important point attributes, displacement measures, and the maintenance of spatial patterns.

6.7.2 Data and Conflict Reduction

The data reduction curves for the quadtree-based generalisation algorithms between zoom level, as well as the Radical Law (Töpfer and Pillewizer 1966) with its initial scale at zoom level 20, are shown in Fig. 6.8. The scale for the zoom levels in Fig. 6.8 spans from a small scale of ~1:500,000,000 for zoom level 0 to a large scale of ~1:500 for zoom level 20, with a scale change of factor two between each consecutive zoom level. It shows that quadtree-based operators retain more points than the Radical Law would suggest, mainly due the selected symbol size and the fact that the Radical Law ignores the spatial configuration of the input data. Quadtree-based generalisation operators account for proximity and density of point symbols and therefore they remove points only where conflicts arise. Once the average distance between data points reaches the size of the quad

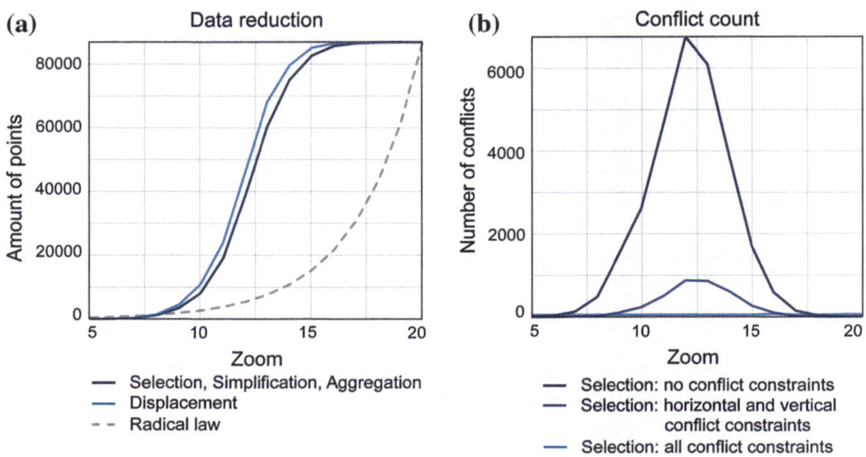

Fig. 6.8 **a** Global data reduction per zoom level, for point reduction algorithms and displacement. **b** Global conflict count per zoom level with different conflict constraints applied

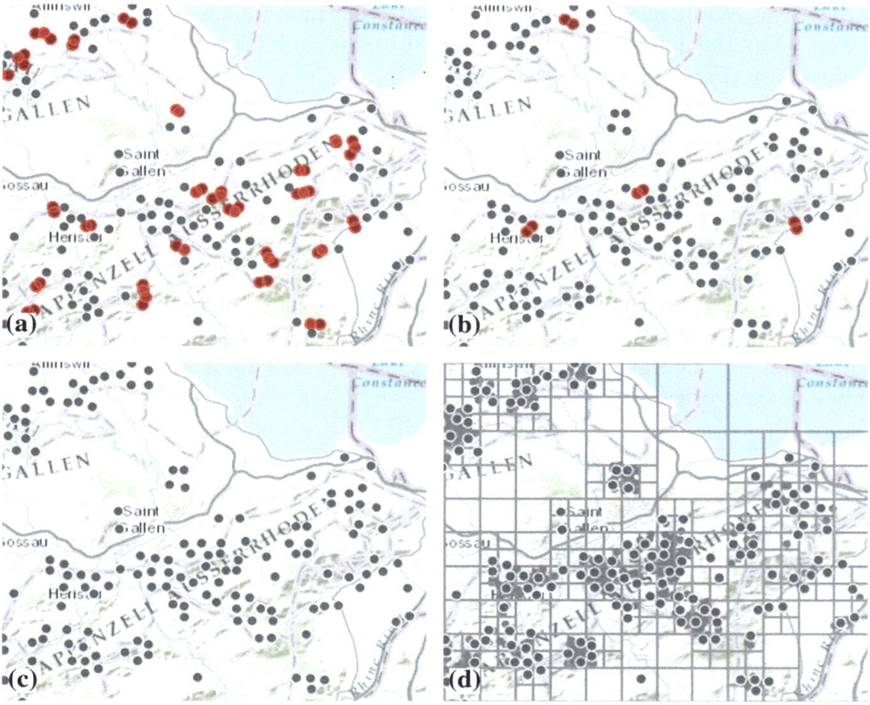

Fig. 6.9 **a–c** Cartographic conflicts (*red dots*) for quadtree-based selection with: **a** no conflict constraints, **b** *horizontal* and *vertical* conflict constraints, **c** including diagonal conflicts constraints, and **d** debug view with the underlying quadtree data structure

cells corresponding to the zoom level, the point reduction rate rapidly increases. Not surprisingly, a comparison between point reduction operators (selection, simplification and aggregation) and displacement illustrates that for those zoom levels where most conflicts arise, a process combining selection and displacement retains more points than with an elimination algorithm only.

Cartographic conflict—overlapping symbols—are not entirely removed by solely retaining one point per quadnode at the target LOD, as they do not consider *per se* potential overlaps from generalisation results residing in neighbouring quadnodes (Fig. 6.8b dark blue curve). Two points may lay across the border between two neighbouring quadnodes, separated by a distance less than the symbol size. This can be alleviated by checking for collisions in the adjacent quadnode neighbours by performing a further elimination or by constraining all points inside the quadnode not allowing for any overlap. Figure 6.8b shows the evolution and reduction of cartographic conflicts over the different zoom levels for the selection algorithm, with the different variations of collision checks applied.

Figure 6.9 shows conflict reduction applying the different variations of conflict checks in the case of value-based selection on lichens data for the Eastern part of

Switzerland. While pure selection still shows some cartographic conflicts (red dots in Fig. 6.9a), the application of conflict constraints in the algorithm reduces them significantly (Fig. 6.9b, c).

6.7.3 Data Enhancement

A variant of value-based selection (Fig. 6.10) illustrates how a particular attribute is retained and enhanced through a set of scales. It can be applied if comparatively rare point features need to be retained and are not evened out over the course of scales. Figure 6.10 highlights how the most endangered species are retained throughout the different zoom levels, rather than maintaining the overall distribution of categories.

6.7.4 Displacement Measures

To further resolve spatial conflicts and retain more elements than solely with point reduction operators, a displacement algorithm can be applied. The displacement algorithm tries to accommodate as many points as possible keeping at most one point per quadnode, and displacing points if the directly neighbouring quadnodes provide sufficient holding capacity for displacement. Remaining overlaps can be removed by further resolving boundary constraints as illustrated in Fig. 6.9. A comparison between the two algorithms (Fig. 6.11a, b) illustrates that displacement retains more points for the displayed zoom level. On the other hand, it shows that displacement has the effect of homogenising dense clusters and thus affecting the overall distribution pattern.

The characteristics of an applied displacement operator can be highlighted by plotting cumulated displacement vectors (in pixels) for each angle of displacement. Figure 6.11c shows nicely that the algorithm (considering in this example only horizontal and vertical neighbours) did not displace points to diagonal neighbours with the majority of displacement angles in horizontal and vertical direction.

6.7.5 Preservation of Spatial Patterns

How well a generalisation algorithm preserves the underlying spatial pattern can be investigated by visually comparing the kernel density estimation (KDE) of a point pattern or by calculating the difference between two kernel density estimations. The kernel density map in Fig. 6.12a for zoom level 9 and the KDE density difference map in Fig. 6.12b between zoom level 9 and 10, show the density distribution of the point pattern and where it changes most, respectively. It shows

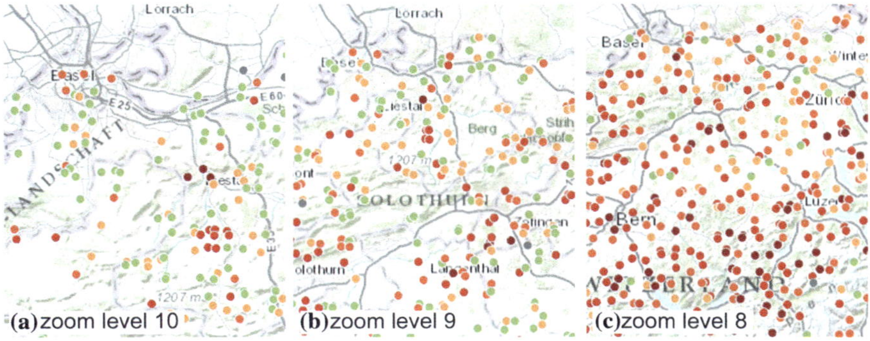

Fig. 6.10 Value-based selection, retaining most endangered lichens in the region of Basel with decreasing zoom level. Color codes range from least endangered (*green*) to most endangered (*red*)

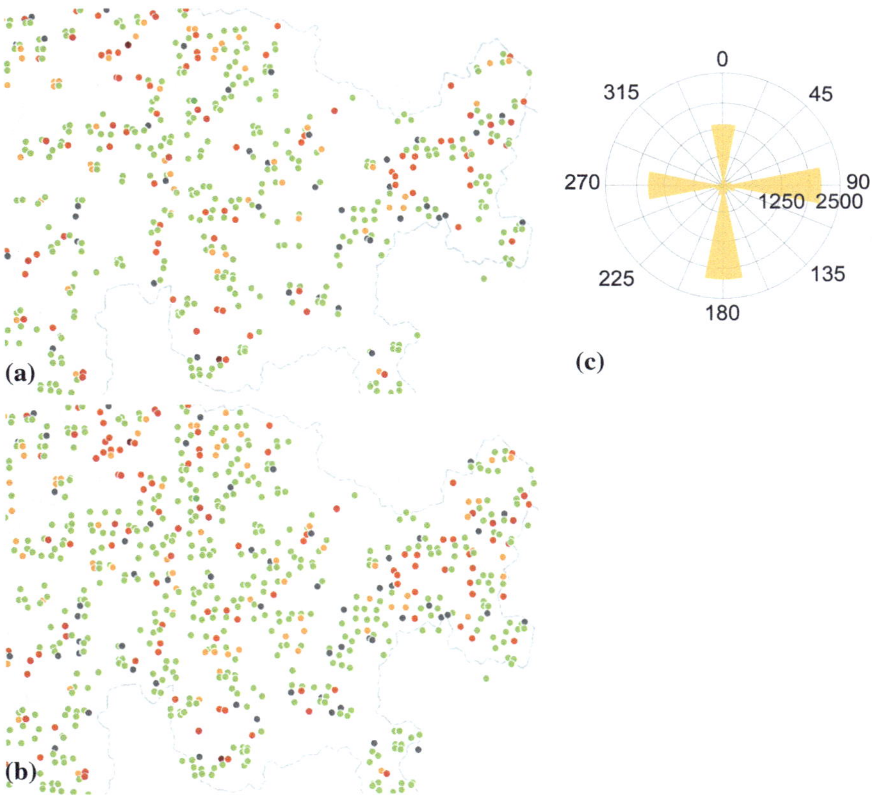

Fig. 6.11 **a** Centrality-based simplification (724 point), **b** displacement applied after centrality-based simplification (974 points) **c** cumulated displacement vectors for zoom level 9

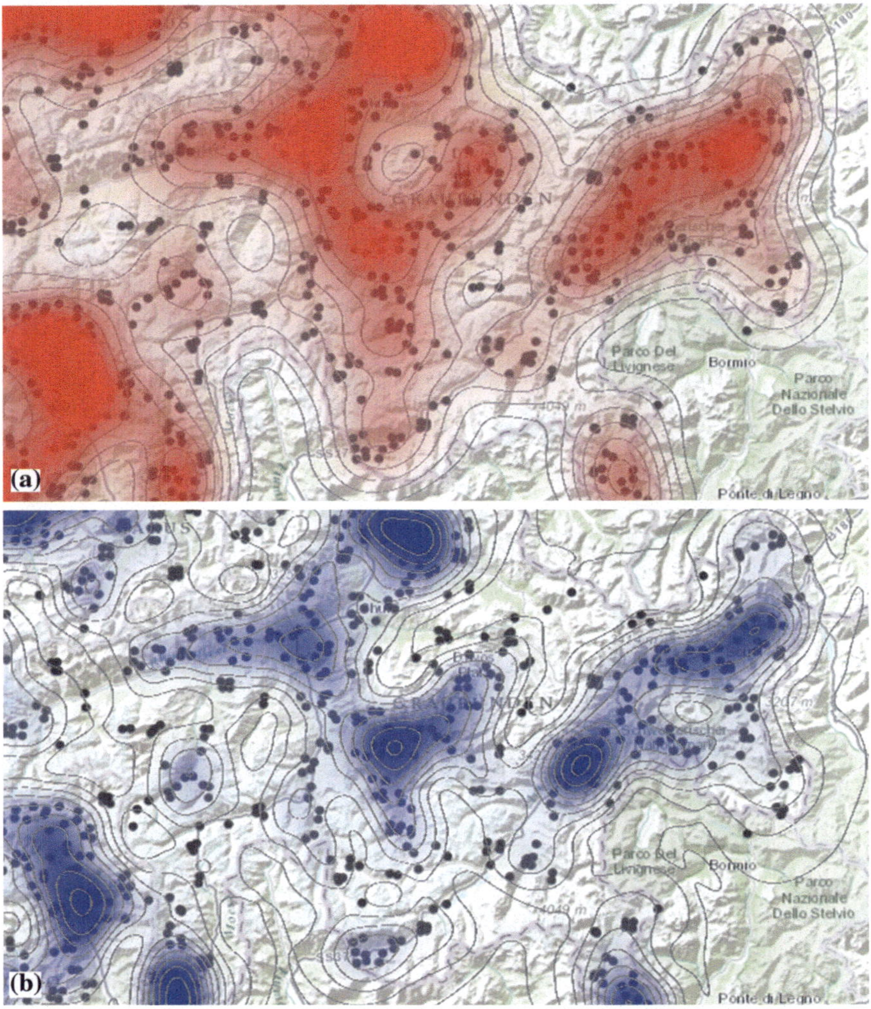

Fig. 6.12 **a** Kernel density estimation for centrality-based simplification at zoom level 9, **b** Kernel density difference between zoom level 9 and 10

that the applied algorithm reduces the point density most at local peaks, where the highest densities are located, and how much the density is decreased around local density peaks. The decrease around density peaks is however less pronounced than with the displacement operator (Fig. 6.11a, b).

6.8 Conclusions

This chapter reviewed generalisation operators, defined as a generic descriptor for the type of spatial or attribute modification to be achieved on some set of geospatial data. Examples of operator typologies demonstrate the wide variety of methods to organise types of operators for paper mapping, digital mapping, and for multi-scale data modelling. The development of methods by which to formalise differing requirements for topographic and thematic mapping across a range of scales has caused a shift from application of individual operators in isolation, to generalisation strategies encompassing integrated sequences of operators.

Recent advances in the design of generalisation strategies increase processing requirements and in some cases mandate data enrichment, which supports generalisation that can be tailored to specific landscape and settlement contexts. Improved methods for reasoning about relative feature priorities permit advanced processing of road networks, stream networks, and very large point data sets, as documented in the three case studies. Additional research on networks should assess whether algorithms are suitable for more complex cyclic, anthropogenic networks (i.e., road and rail networks, or stream and canal networks) or whether refinements or a different set of algorithms are required. Much work in the realm of generalisation operators has been devoted to methods that preserve local density variations of feature distributions. Feature density variations are often related to changing anthropogenic or geophysical conditions, and the case studies all show how maintenance of spatial patterns for these variations enhances map displays with more realistic context.

In the future, one can expect to see continued work to formally characterise feature geometry, spatial context, and to identify and resolve spatial conflicts arising from generalisation for mapping at reduced scales. The extent to which multi-scale characterisation can be articulated will continue to expand the extent to which generalisation can be fully automated; and the foundations for such advances will be initiated in development and refinement of generalisation operators.

Acknowledgments Work for this chapter was funded by several federal agencies in several countries. The work of Dr. Brewer is funded by the U.S. Geological Survey's Center of Excellence for Geospatial Information Science (USGS-CEGIS) grant #06HQAG0131. The work of Dr. Buttenfield is supported by USGS-CEGIS grant #04121HS029. The research reported in the third Case study represents part of the PhD project of Pia Bereuter, funded by the Swiss National Science Foundation through project GenW2 + (SNF no. 200020-138109). Particular thanks go to Silvia Stofer of the National Data and Information Center for Swiss lichens and all the contributors for collecting and providing the data of the SwissLichens dataset.

References

Ai T, Liu Y, Chen J (2006) The hierarchical watershed partitioning and data simplification of river network. Prog Spat Data Handling Part 11:617–632. doi:10.1007/3-540-35589-8_39

Aslan S, Bildirici IO, Simav O, Cetinkaya B (2012) An incremental displacement approach applied to building objects in topographic mapping. In: 13th ICA workshop on generalisation and multiple representation, Istanbul

Balboa JLG, Lopez FJA (2008) Generalization-oriented road line classification by means of an artificial neural network. Geoinformatica 12:289–312

Beard K (1991) Constraints on rule formation. In: Buttenfield BP, McMaster RB (eds) Map generalisation: making rules for knowledge representation. Longman Group, Harlow, pp 121–135

Benz SA, Weibel R (2013) Road network selection using an extended Stroke-Mesh combination algorithm. In: 16th ICA workshop on generalisation and multiple representation, Dresden, Germany, 23–24 Aug 2013

Bereuter P, Weibel R, Burghardt D (2012) Content zooming and exploration for mobile maps. In: Proceedings of the 15th AGILE international conference on geographic information science, pp 74–80

Bereuter P, Weibel R (2013) Real-time generalisation of point data in mobile and web mapping using quadtrees. Cartogr Geogr Inf Sci 40(4):1–11

Bildirici IO, Aslan S, Simav O, Cobankaya ON (2011) A generic approach to building typification. In: 14th ICA workshop on generalisation and multiple representation, Paris, 30 June–1 July

Bobzien M, Burghardt D, Petzold I, Neun M, Weibel R (2008) Multi-representation databases with explicitly modeled horizontal, vertical, and update relations. Cartography Geogr Inf Sci 35(1):3–16

Brewer CA, Stanislawski LV, Buttenfield BP, Raposo P, Sparks KA, Howard MA (2012) Multiscale design for the national map of the United States: road thinning for topographic mapping. In: AutoCarto 2012 Proceedings, Columbus, Ohio, 16–18 Sept 2012

Brewer CA, Stanislawski LV, Buttenfield BP, Sparks KA, McGilloway J, Howard MA (2013a) Automated thinning of road networks and road labels for multiscale design of the national map of the United States. Cartogr Geogr Inf Sci

Brewer CA, Guidero EM, Stanislawski LV, Buttenfield BP, Raposo P (2013b) Labeling through scale using hierarchies of thinned road networks for design of the national map of the United States. In: Proceedings 26th international cartographic conference, Dresden, Germany, Aug 25–30

Briat M-O, Monnot J-L, Punt EM (2011) Scalability of contextual generalisation processing using partitioning and parallelization. In: 14th ICA/ISPRS workshop on generalisation and multiple representation, Paris, 30 June–1 July

Buttenfield BP, Mark DM (1991) Expert systems in cartographic design Chapter 7. In: Taylor DRF (ed) Geographic information systems. New York: Pergamon: pp. 129–150

Buttenfield BP, Stanislawski LV, Brewer CA (2011) Adapting generalisation tools to physiographic diversity for the United States national hydrography dataset. Cartography Geogr Inf Sci 38(3):289–301

Chaudhry O, Mackaness WA (2005) Rural and urban road network generalisation deriving 1:250,000 from OS mastermap. In: Proceedings of the 22nd international cartographic conference, A Coruña, Spain, 11–16 July

Chaudhry O, Mackaness WA (2008a) Partitioning to make manageable the generalisation of national spatial datasets. In: 11th ICA workshop on generalisation and multiple representation, Montpellier, France, 20–21 June 2008

Chaudhry O, Mackaness WA (2008b) Automatic identification of urban settlement boundaries for multiple representation databases. Comput Environ Urban Syst 32(2):95–109

Chen J, Hu Y, Li Z, Zhao R, Meng L (2009) Selective omission of road features based on mesh density for automatic map generalisation. Int J Geogr Inf Sci 23(8):1013–1032

Christophe S, Ruas A (2002) Detecting building alignments for generalisation purposes, advances in spatial data handling. Springer, Berlin, pp 419–432

Esri (2011) ArcGIS desktop 10 help documentation: thin road network (cartography). Environ Syst Res Inst, 11 Sept 2011 http://help.arcgis.com/en/arcgisdesktop/10.0/help/index.html#//007000000014000000. Accessed 6 Feb 2013

Esri (2012) ArcGIS 10.1 help: cartographic partitions (environment setting). Environ Syst Res Ins, http://resources.arcgis.com/en/help/main/10.1/index.html#/Welcome_to_the_ArcGIS_Help_Library/00qn0000001p000000/. Accessed 18 Feb 2013

Foerster TJ, Stoter J, Kobben B (2007) Towards a formal classification of generalisation operators. In: Proceedings of the 23rd international cartographic conference. Moscow, Russia

Fotheringham AS, Brunsdon C, Charlton M (2000) Quantitative geography: perspectives on spatial data analysis. Sage, London

Gülgen F, Gökgöz T (2008) Selection of roads for cartographic generalization. the international archives of the photogrammetry. Remote Sens Spat Inf Sci 37(B4):615–620

Gülgen F, Gökgöz T (2011) A block-based selection method for road network generalisation. Int J Digital Earth 4(2):133–153

Gutman N (2012) Continuous and adaptive cartographic generalization of river networks. Doctoral dissertation. School of Computer Science, University of Oklahoma

Heinzle F, Anders KH, Sester M (2005) Graph based approaches for recognition of patterns and implicit information in road networks. In: Proceedings of the 22nd international cartographic conference, A Coruna, Spain, 9–16 July

Heinzle F, Anders K-H, Sester M (2007) Automatic detection of patterns in road networks methods and evaluation. In: Proceedings of joint workshop visualization and exploration of geospatial data 26(4)

Horton RE (1945) Erosional development of streams and their drainage basins: hydrophysical approach to quantitative morphology. Geol Soc Am Bull 56(3):275–370

Jenny B, Jenny H, Hurni L (2011) Terrain generalisation with multi-scale pyramids constrained by curvature. Cartography Geogr Inf Sci 38(1):110–116

Jiang B, Claramunt C (2004) A structrual approach to the model generalisation of an urban street network. GeoInformatica 8(2):157–171

Jiang B, Liu X, Jia T (2013) Scaling of geographic space as a universal rule for map generalization. Ann Assoc Am Geogr 103(4):844–855

Kilpeläinen T (1997) Multiple representation and generalisation of geo-databases for topographic maps, Doctoral dissertation. Publications of the Finnish Geodetic Institute, Kirkonummi, (124)

Leonowicz AM, Jenny B, Hurni L (2010) Terrain sculptor: generalizing terrain models for relief shading. Cartographic Perspect 67:51–60

Li Z, Zhou Q (2012) Integration of linear and areal hierarchies for continuous multi-scale representation of road networks. Int J Geogr Info Sci 26(5):855–880

Long S, Shoberg T, Corns S, Carlo H, Ramachandran V (2012) Integrating geospatial data from the national map with supply chain networks to improve resiliency following large-scale urban disasters. American association of geographers national conference, New York

Mackaness WA, Ruas A, Sarjakoski LT (2007) Generalisation of geographic information: cartographic modelling and applications, Elsevier, Amsterdam pp 370

McMaster RB (1983) A quantitative analysis of mathematical measures in linear simplification. Unpublished Ph.D. dissertation, Department of Geography-Meteorology, University of Kansas

McMaster RB (1989) Numerical generalisation in cartography. In: McMaster, RB (ed) Monograph 40, numerical generalisaton in cartography. Cartographica 26(1):1–6

McMaster RB, Monmonier M (1989) A conceptual framework for quantitative and qualitative raster-mode generalisation. In: Proceedings of GIS/LIS'89, Orlando. American society for photogrammetry and remote sensing, Maryland, pp 390–403

McMaster RB and Shea KS (1992) Generalisation in digital cartography. In: Association of american geographers resource publications in college geography, Washington, DC, pp 134

Morrison JL (1974) A theoretical framework for cartographic generalisation with emphasis on the process of symbolization. Int Yearb Cartography 14:115–127

Neun M, Burghardt D, Weibel R (2008) Web service approaches for providing enriched data structures to generalisation. Int J Geogr Inf Sci 22(2):133–165

Nickerson BG, Freeman HR (1986) Development of a rule-based system for automatic map generalisation. In: Proceedings of the 2nd international symposium on spatial data handling, Seattle, Washington. International Geographical Union Commission on Geographical Data Sensing and Processing, Williamsville, NY, pp 537–556

Punt EM, Watkins D (2010) User-directed generalisation of roads and buildings for multi-scale cartography. In: 13th ICA workshop on generalisation and multiple representation, Zurich, 12–13 Sept 2010

Raposo P, Brewer CA, Stanislawski LV (2013) Label and attribute-based topographic point thinning. In: 16th ICA workshop on generalisation and multiple representation, Dresden, Germany, 23–24 Aug 2013

Ratajski L (1967) Phenomenes des points de generalisation. Int Yearb Cartography 7:143–151

Regnauld N, McMaster RB (2006) Creating a hydrographic network from its cartographic representation: a case study using Ordnance Survey Mastermap data. Int J Geogr Inf Sci 20(6):611–631

Regnauld N, McMaster RB (2007) A synoptic view of generalisation operators. In: Mackaness WA, Ruas A, Sarjakoski LT (eds) Generalisation of geographic information: cartographic modelling and applications, Elsevier Ltd, Amsterdam pp 37–66

Robinson AH, Sale RD (1969) Elements of cartography, 3rd edn. Wiley, New York

Roth RE, Brewer CA, Stryker MS (2011) A typology of operators for maintaining legible map designs at multiple scales. Cartographic Perspect 68:29–64

Ruas A (1995) Multiple paradigms for automated map generalisation: geometry, topology, hierarchical partitioning and local triangulation. In: Proceedings AutoCarto, Charlotte, USA, 12, pp 69–78

Ruas A (1999) Modele de generalisation de donnees geographiques a base de constraints et d'autonomie, These de doctorat. L'universite de Marne La Vallee, Paris

Ruas A, Duchêne C (2007) A prototype generalisation system based on the multi-agent system paradigm. In: Mackaness WA, Ruas A, Sarjakoski LT (eds) Generalisation of geographic information: cartographic modelling and applications, Elsevier Ltd, pp 269–284

Sarjakoski LT, (2007) Conceptual Models of Generalisation and Multiple Representation, In Mackaness WA, Ruas A, Sarjakoski LT (eds), Generalisation of Geographic Information: Cartographic Modelling and Applications, Elsevier Ltd, pp 11–35

Savino S, Rumor M, Zanon M, Lissandron I (2010) Data enrichment for road generalization through analysis of morphology in the CARGEN project. In: 13th ICA workshop on generalisation and multiple representation, Zurich, Switzerland, 12–13 Sept 2010

Savino S, Rumor M, Canton F, Langiu G, Reineri M (2011a) Model generalisation of the hydrography network in the CARGEN project. In: Ruas A (eds) Advances in cartography and GIScience (1):439–457, Lecture Notes in Geoinformation and Cartography, Springer, Berlin

Savino S, Rumor M, Zanon M (2011b) Pattern recognition and typification of ditches. In: Ruas A (eds), Advances in cartography and GIscience, Vol 1. Springer, pp 425–437 ISBN 9783642191428

Schylberg L (1993) Computational methods for generalisation of cartographic data in a raster environment. Doctoral Thesis, Department of Geodesy and Photogrammetry, Royal Institue of Technology Stockholm, Sweden, Photogrammetric Reports (60) TRITA-FMI Report, (1993:7)

Simley J, Doumbouya A (2012) National hydrography dataset—Linear Referencing. USGS Fact Sheet: 2012–3068

Stanislawski LV (2009) Feature pruning by upstream drainage area to support automated generalisation of the united states national hydrography dataset. Comput Environ Urban Syst 33:325–333

Stanislawski LV, Buttenfield BP (2011) A raster alternative for partitioning line densities to support automated cartographic generalisation. In: 25th International cartography conference, July 3–8, Paris

Stanislawski LV, Briat M, Punt E, Howard M, Brewer CA, Buttenfield BP (2012a) Density-stratified thinning of road networks to support automated generalisation for the na-tional map. In: 15th ICA workshop on generalisation and multiple representation, Istanbul, Turkey, 13–14 Sept 2012

Stanislawski LV, Doumbouya AT, Miller-Corbett CD, Buttenfield BP, Arundel-Murin ST (2012b) Scaling stream densities for hydrologic generalisations. Seventh international conference on GIScience, Columbus, Ohio, 18–21 Sept 2012

Stanislawski LV, Savino S (2011) Pruning of hydrographic networks: a comparison of two approaches. In: 14th ICA/ISPRS workshop on generalisation and multiple representation, Paris, France, 30 June–1 July 2011

Steiniger S, Burghardt D, Weibel R (2006) Recognition of island structures for map generalisation. In: International symposium on advances in geographic information systems, ACM-GIS 2006 pp 67–74

Steiniger S, Lange T, Burghardt D, Weibel R (2008) An approachfor the classification of urban building structures based on discriminant analysis techniques. Trans GIS 12(1):31–59

Steiniger S, Wiebel R (2007) Relations among map objects in cartographic generalisation. Cartography Geog Inf Sci 34(3):175–197

Stoter J, Baella B, Blok C, Burghardt D, Duchêne C, Pla M, Regnauld N, Touya G (2010) State-of-the-art of automated generalisation in commercial software. European spatial data research report, p 270 (2010)

Stofer S, Scheidegger C, Clerc P, Dietrich M, Frei M, Groner U, Jakob P, Keller C, Roth I, Vust M, Zimmermann E (2012) Nationales Daten- und Informationszentrum der Schweizer Flechten—SwissLichens. Datenbankauszug vom 25. September 2012 (SST_20120925). Eidgenössische Forschungsanstalt WSL, Birmensdorf

Strahler AN (1952) Hypsometric (area-altitude) analysis of erosional topology. Geol Soc Am Bull 63(11):1117–1142

Sugarbaker LJ, Carswell WJ (2011) The national map: U.S. geological survey fact sheet 2011–3042. pp 4 http://pubs.usgs.gov/fs/2011/3042. Accessed 29 Jan 2013

Thomson RC, Brooks R (2002) Exploiting perceptual grouping for map analysis, understanding and generalization: the case of road and river networks. Graphics recognition algorithms and applications, Lecture notes in computer science 2390: 148–157

Thomson RC, Brooks R (2007) Generalisation of geographical networks. In: Mackaness WA, Ruas A, Sarjakoski LT (eds) Generalisation of geographic information: cartographic modelling and applications, Elsevier Ltd, Boston pp 255–267

Thomson RC, Richardson DE (1995) A graph theory approach to road network generalisation. In: Proceedings 17th international cartographic conference, Barcelona, Spain, 3–9 Sept 1995

Thompson RC, Richardson DE (1999) The 'Good Continuity' principle of perceptual organisation applied to the generalisation of road networks. In: Proceeding of the 19th international cartographic conference, Ottawa pp 1215–1225

Töpfer F, Pillewizer W (1966) The principles of selection. Cartographic J 3(1):10–16

Touya G (2007) River network selection based on structure and pattern recognition. In: Proceedings 23rd International cartographic conference, Moscow, Russia

Touya G (2010) A road network selection process based on data enrichment and structure detection. Trans GIS 14(5):595–614

Touya G, Duchêne C, Ruas A (2010) Collaborative generalization: formalization of generalization knowledge to orchestrate different cartographic generalization processes. GIScience 2010, Zurich, Switzerland, 14–17, Sept 2010

USGS (1985) Standards for 1:100,000-scale quadrangle maps. National mapping program technical instructions. U.S. Geological survey, Department of the Interior

USGS (2006) The best practices data model—The national map, March 1 2006, http://services.nationalmap.gov/bestpractices/model/acrodocs/Poster_BPTrans_03_01_2006.pdf. Accessed 5 Feb 2013

van Oosterom P (2005) Variable-scale topological data structures suitable for progressive data transfer: the GAP-face tree and GAP-edge forest. Cartography Geogr Inf Sci 32(4):331–346

van Oosterom P, Meijers M (2011) Towards a true vario-scale structure supporting smooth-zoom. 14th ICA/ISPRS workshop on generalisation and multiple representation, Paris, 30 June–1 July

Verdin KL, Verdin JP (1999) A topological system for delineation and codification of the earth's river basins. J Hydrol 218:1–12

Wallace RM, Tarboton DG,Watson DW, Schreuders KAT, Tesfa TK (2010) Parallel algorithms for processing hydrologic properties for digital terrain. GIScience, Zurich, 14–17, Sept 2010

Weibel R (1991) Amplified intelligence and rule-based systems. In: Buttenfield BP, McMaster RB (eds) Map generalisation: making rules for knowledge generation. Longman, London

Weibel R (1992) Models and experiments for adaptive computer assisted terrain generalization. Cartography Geogr Inf Syst 19(3):133–153

Weibel R, Burghardt D (2008) Generalisation on-the-fly. In: Shekhar S, Xiong H (eds) Encyclopedia of GIS. Springer, New York, pp 339–344

Weiss R, Weibel R (2013) Road network selection for small-scale maps using an improved centrality approach. In: 16th ICA workshop on generalisation and multiple representation, Dresden, Germany, 23–24 Aug 2013

Werder S, Kieler B, Sester M (2010) Semi-automatic interpretation of buildings and settlement areas in user-generated spatial data. In: Proceedings of the 18th SIGSPATIAL international conference on advances in geographic information systems, pp 330–339. doi: 10.1145/1869790.1869836

Wu H, Kimball JS, Li H, Huang M, Leung LR, Adler RF (2012) A new global river network database for macroscale hydrologic modeling. Water resources research, vol. 48, W09701, pp 5 doi:10.1029/2012WR012313

Yager DB, Hofstra AH, Granitto M (2012) Analyzing legacy U.S. geological survey geochemical databases using GIS—Applications for a national mineral resource assessment: U.S. Geological survey techniques and methods 11–C5, pp 28

Zhang X, Ai T, Stoter J (2010) Characterization and detection of building patterns in cartographic data: two algorithms. The international archives of the photogrammetry, remote sensing and spatial information sciences 38(II):261–266

Zhou Q, Li Z (2012) A comparative study of various strategies to concatenate road segments into strokes for map generalization. Int J Geogr Inf Sci 26(4):691–715

Chapter 7
Process Modelling, Web Services and Geoprocessing

Nicolas Regnauld, Guillaume Touya, Nicholas Gould and Theodor Foerster

Abstract Process modelling has always been an important part of research in generalisation. In the early days this would take the form of a static sequence of generalisation actions, but currently the focus is on modelling much more complex processes, capable of generalising geographic data into various maps according to specific user requirements. To channel the growing complexity of the processes required, better process models had to be developed. This chapter discusses several aspects of the problem of building such systems. As the system gets more complex, it becomes important to be able to reuse components which already exist. Web services have been used to encapsulate generalisation processes in a way that maximises their interoperability and therefore reusability. However, for a system to discover and trigger such a service, it needs to be formalised and described in a machine understandable way, and the system needs to have the knowledge about where and when to use such tools. This chapter therefore explores the requirements and potential approaches to the design and building of such systems.

N. Regnauld (✉)
1Spatial, Tennyson House, Cambridge Business Park, Cambridge, UK
e-mail: Nicolas.regnauld@1Spatial.com

G. Touya
Laboratoire COGIT, IGN, Saint-Mandé, France
e-mail: guillaume.touya@ign.fr

N. Gould
Manchester Metropolitan University, Manchester, UK
e-mail: nicholas.m.gould@stu.mmu.ac.uk

T. Foerster
Institute for Geoinformatics, University of Muenster, Münster, Germany
e-mail: theodor.foerster@uni-muenster.de

D. Burghardt et al. (eds.), *Abstracting Geographic Information in a Data Rich World*, Lecture Notes in Geoinformation and Cartography, DOI: 10.1007/978-3-319-00203-3_7,

7.1 Introduction and State of the Art

In recent years, the ever increasing speed of computers coupled with exploding huge growth in numbers, and that they are better connected than ever before, has led to a revolution in the way software is designed and used. Service Oriented Architectures have emerged to take advantage of the distributed power provided by multiple computers connected via a network. The increase in computer power has also opened the door for new advances in Artificial Intelligence. All this has benefited research in generalisation which has embraced some of these new concepts and developments.
Early research on automated generalisation focused on developing algorithms to automate single generalisation operations (Jenks 1989; Douglas and Peucker 1973; Jäger 1991). After a few of these emerged, the next logical question was how to combine these atomic processes into more complex ones capable of generalising an entire map? In order to control the growing complexity of these processes, the research community started to focus on process modelling. The first studies proposed static sequences of algorithms to derive a specific map, this is often referred to as batch processing. Weibel et al. (2007) explain how this domain has evolved into more advanced types of modelling, which the authors have classified into three main categories: condition-action modelling (rule based system), human interaction modelling, and constraint-based modelling. The rule based system, which has sometimes been implemented as an expert system, was very popular at first, but proved difficult to extend beyond a certain point, due to the difficulty of formalising and capturing all the rules required, and to avoid inconsistencies when the number of rules grows. While inconsistencies in a rule set can be tracked down (Brenner and Sester 2005), resolving them can be very difficult. The human interaction model was letting the human make most decision of what tool to apply where, but such systems offer a limited increase in productivity. In order to overcome the limitations of these two models, the constraint-based model was developed. The principle is to express the requirements for the target map in terms of constraints (see Chap. 2), and to use a mechanism to trigger a combination of algorithms to maximise the satisfaction of this set of constraints. Different models have been studied to achieve an optimum:

- Multi agent models;
- Combinatorial optimisation models;
- Continuous optimisation model.

Since the review of Weibel et al. (2007), no ground breaking advances have been made in this domain. Instead, the existing techniques have been refined, and applied to real world use cases with very positive results. The most interesting advance on the methodology front has been to add another layer to the decision tree, and apply different subsystems to different tasks, or different geographic configurations (Sect. 7.2).

Chapter 11 provides an overview of the most recent success in applying automated generalisation to real production systems, all of them implemented in National Mapping Agencies, using different technologies, and achieving different level of automation on different types of products. Systems described in Sects. 11.3, 11.4 and 11.8 produce paper maps at scales between 1:25 and 1:100 k from topographic data, respectively at IGN France, at SwissTopo and in a group of German federal states (Bundesländer). In these three cases a first pass applies automatic generalisation to the data, before manual editing takes place to finish the maps. IGN uses Radius Clarity (1Spatial) for the automated generalisation followed by GeoConcept Publisher for the manual editing. SwissTopo uses Axpand (Axis Systems) in combination with ArcGIS (ESRI). The German solution is also based on Radius Clarity for the automated generalisation part. Section 11.5 describes another system using automated generalisation, but this time the generalisation system is fully automated, and manual finishing is not required. The product obtained is not the usual high quality topographic map, but a lighter backdrop map, designed to be used at scales around 1:25 k for overlaying other data onto it. The generalisation system is based on 1Spatial software (Radius Clarity and Radius Studio). Sections 11.6 and 11.7 present automated generalisation processes developed respectively at USGS and Kadaster NL using ArcGIS (ESRI). All these systems have been built by heavily customising the original platform. In some cases the customisation and extensions have been made by the software company itself (1Spatial for the German federal states, Axis Systems with SwissTopo), and sometimes by the customer (IGN and OS). So platforms exist for developing complex automated generalisation processes that bring real benefits to production systems. What we have not seen yet is a platform that can be used out of the box by simple configuration (importing the input data into the system and expressing the requirements). This is one of the main objectives of the current research on on-demand mapping. The general idea is to build a system that can easily integrate data from different sources, and generalise them together to obtain a map as specified by the user. The general idea has been described by Regnauld (2007), and refined through Foerster et al. (2012b). Balley and Regnauld (2012) proposes an architecture for such a system which decouples its main components. These are summarised in Fig. 7.1. They include

- The product specifications, to formalise the requirements (geometric, topological, contextual, etc.);
- The data access manager, to link the data with the system's internal schema;
- Web services, to provide algorithms (generalisation algorithms, spatial analysis tools, etc.), which should be formally described;
- The knowledge base, which contains all the knowledge required by the system (procedural knowledge, cartographic knowledge, geographic knowledge) (Armstrong 1991). This knowledge can take several forms, such as rules, constraints or facts;

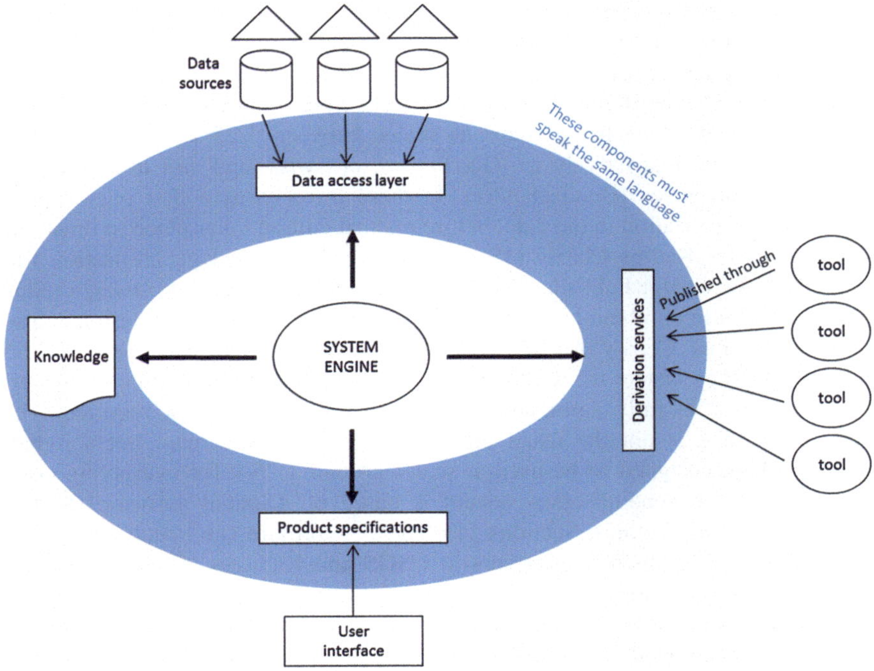

Fig. 7.1 High-level architecture deriving on-demand products

- The system engine is basically an inference engine, which is capable of using the knowledge to interpret the specifications and build and apply the workflow that will derive the required map from the given data, using the tools available.

Such a system, in order to deliver true on-demand mapping, would be very big and complex, include large libraries of well described tools and rich knowledge of many different kinds. Building such a system needs to be done incrementally, focusing on a few types of requirement at a time, and enriching the system with the components required. For this to happen we need to decide upon the essential components of the generalisation process.

7.2 Deciding on the Components of the Generalisation Process

Decomposing a complex system into simpler components helps ensure a modular and robust design. The ambition is to decompose a generalisation system into components that can be developed independently and shared. These components can be made up of simple algorithms, or perform much more complex tasks. The

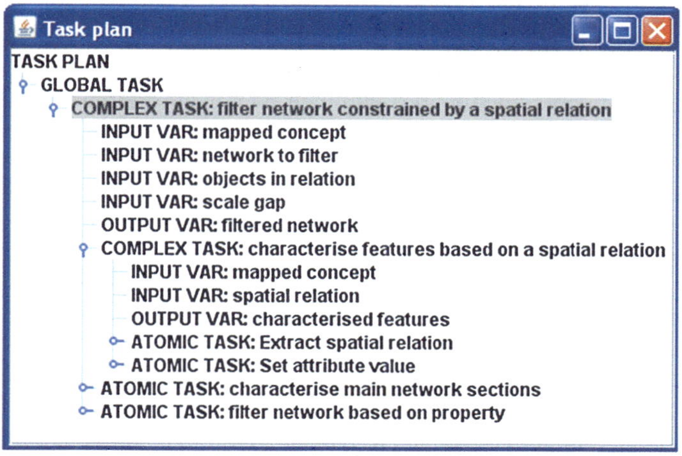

Fig. 7.2 Example of a complex task

benefit of having complex tasks is that they can encapsulate the knowledge specific to the task. This can reduce the amount of knowledge required in the top level system. This is a way of organising the knowledge to avoid having a huge flat set of rules and constraints which would quickly become unmanageable and impossible to extend. Figure 7.2 shows an example of complex task from Balley et al. (2012). This task was instantiated in order to prune a road network (*network to filter*) while making sure that roads supporting (*spatial relation* alignment) cycle routes (*mapped concept*) were kept. This task is made of three subtasks. The first one is a complex task responsible for marking the sections of the road network which are associated with a cycle route. The second task is atomic and identifies the main roads in the road network. The final atomic subtask performs the filtering based on the attributes set during the two previous subtasks.

This is also a good way of reusing existing processes to incorporate them into bigger, more complete systems. For example Touya and Duchêne (2011) combined existing processes to make a generalisation system capable of handling different complex situations. Balley and Regnauld (2012) have proposed a model using tasks and subtasks for organising the knowledge in their on-demand mapping system. Goals are chosen depending on the user requirements, and then each goal is associated with a task designed to achieve it. Each task encapsulates its own procedural knowledge, used to trigger the appropriate actions.

To ensure interoperability between these tasks, there is a need for them to conform to some standard. The requirements for a shared development platform for generalisation has been identified by Edwardes et al. (2003). They had proposed a number of possible approaches, and this led to the development of WebGen, a client–server platform for sharing processes using Web services (Neun and Burghardt 2005). Here we review the different ways of sharing processes:

- Open source. There are a number of open source projects (Open JUMP, GeoTools, QGIS, GeOxygene, 52°North WPS, GRASS, sextante) that propose a platform for developing geospatial processes. Users can either develop on the platform, or import some libraries into their own platform to use existing tools. This approach has limitations though, as the integration can be cumbersome (incompatible open source license models, linking libraries and writing translators to cope with different data models), and needs to be done every time a new library is required. There are also potential compatibility issues. Users are often already using their own development platform, integrated with their own systems. Switching to a new one is not often possible. However, using libraries of generic tools is a common way of reusing existing open source software such as JTS, JCS, and CGAL.
- Web Services. The Web service approach is a way to package processes in a standard way. Web services are hosted on servers and can be accessed by remote clients through the web. The main advantage of the web service is that the client does not need to run on the same platform as the service it is calling. This provides true platform interoperability. It is also very versatile, as there is no limit to what can be encapsulated in a service. It could be a simple generalisation operator, or even a simple measure, or it could be a full generalisation system. A disadvantage of the web service approach is that it involves moving the data from the client to the server, doing the processing remotely and downloading the results; this can involve transferring large amounts of data. Web services for generalisation have been studied. An initial platform called WebGen was developed at the University of Zurich (Burghardt et al. 2005; Neun and Burghardt 2005). It was later adapted to the OGC standard WPS, at the request of several members of the ICA Commission on Generalisation and Multiple Representation (Foerster et al. 2008). Tests have been done at Ordnance Survey to demonstrate the cross platform interoperability benefits of the approach. A server capable of hosting services that rely on the platform Radius Clarity has been setup. These services have been successfully called by an OpenJump (http://openjump.org/) client. A client for ArcGIS is currently under development at 52°North, funded by the Ordnance Survey (GB). More information and updates on WebGen-WPS can be found on the commission website (http://generalisation.icaci.org/index.php/web-services).
- Code moving. In order to avoid the issue of transferring large amount of data between the client and the server, an alternative approach to web services would be to send the executable code to the client, where it can execute the process locally. This code moving concept and its application to geoprocessing is described by Müller et al. (2010, 2012). It has the advantage over web services in that it requires less data transfer, as the size of the code is often much smaller than the size of the geographic data being processed. This could therefore result in much faster overall execution time. It also reduces the risks related to data security, as the data does not leave the client. The main problem with this

method is that it requires the client to provide a compatible runtime environment, compatible hardware, and there may also be licensing issues if the code includes third party libraries. This approach also allows organisations to publish their tools without having to maintain a powerful server capable of running the process locally. This technology is currently less mature, less readily available than the web service approach, but the concept is interesting.

Both the web service and the code moving approaches require a formal description of the process proposed. This includes a high level description of what the service does, and also the type of all the parameters required. This is often referred to as the service contract, which formalises the requirement of the service to ensure its successful execution. This should be enough for a human to choose what service to use. In a context of automatic discovery of services, it becomes essential that these descriptions are formalised and use a standard vocabulary. This has been discussed by Balley and Regnauld (2011) and led to the definition of the *semantic referential*, which defines all the concepts that need to be shared by all the components of their on-demand mapping system. Touya et al. (2010) also identified similar requirements, and proposed generalisation domain ontology to define these concepts. Required geographic concepts relate to real world objects and can be easily identified (for example the feature types defined by INSPIRE), even if reaching a consensus is often difficult. Many classifications of generalisation operators have also already been proposed (McMaster and Shea 1992; Regnauld and McMaster 2007; Foerster et al. 2007), so again, the challenge is to adopt a common one. It becomes more difficult when it comes to formalising the description of the parameters, as these are sometimes very specific to a particular process (Fig. 7.4). The best way to overcome this seems to be to rely on translators that derive the values of the parameters required by a service from meaningful formalised user requirements.

7.3 Formalising the Procedural Knowledge

Procedural knowledge is used to guide the selection and application of generalisation operators (Armstrong 1991). These can take various forms such as rules, constraints or ontologies. Technically this is declarative knowledge, but we often associate this to the domain expert's savoir-faire, which we include in the procedural knowledge. This selection mechanism depends on the input data, the output required and of course the operators available. Formalising this knowledge therefore requires an existing formalism to describe the types of geographic features handled, the types of operators required, and the requirements, often referred as map specifications (Chap. 2).

The procedural knowledge can take many forms. In its "unformalised form", it can be found in the implementation of batch processes, as a static sequence of operations, possibly enhanced by the use of conditional statements to adapt the sequence to the conditions. In more advances system, such as those based on optimisation techniques, the procedural knowledge provides the heuristics used by the system to explore the space of solutions. This is rarely well formalised. Taillandier et al. (2008, 2011) formalises part of the procedural knowledge in the AGENT system in order to revise this knowledge automatically, and to improve its performance.

Formalising the procedural knowledge is part of formalising all the knowledge required by the system. Concepts used to describe the procedural knowledge must match those used to describe the map specifications (Chap. 2). The key components of the system that need to be formally described are the basic tools (generalisation operators, measures, spatial structures, etc.). Some elements of the description are purely functional, while others are dependent on how they are made available (web service, moving code, library). Items required for describing tools include:

- Type of operation performed (Sect. 7.7)
- Type of data processed. Many data models for MRDB have been developed in the past, for example MADS (Parent et al. 1998) proposes a hierarchy of spatial abstract types that can be used to categorise the geographic classes. Ontologies have also been used to organise and describe geographic data and their interactions (Sect. 2.3)
- Geographic context (scale, type of geographic area, extent)
- Parameters
- Software dependencies (for code moving)
- Hardware dependencies (for code moving)

Knowledge is also required to link the tools to the conditions in which they can be used. These conditions are influenced by various factors, some coming from the requirements (target map specifications), others from local context. Several models have been defined to try to automatically choose the tools based on the requirements and context:

- Model developed by Balley (2012) (goals and tasks)
- Model developed by Touya (Sect. 7.6).

7.4 Chaining Processes

Once a library of tools to analyse the data and transform them is available and the specifications of the required output are available, the next challenge is to trigger the appropriate actions in the right order. In this section we discuss the different ways that processes (available as services) can be chained together. All these techniques rely on some kind of procedural knowledge (as described in Sect. 7.3), with different requirements with regard to its formalisation.

- Workflows. A workflow is a predefined sequence of steps. Using workflows to link existing processes is a typical way of chaining. Workflows are in general very static, built on top of the existing tools (Burghardt et al. 2010).
- Service chaining is the process of building a sequence of individual services to perform a more complex task. Depending on how the services are described, this chaining can be done in different ways. When the services are poorly described, the chaining has to be done by an expert who knows exactly what task the service is performing. For services well described using natural language, the service can be used by someone who does not need to know how the service works, but understands what it delivers. When the service description is formalised, it opens the door for chaining being automatically done. A chain of services can be encapsulated as a new service. Services can therefore be made available for all sorts of tasks, from the atomic operations to the full generalisation process.

Standards such as BPEL (Business Process Execution Language) have been developed to assemble such services. Schaeffer and Foerster (2008) present an approach that uses it for chaining OGC services for example. Service chaining is to some extent similar to creating a workflow using services. However it adds the interoperability aspect, and as the service becomes described with more formal languages, the chaining should soon include aspects of automatic service discovery which provides a much more dynamic way of chaining services.

- AGENT System: In the AGENT system (Ruas and Duchêne 2007), constraints are used to guide the choice of actions. Each constraint proposes a list of actions that can be triggered in order to reach a better solution. The agent system has an optimisation engine that looks for an optimum global satisfaction of all the constraints. While the current implementations of AGENT use actions available on the same platform, they could easily be replaced by service calls. The interesting aspect of the system is that the chaining of actions is built dynamically.
- Hybrids. A hybrid system combines different approaches to chain processes or services. These systems are usually designed using a very pragmatic approach, trying to reuse what is already available and combine them. For example, CollaGen (Sect. 7.2) uses a workflow to control different generalisation subsystems responsible for generalising a specific type of geographic context. These subsystems are already complex systems, some based on the AGENT paradigm as described above. The model proposed by Balley and Regnauld (2012) uses goals and tasks to dynamically build the sequence of high level tasks to derive the map specified by the user. Each task contains its own knowledge base that allows it to build its own chain of subprocesses (in this case implemented as services). The interest in Hybrid systems is expected to grow and such systems will become more and more powerful. They could use computational

intelligence engines of different types. Some such as neural nets would use a learning approach based on examples to build their own implicit knowledge base. Others could be rule based and rely on explicit knowledge bases.

7.5 Future Opportunities

Cloud computing is one development that is highly relevant to map generalisation yet has seen little application to date. With the current trend to componentise the process of generalisation, and allow interoperability between different platforms, exploiting the power offered by parallel processing is becoming a very realistic prospect. Frameworks such as MapReduce (Dean and Ghemawat 2004) or Hadoop (http://hadoop.apache.org/) have been designed for processing large datasets efficiently. They rely on mechanisms that split the task and distribute the processing over a cluster of processing nodes. While this is already widely used in other application domains, we still haven't seen it used for generalisation. However, we can easily imagine that once the tasks of a workflow have been identified as being independent of one another, that they could be computed simultaneously in the cloud. Computing in the cloud also offers the benefit to the user of getting the processing power they need when they need it, without having to maintain expensive hardware and software in house.

The other benefit of a component based approach is that it provides a basis by which components can be described in a standard machine readable way. Once this happens, there will be no need to put into the procedural knowledge any direct reference to specific algorithm implementation (or services). The procedural knowledge will only mention the abstract operations required. Then depending on circumstances (input data, geographic context, requirements), the right service could be dynamically chosen. In the absence of precise definitions, there will be a high degree of failure (algorithm not stable, description misleading, lack of requirements). This is where optimisation systems such as AGENT, or simulated annealing engines (Swan et al. 2007) can be used. They would be able to find the best (or at least an acceptable) solution among potential candidates. This of course will add a considerable overhead, but this can be mitigated in two ways. First trying alternative solutions can be done in parallel. In addition, we can use automatic learning to record past experiences and use it to refine the knowledge used by the optimisation engine, so that it gradually stops trying options that tend to never deliver good results, and try those which are likely to provide satisfying results first. These self-optimising techniques have already been studied for an agent system performing generalisation, and proved very successful (Taillandier et al. 2011; Taillandier and Gaffuri 2012). With more distributed systems on the horizon, greater optimisation of this type will be required to efficiently harvest the power of Web based processing.

7.6 Case Study I: Collaborative Generalisation

Guillaume Touya

7.6.1 Principles of Collaborative Generalisation

Past and current research shows that existing automatic generalisation processes are not able to correctly generalise a complete topographic map, despite some very good results for some specific parts of the map (e.g. landscapes such as urban or rural, or themes such as roads or land use) (Touya 2008). As a consequence, generalising a complete map requires the optimal use of available processes (on a software platform or via web services). Collaborative generalisation is a hybrid kind of chaining that follows from the previous classification (Sect. 7.4).

7.6.1.1 Collaboration Within Automated Generalisation Processes

According to McMaster and Shea (1988), an automatic generalisation process has to be able to know how, where and when to apply a generalisation operation. Collaborative Generalisation (Touya et al. 2010) seeks to answer the same 'how, where, when' but at the upper level of processes. This is where processes collaborate in order to decide which part of space they are best suited for. Figure 7.3 shows a schematic view of collaborative generalisation principles. The map is partitioned into spaces representing a landscape (e.g. urban, rural areas) or a data theme (e.g. the road network) (Touya 2010). Each of the spaces that needs generalisation is matched to the best suited available process, then the potential side effects at the space boundary are automatically corrected.

7.6.1.2 Interoperability Problems with Collaborative Generalisation

The collaborative generalisation mechanism is intended to make processes collaborate in the creation of an overall solution, even though these processes were never designed to work alongside one another. Interoperability issues arise from different kinds of heterogeneities in the collaborative generalisations (Sect. 7.2):

- *Process capabilities description*: the capabilities of generalisation processes have to be specifically formalised to be able to choose the solution that is best suited to a given region of the map.
- *Parameter heterogeneity*: each process requires its own set of parameters while the overall generalisation should have a single standard way of parameterisation. Figure 7.4 shows that a single way of parameterisation can be achieved by adding a translator to convert each process specific parameter into the standard parameter format.

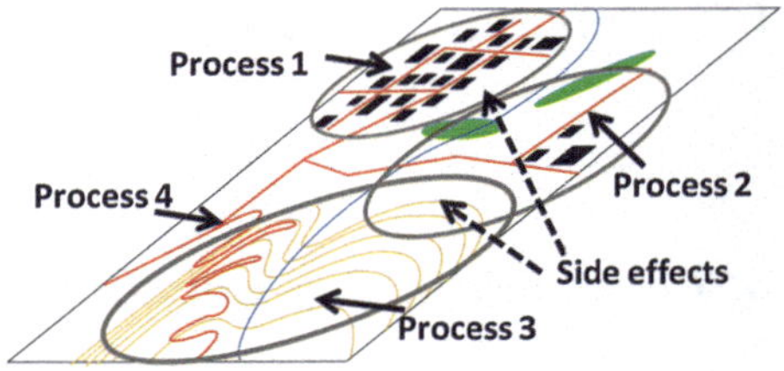

Fig. 7.3 The collaboration principle between generalisation processes: *process 1* is carried out on the town area, *process 2* on the rural area, then *process 3* on the mountain area and finally *process 4* on the road network. Side effects are corrected for the spaces at the boundaries between generalised regions

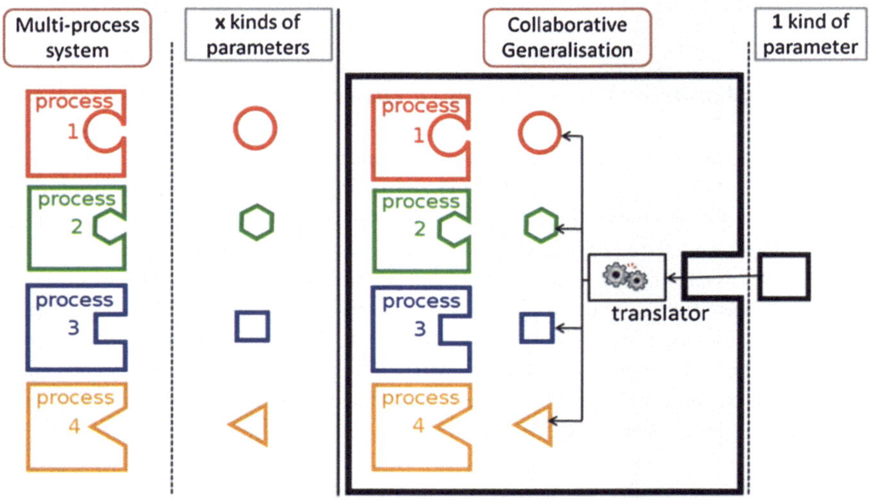

Fig. 7.4 Parameter heterogeneity requires a standard kind of parameter and a translator

- *Evaluation heterogeneity*: collaborative generalisation is an iterative process that requires a generic self-evaluation model, independent from the processes.
- *Global syntactic and semantic heterogeneity*: process capabilities, parameters or evaluation rely on shared generalisation knowledge to avoid syntactic and semantic heterogeneity.

7.6.2 Knowledge Formalisation for Collaborative Generalisation

The CollaGen model (Touya and Duchêne 2011) supports collaborative generalisation. Some CollaGen components were developed to overcome the interoperability problems previously described. This section briefly describes each of these components with some experimental results.

7.6.2.1 Generalisation Knowledge Ontology

In order to solve syntactic and semantic heterogeneity, CollaGen is fitted with a generalisation knowledge ontology (part of the semantic referential introduced in Sect. 7.2). It contains a shared vocabulary of geographic entities (e.g. building), generalisation concepts (e.g. meso object) and operations (e.g. typification), or spatial relations (e.g. rivers flow into talwegs) (Touya et al. 2010). CollaGen's formalised knowledge described in the next section, shares this vocabulary by always referring to this ontology.

7.6.2.2 Specifications Formalisation by Constraints and Rules

The chosen pivot format for specifications (i.e. exchange format for specifications/ parameters) in CollaGen is a formal model of generalisation constraints (Chap. 2). The formal model is visualised in Fig. 3.8: a generalisation constraint constrains a geographic entity (e.g. building), about a character (e.g. area) with a type of expression (e.g. character value < threshold).

Stoter et al. (2009) noticed that, sometimes, specifying the operation to do or not do was useful (e.g. small buildings shouldn't be aggregated), so it is possible to define such rules instead of constraints in the model. Eighty formal constraints or rules have been defined in the CollaGen prototype.

A translator function (Fig. 7.4) is associated with each available process to transform the formal constraints and rules into the specific parameters of the process. For instance, the translator function for the Least Squares generalisation process (Harrie and Sarjakoski 2002) transforms the formal constraints and rules into a system of linear equations.

7.6.2.3 Constraint Monitors for Interoperable Evaluation

The CollaGen model needs to iteratively (after each process is triggered) assess the satisfaction of the constraints in the map, in a process-independent way. Constraint *monitors* are created (i.e. database objects) to monitor each formal constraint for each object concerned by the constraint (Touya and Duchêne 2011). For instance, n monitors are created to monitor the minimum building size for the n buildings of the

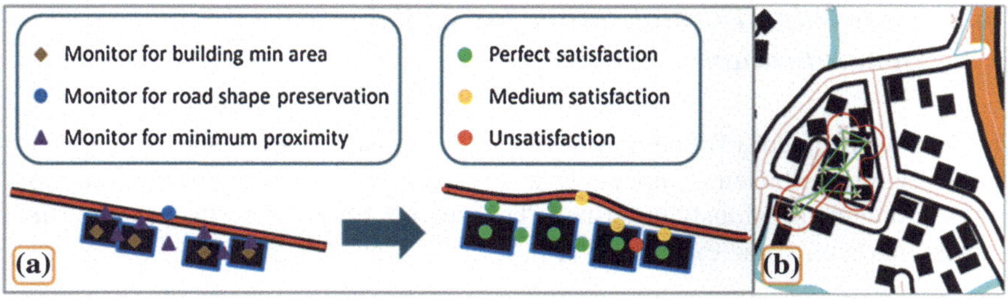

Fig. 7.5 **a** Constraints monitors for generic evaluation of generalisation. **b** A group of unsatisfied monitors identified by CollaGen prototype

dataset. Constraint monitors allow us to identify where the map is badly generalised (Fig. 7.5). Generalising a complete map implies the handling of a huge number of monitors and therefore it is a challenge to evaluate the map overall by summarising the evaluation of each monitor into a single global evaluation (Touya 2012).

7.6.2.4 Generalisation Process Capabilities Description

Formalising the description of capabilities is a key issue in the quest for web services interoperability. Following the ideas of Lutz (2007) for geo-services, CollaGen describes generalisation processes with pre-conditions (i.e. the conditions the input data have to meet to be properly processed) and post-conditions (i.e. what the data are expected to conform to after the process has been applied). For generalisation processes, pre-conditions are the type of situation the process is able to treat [e.g. CartACom process is able to generalise rural spaces (Ruas and Duchêne 2007)] and post-conditions are the constraints that are expected to be satisfied after generalisation (Touya et al. 2010). In addition to the conditions, the formal description contains an optimal scale range and a list of required enrichments (Mackaness and Edwards 2002). These pre and post conditions should help selecting the best suited algorithm for each situation. In a situation like the one shown on Fig. 7.6, CartACom would be expected to deliver better results than AGENT, because while they both have preconditions stating that they can be used in rural environments, only CartACom has a post condition stating that it preserves building/road parallelism.

7.6.2.5 Formalising Orchestration Knowledge

In order to orchestrate the processes in charge of generalising the different parts of the space, CollaGen allows the definition of orchestrating rules (e.g. "generalise urban areas before rural ones") that play Ruas and Plazanet's (1996) *Global Master Plan* role. This enables us to control the sequence such that 'the road network should be generalised before the rural areas'. In Fig. 7.6, result (d), where the road network has been generalised before the rural area, is better than result (c), as the dead-end road symbol does not overlap buildings. This shows the value in specifying such rules.

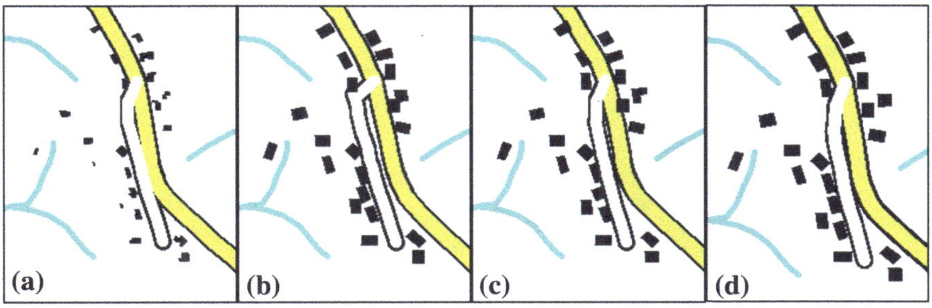

Fig. 7.6 **a** before generalisation. **b** Generalisation with AGENT then Least Squares process (Harrie and Sarjakoski 2002). **c** CartACom then the Beams (Bader et al. 2005). **d** The Beams then CartACom

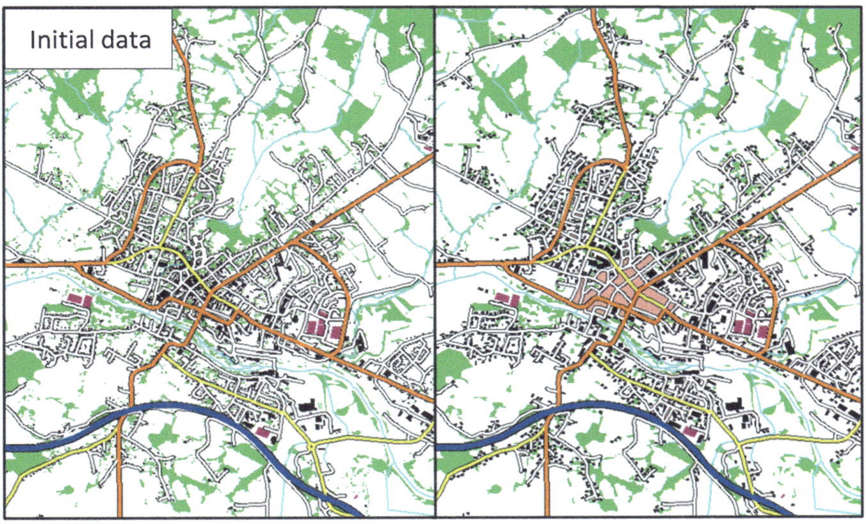

Fig. 7.7 Extract of a CollaGen generalisation on a large area at the 1:50 k scale, where nine processes collaborated

7.6.3 CollaGen Results

The CollaGen model is based on formalised knowledge that enables us to carry out automatic collaborative generalisations (Touya and Duchêne 2011). A prototype was developed with access to nine automatic processes, all dedicated to topographic maps. Figure 7.7 shows results for a 1:50 k map that contains urban and rural landscapes. These results have been evaluated as being much better than any automatic process used alone. Figure 7.8 shows that CollaGen performs better

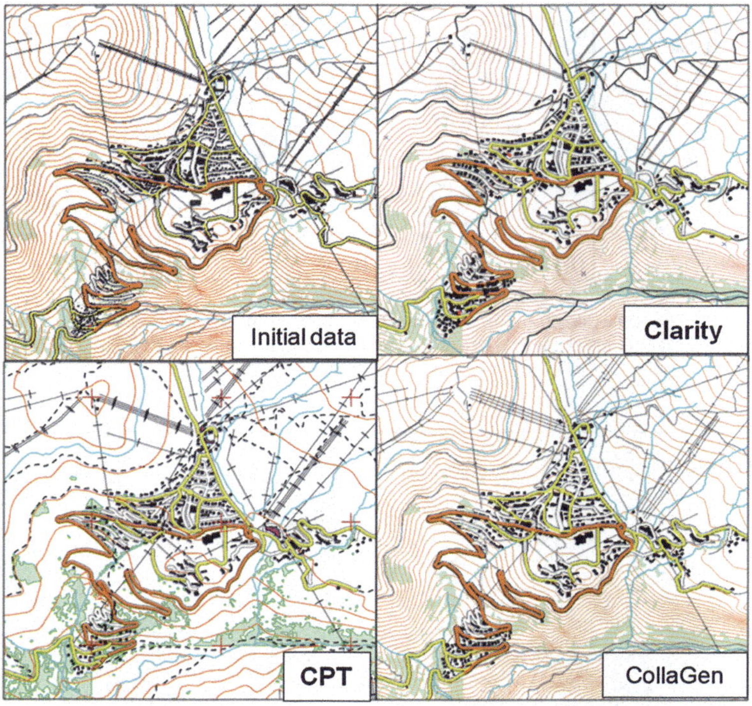

Fig. 7.8 A mountainous French dataset from EuroSDR tests (Stoter et al. 2009) generalised with CollaGen compared to the best results

than the best (non automatic) benchmark tests with commercial software from the EuroSDR tests (Stoter et al. 2009).

7.7 Case Study II: An Ontological Approach to On-Demand Mapping and Generalisation

Nicholas Gould

7.7.1 The Case for Ontology-Driven Generalisation

This Case study relates to on-demand mapping and in particular focusses on knowledge formalisation and how it can be used to aid the automatic selection of generalisation operators and algorithms. If we wish to automate any process then we must *formalise* the knowledge for that particular domain. The knowledge needs to be machine understandable not merely machine readable. In that way the system can make decisions based on the knowledge it has of the process. The acquisition and formalisation of cartographic generalisation knowledge has not proved easy (Rieger and Coulson 1993; Kilpelainen 2000). Consider, for example, the naming and classification of generalisation operators.

As discussed in Sect. 7.2, there have been numerous attempts to classify and describe generalisation operators but the problems highlighted by Rieger and Coulson (1993) remain. As well as differences between the proposed categories of operators there are also problems where different terms are used for the same concept (Aggregation or Combine?) and in granularity; McMaster and Shea (1992) define Smoothing, Enhancement and Exaggeration where Foerster et al. (2007) simply define Enhancement. There is also disagreement as to what functions can be regarded as generalisation operators. For example, is Symbolisation a generalisation operator (McMaster and Shea 1992) or a pre-processing step (Foerster et al. 2007)?

The use of different operator taxonomies in closed systems does not matter, but, if we are to develop an interoperable on-demand system, an agreed taxonomy as well as the semantic description of the operators is required. This is because we cannot simply ask for a web service that performs Smoothing, say, since that operation can be performed by a number of different algorithms (Gaussian, Cubic Spline, Fourier transform etc.), each with their own parameters and often with different results. Similarly, some operators apply to multiple geometry types and will need to be implemented by different algorithms. Likewise some algorithms specialise in different feature types such as buildings (Guercke and Sester 2011). Thus these details need to be formally defined so that automatic selection and execution is possible by the on-demand system.

Formalisation of knowledge can lead to the discovery of new knowledge as long as appropriate formalisation tools are available (Kilpelainen 2000). One such tool is the ontology: the explicit specification of the objects, concepts and the relationships in a body of knowledge concerning a particular subject or domain (Gruber 1993). Ontologies have the advantage of allowing the sharing and reuse of formalised knowledge (Gruber 1993). Rule-based systems hold *procedural knowledge* that describes explicitly how a process is to be performed. As described earlier (Sect. 7.1), rule-based systems are likely to suffer from rule explosion. Ontologies can hold *declarative knowledge*. One advantage of declarative knowledge is that it can be extended by means of reasoning which can be used to derive additional knowledge (Genesererh and Nilsson 1998).

The application of ontologies to generalisation is not new. However, to date, their use has been restricted to aiding the process of generalisation, for example by pattern identification (Lüscher et al. 2007); by describing geographical relationships (Dutton and Edwardes 2006); and by semantically enhancing a line simplification algorithm (Kulik et al. 2005). However, what is proposed is using ontologies to describe the complete process of generalisation. The intention is to formalise the *why*, *when* and *how* of generalisation (McMaster and Shea 1992).

7.7.2 Designing the Ontology

The first stage in designing an ontology is to determine its scope by defining a set of competency questions that the ontology is expected to answer (Noy and McGuinness 2001). In the domain of on-demand mapping the competency questions include: Under what conditions is generalisation required? Which generalisation operators should be applied? What algorithms should be applied to implement the selected operators? The next step is to enumerate the important terms in the domain.

There are a number of reasons *why* a set of geographic features should be generalised but, if we consider *legibility* in the first instance, we can define a number of *geometric conditions*, such as *congestion* and *imperceptibility*, which are the result of a change from large (detailed) scale to a small scale, and govern legibility. These conditions can be evaluated by applying a number of *measures* (Stigmar and Harrie 2011). For example, the existence of congestion can be determined by applying a feature *density* measure and will determine *when* generalisation is necessary.

The *how* of generalisation is answered by generalisation operators. But how are they to be defined and classified given the disparate taxonomies described earlier? A (loose) analogy with medicine can be applied. Consider the congestion of features. Congestion can be regarded as a *condition* and a *symptom* of that condition is a high feature density. To check whether the data has that condition a *measure* algorithm can be applied (where a measure algorithm is analogous to a thermometer, say). If the condition is present then a *remedy* such as a reduction in feature size or in feature count is appropriate. Generalisation operators are defined by the remedies they implement. For example, if a set of buildings features is determined to be congested then a reduction in feature count can be applied by the *SelectionByAttribute* of the more important buildings only or by the *Amalgamation* of buildings into single features. The mechanism by which an algorithm can be used to resolve a condition is shown, in a much simplified manner, in Fig. 7.9. That mechanism applied to the particular case of congestion in point data, is shown in Fig. 7.10. In this case the ontology identified two operations, Amalgamation and Selection, as candidates for resolving the congestion. The ontology uses the term *transformation* algorithm instead of generalisation algorithm since this is the only distinction made between algorithms that measure and those that transform be it by generalisation or other means.

Operators may also have specific requirements. For example, *SelectionByAttribute* requires the source dataset to have an attribute that is used to rank the importance of features (such as the severity of road accidents). If this attribute is not present then the operator can be ruled out. All of this knowledge, in combination, will aid the automatic selection of appropriate operators.

However, the domain of the ontology does not stop at the level of the operator. As discussed earlier there is no one-to-one mapping between an operator and the algorithm it implements. Quite different algorithms are required to *Simplify* line

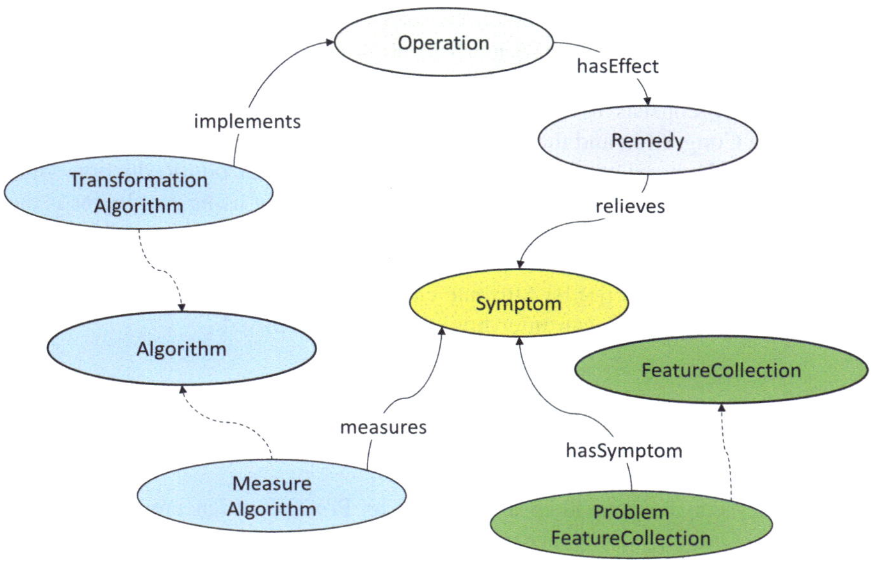

Fig. 7.9 Selecting generalisation algorithms—general case (dashed lines represent "is a kind of" relations)

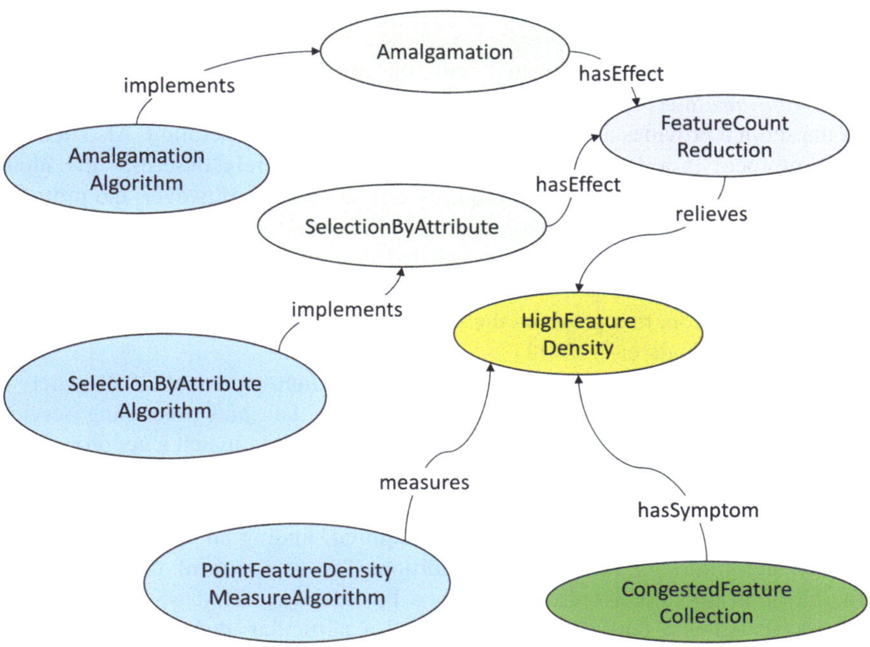

Fig. 7.10 Selecting generalisation algorithms—particular case

features, such as roads, and area features, such as buildings. The ontology needs a sufficiently detailed description of generalisation algorithms to allow the relevant algorithm to be selected automatically.

An ontology consists of assertions. We assert that HighFeatureDensity is a Condition of Congestion and that Congestion is a barrier to Legibility. We can also assert that HighFeatureDensity can be relieved by a FeatureCountReduction which is an effect of Amalgamation (Fig. 7.10). The advantage of using ontologies is that we can use *reasoning* to *infer* further knowledge. This is the additional knowledge described earlier. For example, we do not have to explicitly state that both Amalgamation and SelectionByAttribute can resolve congestion. If we sufficiently describe the operators we can infer how they can be utilised.

7.7.3 Applying the Ontology

The ontology can be created in a tool such as Protégé which creates and edits OWL (Web Ontology Language) files. Protégé also employs reasoners such as HermiT that can be used to derive inferences. But once developed, how can the ontology be applied in a distributed web service-based system? The application of semantically described geoprocessing services in Spatial Data Infrastructures (Brauner 2012) provides a parallel.

The standard for implementing geospatial web services is the OGC's Web Processing Service (WPS) protocol and, as described earlier, generalisation algorithms have been implemented with this protocol. The protocol defines a *GetCapabilities* interface that will return a list of each individual spatial operation that the service provides and a free text description of each operation. Also defined in the protocol is a *DescribeProcess* interface that merely describes the input parameters the specified operation requires and its outputs. However, the protocol does not provide for semantic interoperability (Janowicz et al. 2010); that is, there is no method of providing machine readable descriptions of the operations that allow the operation to be selected automatically. What is required is a technique, semantic annotation, that provides these descriptions (Lemmens et al. 2007; Maue et al. 2009; Mladenic et al. 2011).

One solution is the *Semantic Enablement Layer* (Janowicz et al. 2010) where a Web Ontology Service injects semantics into both data and processing service descriptions. A Web Reasoning Service can then be used to match a geoprocessing service to a dataset. Their architecture is aimed at geoprocessing in general rather than generalisation but can be expanded to firstly finding an appropriate measure algorithm for the source data and then, if required, finding an appropriate generalisation algorithm for the existing condition. The selection of the appropriate generalisation operator would be bypassed by inference. That is, if a particular operator remedies a particular condition and a particular algorithm implements that operator then we can infer that the algorithm will remedy the condition.

The ontology can be regarded as component of the semantic referential described earlier (Sect. 7.2) and it extends the Generalisation Knowledge Ontology of Collagen (Sect. 7.6.2.1) by expanding the description of the operations. In summary, the use of ontologies to aid the selection of geoprocessing services that implement a specified operation (e.g. buffer) is a well researched area; our contention is that the use of ontologies can be extended to the selection of the (generalisation) operation itself.

7.8 Case Study III: Live Geoinformation with Standardised Geoprocessing Services

Theodor Foerster

Live geoinformation (available without significant delay) is considered to be crucial for applications in which decisions are (a) based on massive volumes of data and (b) need to be carried out in real-time. For instance in risk management scenarios, live geoinformation can directly support time critical decision making for saving human lives and infrastructure. Other examples are near real-time analysis of crowd-sourced geodata. All these applications are framed by the idea of the Digital Earth (Gore 1998) which provides an integrated platform for accessing different kinds of distributed data in near-real time.

Providing such information and transforming raw data into value-added information is supported by geoprocessing. Generalisation is involved in any task, in which the scale of the information is affected and is thereby chosen as a representative example. Currently, generalisation processes as well as their output (maps, raw data) become available through web service interfaces. These web service interfaces are currently designed along a sequential request-response mechanism, in which the data are sent to the service, processed and then sent back. These different phases are handled sequentially, which means, that the service and the client remain idle in the meantime and wait for the other party to complete. This is not sufficient for live geoinformation and its emerging requirements:

- Performance—Using the idle time of the service, while transferring data.
- Handling, processing, creating of geodata streams—Streams of geodata such as provided by sensors become a valuable source of information for GIS and Digital Earth.
- Loss-less reliable encoding and transfer of data (in contrast to existing lossy unreliable multi media encodings)—The data needs to be transferred in a reliable manner, guaranteeing data completeness.

- Interoperability and portability—The data needs to be transferred in an interoperable way by reusing existing standards and agnostic of the technical setup.

To meet these requirements and realise live geoinformation, HTTP Live Streaming as a loss-less format for real-time data streaming has been combined with the OGC Web Processing Service, which is an established web service interface and de-facto standard for processing geodata on the web. The presented approach is applied to automated generalisation of OpenStreetMap data.

7.8.1 Related Work

The building blocks of live geoinformation require the efficient creation and handling of live geodata streams. In the context of geoprocessing services, the real-time processing of live geodata streams and publishing such streams is required. Detecting and extracting events from such geodata streams is highly relevant in the context of Complex Event Processing (Everding et al. 2009). Live geoinformation requires a scalable event- and streaming-based architecture in order to support concepts such as Digital Earth. Regarding the communication within the architecture, we envision a fully push-based architecture, in which the processes are triggered from the sources (e.g. sensors or created by events). This will limit the communication overhead to a minimum. Technically, this is realised through notification and call-back methods.

Several approaches for improving the scalability and performance of geoprocessing services have been described, e.g. applying Cloud and Grid Computing infrastructures (Baranski et al. 2011; Baranski 2008; Di et al. 2003; Lanig et al. 2008) or the mobile code paradigm (Müller et al. 2010). Scholten et al. (2006) identify caching, network adaptation, data granularity and communication modes (synchronous vs. asynchronous) as performance criteria.

From a generalisation perspective, the work of Bertolotto and Egenhofer (2001), van Oosterom (2005), Buttenfield (2002) concerning progressive transfer addressed a related problem from a users' perspective, namely that users want to receive the most important data first. This is handled by extracting the most important aspects of the data by automated generalisation and providing it successively. A streaming based approach could be designed to behave in a similar way.

7.8.2 Approaches to Stream-Based Processing

When the WPS receives an asynchronous Execute request, an Execute response is instantly returned to the client and the process execution is scheduled in the background. The Execute response includes a 'Status' element that contains information about the overall status of the process ('accepted', 'started', 'paused', 'succeeded' or 'failed') and an (optional) progress indicator showing the percentage rate of process completion. The Execute response includes a 'statusLocation' element that links another Execute response, which always contains the latest status information about a process. As soon as a process has completed, this Execute response contains the process result(s). The client can constantly pull this Execute response until the final result is available.

In the proposed approach, the body of the 'Status' element does not indicate detailed information about the progress of the process, as in current WPS implementations (e.g. the amount of features that have been processed). Instead it includes a URL to a playlist file as specified by the HTTP Live Streaming draft specification. Table 7.1 demonstrates an example of an Execute response containing a reference to a playlist file.

The playlist file contains a sorted list of URLs that represents previous and current intermediate results. When an intermediate result is created and stored by the service, the service also updates the playlist file (a URL returning the latest intermediate result is attached). Therefore, by frequently calling the playlist file URL the client receives the latest intermediate results. As soon as a process is completed, the service adds a special tag to the playlist file accordingly. By not adding such a tag, the client knows that the process might run continuously. Further details such as the playlist format encoding and the implementation are described in Foerster et al. (2012a).

In Fig. 7.11 the effect of generalising road segments (taken from an Open Street Map dataset) based on the streaming-based approach is portrayed. Continuously (t1,..., t4 in Fig. 7.11), the different processed road segments are transferred and displayed as processed by the streaming-based WPS.

7.8.3 Discussion and Related Challenges for Generalisation

As described, Web Services face a challenge of providing the most current data as soon as it is available. Mostly, this data needs to be adopted regarding scale through processing, so generalisation is a challenge. To tackle this challenge, a loss-less, asynchronous stream-based and interoperable approach towards web service interfaces is required. The presented approach is based on HTTP Live Streaming and is applied to the generalisation of Open Street Map data. From the research it becomes evident, that not all generalisation processes are suitable for this approach. If the spatial context of the objects and their topology play a significant role in a specific process, this process cannot start until all the contextual data has been received. This means that processes requiring a large or unknown amount of contextual information are not suitable for streaming-based processing. However, tasks of simplification (for instance Douglas Peucker, which are still heavily used in generalisation batch jobs) appear to be suitable candidates.

Table 7.1 Example of an execute response with a URL of a playlist that contains real-time intermediate results

<ExecuteResponse service = "WPS" version = "1.0.0" statusLocation = "…">
<Process ns:processVersion = "1.0.0">
<Identifier>StreamDouglasPeuckerAlgorithm</Identifier>
</Process>
<Status creationTime = "…">
<ProcessStarted>
http://host:port/wps/playlist?id=123&pollingRate=1
</ProcessStarted>

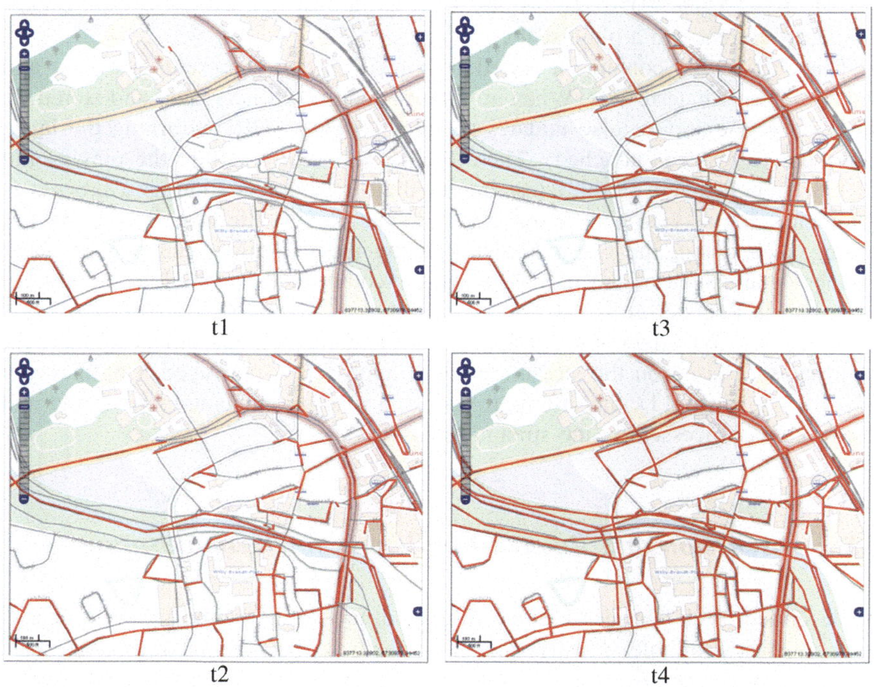

Fig. 7.11 The evolution (from t1 to t4) of transferred road segments based on the streaming-based approach

Future research needs to address these issues in a more detailed manner and we need to find intelligent ways to enable streaming-based processing for a broader range of generalisation functionality. This can be done by automatically detecting the partition requirement and adjust the size of transferred data chunks accordingly. As soon as this is achieved, the chaining and orchestration of these streaming-enabled generalisation processes becomes an interesting application, to deliver highly customised datasets in a timely fashion.

7.9 Conclusions

We have seen in this chapter that modelling the process of generalisation is still a very open question, and continues to generate interest in the research community. While a lot of generalisation tools have been studied and implemented, the challenge of creating systems capable of using them automatically is still alive. We have seen that different models have been studied and automated solutions have been successfully implemented, but these successes have been limited to building systems that provide a static solution to derive a specific product from a specific set of data.

In order to overcome this limitation, it has long been recognised that studies are needed that concentrate on the process modelling side. But in order to do this successfully, and to be able to test the concepts, we had to have the basic tools (generalisation operators, measures) readily available. Web services have been proposed as a way to encapsulate and publish these tools, so that they can be easily reused.

Several studies now focus on "on-demand mapping", which looks into designing systems capable of interpreting user requirements (in the form of formal machine readable specifications (Chap. 2)), and deriving the appropriate map. This requires a significant effort in formalising the description of all the components of the system: specifications, data, tools, knowledge. The Collaborative Generalisation approach described in the first Case study of this chapter, shows how to apply different existing models in different situations occurring on the same map. The second Case study presented an ontology based approach to resolve cartographic conflicts as they occur during the automatic creation of a map. Both these approaches rely on formalising aspects of the generalisation knowledge. Both also rely on existing software components providing the lower functionalities of the system (operations, measures, or a particular subsystem to perform a specific generalisation task). As it is often suggested that such decoupled system should use Web Services to access the lower functionalities, we have included a third Case study focusing on improving the performance of web processing services, using the concept of streaming based processing.

The next challenges in the area of process modelling related to generalisation are related to the design of systems able to perform on-demand mapping, which includes generalisation, but also the collection of user requirements, data integration and automatic cartographic design (to style the resulting map). As prototypes of these systems emerge, we know that we will also need to address performance issues. This is due to the fact that the lack of predefined sequences of actions will be overcome by complex strategies to build the sequence, possibly including expensive trial and error strategies. More research in optimisation techniques will therefore be required. These are likely to include machine learning, parallel computing, and streaming based processing.

References

Alameh N (2003) Chaining geographic information web services. IEEE Internet Comput 07:22–29

Armstrong MP (1991) Knowledge classification and organization. In: Buttenfield B, McMaster RB (eds) Map generalization: making rules for knowledge representation. Longman, London, pp 86–102

Bader M, Barrault M, Weibel R (2005) Building displacement over a ductile truss. Int J Geog Inform Sci 19(8):915–936

Balley S, Regnauld N (2011) Models and standards for on-demand mapping. In: Proceedings of the 25th international cartographic conference. Paris, Jul 2011

Balley S, Jaara K, Regnauld N (2012) Towards a prototype for deriving custom maps from multisource data. 15th workshop of the ICA commission on generalisation and multiple representation. Istanbul, Sep 2012

Baranski B (2008) Grid computing enabled web processing service. In: Pebesma E, Bishr M, and Bartoschek T (eds) Proceedings of the 6th geographic information Days, IfGI prints, vol 32. Presented at the GI-days 2008 in the institute for geoinformatics. Muenster, Germany, pp 243–256. Retrieved from http://www.gi-tage.de/archive/2008/downloads/acceptedPapers/Papers/Baranski.pdf

Baranski B, Foerster T, Schäffer B, Lange K (2011) Matching INSPIRE quality of service requirements with hybrid clouds. Transactions GIS 15(s1):125–142. doi:10.1111/j.1467-9671.2011.01265.x

Bertolotto M, Egenhofer MJ (2001) Progressive transmission of vector map data over the World Wide Web. Geoinformatica 5(4):345–373

Brauner J (2012) Ad-hoc-geoprocessing in spatial data infrastructures—formalizations for geooperators. In: Bernard L, Pundt H (eds) 1st AGILE PhD School. Shaker, Wernigerode, Germany

Brenner C, Sester M (2005) Cartographic generalization using primitives and constraints. In: Proceedings of the 22nd international cartographic conference. A Coruna, Jul 2005

Burghardt D, Neun M, Weibel R (2005) Generalisation services on the web—a classification and an initial prototype implementation. Cartography Geogr Inf Sci 32(4):257–268

Burghardt D, Petzold I, Bobzien M (2010) Relation modelling within multiple representation databases and generalisation services. Cartographic J 47(3):238–249

Buttenfield (2002) Transmitting vector geospatial data across the internet. GIScience 2002: 51–64

Dean J, Ghemawat S (2004) MapReduce: simplified data processing on large clusters. In: Proceedings of the 6th symposium on operating system design and implementation, San Francisco, CA, Dec 2004

Di L, Chen A, Yang W, Zhao P (2003). The integration of grid technology with OGC web services (OWS) in NWGISS for NASA EOS data. Presented at the GGF8 & HPDC12 2003. Science Press, Seattle, pp 24–27

Douglas D, Peucker T (1973) Algorithms for the reduction of the number of points required to represent a digitised line or its caricature. Can Cartographer 10(2):112–122

Dutton G, Edwardes A (2006) Ontological modeling of geographical relationships for map generalization. In: Proceedings of the 9th workshop of the ICA commission on map generalization and multiple representation, Portland, USA

Edwardes A, Burghardt D, Bobzien M, Harrie L, Reichenbacher T, Sester M, Weibel R (2003) Map generalisation technology: addressing the need for a common research platform. In: Proceedings of the 21st international cartographic conference, Durban, pp 170–179

Foerster T, Stoter J, Köbben B (2007) Towards a formal classification of generalization operators. In: Proceedings of the 23rd international cartographic conference, Moscow, Aug 2007

Foerster T, Stoter J, van Oosterom P (2012a) On-demand base maps on the web generalized according to user profiles. Int J Geogr Inf Sci 26(1):99–121

Foerster T, Baranski B, Borsutzky H (2012b) Live geoinformation with standardized geoprocessing services. In: Gensel J, Josselin D, Vandenbroucke D (eds) Bridging the geographic information sciences. Springer, Berlin, pp 99–118

Genesererh MR, Nilsson NJ (1998) Logical foundations of artificial intelligence. Morgan Kaufmann Publishers, Palo Alto

Gore A (1998) The digital earth: understanding our planet in the 21st century. Aust surveyor 43(2):89–91

Gruber T (1993) A translation approach to portable ontology specifications. Knowl Acquisition 5(2):199–220

Guercke R, Sester M (2011) Building footprint simplification based on hough transform and least squares adjustment. In: Proceedings of the 14th workshop of the ICA commission on generalisation and multiple representation, Paris

Harrie LE, Sarjakoski T (2002) Simultaneous graphic generalization of vector data sets. Geoinformatica 6(3):233–261. doi:10.1023/A:1019765902987

Jäger E (1991) Investigations on automated feature displacement for small scale maps in raster format. In: Proceedings of the 15th international cartographic conference, pp 245–256

Janowicz K, Schade S, Broring A, Kessler C, Maue P, Stasch C (2010) Semantic enablement for spatial data infrastructures. Trans GIS 14(2):111

Jenks GF (1989) Geographic logic in line generalisation. Cartographica 26(1):27–42

Kilpelainen T (2000) Knowledge acquisition for generalization rules. Cartography Geogr Inf Sci 27(1):41–50

Kulik L, Duckham M, Egenhofer M (2005) Ontology-driven map generalization. J Vis Lang Comput 16(3):245–267

Lanig S, Schilling A, Stollberg B, Zipf A (2008). Towards standards-based processing of digital elevation models for grid computing through web processing service (WPS). ICCSA, Lecture notes in computer science vol 5073. Presented at the computational science and its applications—ICCSA 2008. Springer, Perugia, pp 191–203 doi:http://dx.doi.org/10.1007/978-3-540-69848-7_17

Lemmens R, de By RA, Gould M, Wytzisk A, Granell C, van Oosterom P (2007) Enhancing geo-service chaining through deep service descriptions. Trans GIS 11(6):849–871

Lüscher P, Burghardt D, Weibel R (2007) Ontology-driven enrichment of spatial databases. In: Proceedings of the 10th workshop of the ICA commission on generalisation and multiple representation. Moscow, Aug 2007

Lutz M (2007) Ontology-based descriptions for semantic discovery and composition of geoprocessing services. GeoInformatica 11(1):1–36. doi:10.1007/s10707-006-7635-9

Mackaness WA, Edwards G (2002) The importance of modelling pattern and structure in automated map generalisation. In: Proceedings of the joint ISPRS/ICA workshop on multi-scale representations of spatial data, pp 7–8

Maue P, Schade S, Duchesne P (2009) Semantic annotations in OGC standards open geospatial consortium

McMaster RB, Shea KS (1992) Generalization in digital cartography. Association of American Geographers, Washington DC

Mladenic D, Moraru A, Skrjanc M (2011) Deliverable D4.2 model annotation component and guidelines. Envision consortium. http://www.envision-project.eu/2011/01/deliverable-4-2-model-annotation-components-and-guidelines. Accessed 15 Mar 2013

Müller M, Bernard L, Brauner J (2010) Moving code in spatial data infrastructures—web service based deployment of geoprocessing algorithms. Trans GIS 14(S1):101–118

Müller M, Kadner D, Bernard L (2012) Moving code—sharing geospatial computation logic on the web. In: Proceedings of the 15th AGILE international conference on geographic information sciences, Avignon

Neun M, Burghardt D (2005) Web services for an open generalisation research platform. In: Proceedings of the 8th ICA workshop on generalisation and multiple representation, A Coruna

Noy NF, McGuinness D (2001) Ontology development 101: a guide to creating your first ontology. Stanford University. http://www.ksl.stanford.edu/people/dlm/papers/ontology-tutorial-noy-mcguinness.pdf. Accessed 15 Mar 2013

Regnauld N, McMaster RB (2007) A synoptic view of generalisation operators. In: Mackaness WA, Ruas A, Sarjakoski LT (eds) Generalisation of geographic information. Elsevier Science B.V, Amsterdam, pp 37–66

Rieger MK, Coulson MRC (1993) Consensus or confusion: cartographers' knowledge of generalization. Cartographica Int J Geogr Inf Geovisualization 30(2):69–80

Ruas A, Duchêne C (2007) A prototype generalisation system based on the multi-agent system paradigm. In: Mackaness WA, Ruas A, Sarjakoski LT (eds) The generalisation of geographic information: models and applications. Elsevier, Amsterdam, pp 269–284 (Chapter 14)

Ruas A, Plazanet C (1996) Strategies for automated generalization. In: 7th international symposium on spatial data handling, Delft, pp 319–336

Schaeffer B, Foerster T (2008) A Client for distributed geo-processing and workflow design. J Location Based Serv 2(3):194–210

Scholten M, Klamma R, Kiehle C (2006) Evaluating performance in spatial data infrastructures for geoprocessing. IEEE Internet Comput 10(5):34–41

Stigmar H, Harrie L (2011) Evaluation of analytical measures of map legibility. Cartographic J 48(1):41–53

Stoter J, Burghardt D, Duchêne C, Baella B, Bakker N, Blok C, Pla M, Regnauld N, Touya G, Schmid S (2009) Methodology for evaluating automated map generalization in commercial software. Comput Environ Urban Syst 3(5):311–324

Swan J, Anand S, Ware M, Jackson M (2007) Automated schematization for web service applications. In: Web and wireless geographical information systems—W2GIS. Cardiff, pp 216–226

Taillandier P, Gaffuri J (2012) Improving map generalisation with new pruning heuristics. Int J Geogr Inf Sci 26(7):1309–1323

Taillandier P, Duchêne C, Drogoul A (2008) Knowledge revision in systems based on an informed tree search strategy: application to cartographic generalisation. In: Proceedings of the international conference on soft computing as transdisciplinary science and technology (CSTST), Cergy-Pontoise, Jul 2008

Taillandier P, Duchêne C, Drogoul A (2011) Automatic revision of rules used to guide the generalisation process in systems based on a trial and error strategy. Int J Geogr Inf Sci 25(12):1971–1999

Touya G (2008) First thoughts for the orchestration of generalisation methods on heterogeneous landscapes. In: Proceedings of 11th ICA workshop on generalisation and multiple representation, Montpellier

Touya G (2010) Relevant space partitioning for collaborative generalisation. In: Proceedings of 12th ICA workshop on generalisation and multiple representation, Zurich

Touya G (2012) Social welfare to assess the global legibility of a generalized map. In: Xiao N, Kwan MP, Goodchild MF, Shekhar S (eds) Geographic information science. 7th international conference, GIScience 2012, Columbus. Lecture notes in computer science, vol 7478. Springer, Berlin, pp 198–211

Touya G, Duchêne C (2011) CollaGen: collaboration between automatic cartographic generalisation processes. In: Ruas A (eds) Advances in cartography and GIScience. Lecture notes in geoinformation and cartography, vol 1. Springer, Berlin, pp 541–558

Touya G, Duchêne C, Ruas A (2010) Collaborative generalisation: formalisation of generalisation knowledge to orchestrate different cartographic generalisation processes. In: Fabrikant S, Reichenbacher T, van Kreveld M, Schlieder C (eds) Geographic information science. Lecture notes in computer science, vol 6292. Springer, Heidelberg, pp 264–278

van Oosterom P (2005) Variable-scale topological data structures suitable for progressive data transfer: the GAP-face tree and GAP-edge forest. Cartography Geogr Inf Sci 32(4):331–346

Chapter 8
Terrain Generalisation

Eric Guilbert, Julien Gaffuri and Bernhard Jenny

Abstract This chapter reviews recent development in terrain generalisation. We focus on issues of aesthetics and legibility in the application of cartographic generalisation. Generalisation methods are relevant to traditional terrain representations (spot heights, contours, hypsometric colours, shaded relief) and to grid and triangulated surface generalisation. First we consider issues related to relief representation at different scales. As generalisation requires knowledge about the terrain morphology, several approaches focusing on the classification of terrain features according to morphometric or topological criteria have been developed. Cartographic generalisation methods are reviewed with consideration given to conflicts between terrain representations and other object type data on the map. In the second part of this chapter, three case studies illustrating previous developments are presented. First, a generalisation method for hypsometric map production is described where important valleys and mountain ridges are accentuated to improve their representation. Second, a method selecting features represented by isobaths and answering specific constraints of nautical charts is presented. The third case study is a generalisation method which models the relationship between terrain and other objects such as buildings and rivers.

E. Guilbert (✉)
Land Surveying and Geo-Informatics Department, The Hong Kong Polytechnic University, Kowloon, Hong Kong
e-mail: lseguil@polyu.edu.hk

J. Gaffuri
Digital Earth and Reference Data unit, Institute for Environment and Sustainability, Joint Research Centre, European Commission, Ispra, Italy
e-mail: julien.gaffuri@gmail.com

J. Gaffuri
Laboratoire COGIT, IGN, Saint-Mandé, France

B. Jenny
College of Earth, Ocean, and Atmospheric Sciences, Oregon State University, Corvallis, USA
e-mail: jennyb@geo.oregonstate.edu

D. Burghardt et al. (eds.), *Abstracting Geographic Information in a Data Rich World*, Lecture Notes in Geoinformation and Cartography, DOI: 10.1007/978-3-319-00203-3_8,

8.1 Introduction

Portrayal of relief and landforms on maps is based on different techniques that provide a stylised representation of the terrain. While the oldest maps mostly provide a qualitative representation of the relief with limited accuracy, the introduction of more rigorous representations starting with the introduction of contour lines at the beginning of the eighteenth century offer a more accurate description of terrain. In addition to contouring, other techniques are variously used depending on the scale and purpose of the map. Maps are now commonly derived from remotely sensed Digital Terrain Models (DTM). They are designed for 2D visualisation on paper and mobile devices but also for representation in a perspective view (where the level of detail depends on the distance from the view point). These are more common on electronic devices.

Depicting terrain has always been a challenge for cartographers. They have always had to strike a balance between effective visual techniques for portraying landforms and methods that yield accurate terrain values. Each technique is applied according to the scale of the map and the level of detail required to be represented and requires a trade-off between visual quality and accuracy (Table 8.1). Relief on large-scale maps is usually portrayed with spot heights, contour lines, and shaded relief. Shaded relief can also be used at smaller scales, although hypsometric colours are preferred for small and very small scale maps (Imhof 1982).

The generalisation process is performed by simplifying the relief and emphasising characteristic landforms from a DTM. Earlier methods mostly focused on adapting the amount of information to the scale of the map by filtering out or smoothing details. Although they efficiently provide simplified terrain representations at a required accuracy, they fall short in highlighting landforms so that relevant features visually stand out from others on the map. More recent developments, in addition to a constant focus on accuracy, give further consideration to integrating knowledge about landforms and surrounding topographic elements in order to better model the relationships between terrain and associated entities. In other words, relief is no longer perceived solely as a field-based phenomenon but can also be considered as being composed of landforms seen as objects. Landforms are then related together and with other objects on the map that can have their own semantic attributes and methods.

This chapter provides a review of recent developments in terrain generalisation. It begins with an overview of the problem with the description of different representation techniques and issues brought by multiple scale representation on maps. The next section addresses the characterisation of terrain features. It addresses their identification from terrain data and their classification in a topological data structure. Section 8.4 reviews recent advances in generalisation techniques. It presents algorithms designed for traditional portrayal on 2D maps and for DTM generalisation, focusing on cartographic generalisation. The following three sections present applications of these techniques. First a method for

Table 8.1 Terrain visualisation types according to map scale (After Imhof 1982)

Scale range		Terrain visualisation types
Very large	1:1,000–1:10,000	Contour lines, spot heights
Large to medium	1:24,000–1:250,000	Contour lines, spot heights, shaded relief
Small	1:500,000–1:75 million	Shaded relief, hypsometric tints
Very small	1:100 million and smaller	Hypsometric tints

generalising hypsometric maps with consideration of terrain features is presented. Second, a method for selecting isobaths with respect to nautical chart constraints where navigation hazards must be emphasised is detailed. The third application presents a model which preserves relationships between the terrain and objects on the map. The last section provides concluding remarks and perspectives on future developments.

8.2 Issues in Terrain Generalisation

8.2.1 Approaches to Terrain Generalisation

DTM generalisation is often considered as an optimisation problem where a representation at a given resolution is required. The objective is mainly to reduce any confusion and to convey the underlying trends of the terrain (Jordan 2007). In cartography, further work is often required to adapt the terrain representation to the map purpose, to avoid conflicts with other topographic elements, or to improve map aesthetics. In generalisation, the first problem is referred to as model generalisation and yields a digital landscape model (DLM) and the second is referred to as cartographic generalisation and produces the digital cartographic model (DCM) (Fig. 8.1). In the latter specific tools are used to highlight or modify terrain features for each representation technique.

Weibel (1992) described three different methodologies that can be combined in DTM generalisation. The first approach, global filtering, is based on resampling and filtering methods such as regular sampling or smoothing as used in image processing for smoothing the surface. Such methods do not take into consideration the terrain morphology and therefore cannot integrate cartographic constraints. They are usually considered for sampling very large datasets or over large changes in scale.

The second approach, selective filtering, eliminates non-significant points on the DTM. It consists mainly of grid and triangulated irregular network (TIN) DTM generalisation methods that preserve morphometric features. Methods were not only developed for terrain generalisation but also for data simplification and compression in computer graphics and for hydrological and geological applications. Two types of method are considered: the first selects critical points based on a distance or error threshold (Fei and He 2009), and the second is based on the

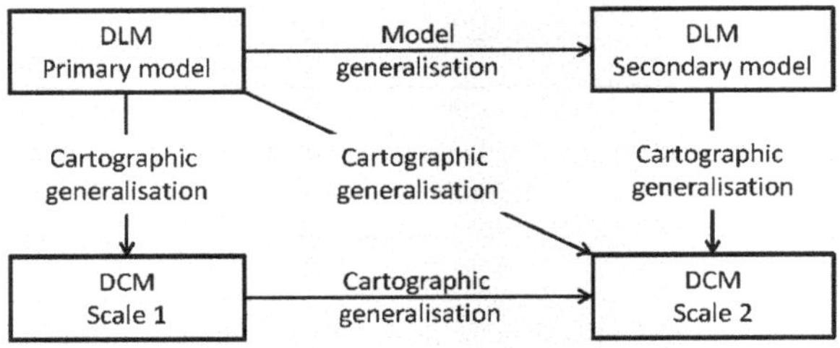

Fig. 8.1 Model and cartographic generalisation (João 1998)

extraction of feature points and lines obtained from the drainage system (Chen et al. 2012). Point selection methods are simple to implement and usually perform fast whereas drainage based methods tend to better preserve the terrain features and derivatives. Global and selective filtering approaches rely on mathematical principles and are mainly used to derive a secondary DLM from a primary DLM.

The third heuristic approach utilises operators that generalise specific terrain elements. It attempts to emulate manual techniques and consists of applying individual operators to various elements (e.g. contours, spot heights) composing landforms for the production of the DCM. Operators are also defined that perform specific tasks such as smoothing, displacement or removal. Each operation can be automated but combining operations is still a difficult task as combinations are not unique and the final result depends on the order in which they are applied. Currently, the most efficient models are based on multi-agent systems which allow the integration of both continuous and discrete operations and can draw up plans of action in order to evaluate different solutions (Ruas and Duchêne 2007).

8.2.2 Representation of Landforms

Weibel's (1992) strategy suggests that generalisation should be structure- and purpose-dependent. The idea is that generalisation procedures should include mechanisms for terrain structure recognition. Landforms should be addressable as objects to allow the possibility of applying certain generalisation operators to specific objects. Landform characterisation depends on the purpose and scale of the map as landforms are generalised according to their meaning and the required level of detail. Landform classification methods fall into two groups (Deng 2007): set theory where components are morphometric points and, category theory where landforms are identified as objects. In the first group, each point of the terrain belongs to one of the six morphometric classes (peak, pit, pass, ridge, channel and plane). The results are scale-dependent and landform delineation may be fuzzy, with multi-scale classification and where fuzziness is also considered in the classification process (Wood 1996; Fisher et al. 2004).

In the second group, landforms are identified as belonging to some categories of objects. Landforms are usually associated with salient terrain features and not to their boundaries which are not always well-defined. For example, the presence of a mountain is easily associated with the existence of a peak significantly higher than its surroundings but there is no consensual definition of the spatial extent of a mountain or of the difference between a hill and a mountain. Uncertainty of landform boundaries is a modelling issue that has been discussed in related works by Frank (1996), Smith and Mark (2003). A landform is considered as a subjectively defined region in a rough part of the Earth's surface. It follows that the objective of qualitative methods is not to explicitly locate the beginning and ending of a landform, but to find out the presence of landforms corresponding to an end-user typology. Therefore, landforms are not restricted to morphometric features but must be classified according to the map requirements.

Although these methods provide a classification of the terrain, landforms are not organised in a data structure describing the surface topology through different scales. The first structure describing the topology of a 2D manifold was the Reeb graph (Rana 2004). Nodes of the graphs are peaks, pits and passes of the surface. A topologically equivalent data structure is the contour tree (Fig. 8.2, right) that can be built from a contour map of the surface (Takahashi 2004). Surface networks (Rana 2004) describe the surface topology in a graph where edges are ridges and channels are those lines that connect critical points (Fig. 8.2, bottom left). Contour trees were used for terrain analysis and identification of landforms (Kweon and Kanade 1994) but multiple scale representation was not yet considered and only features characterised by the tree leaves were identified.

In the computer graphics field, several methods were developed for TIN simplification with preservation of morphometric features based on hierarchical watersheds (Beucher 1994) and on a critical net (Danovaro et al. 2003). Danovaro et al. (2010) also proposed a data structure that gave access to representations at adaptive resolutions. Such approaches provide a terrain representation to multiple

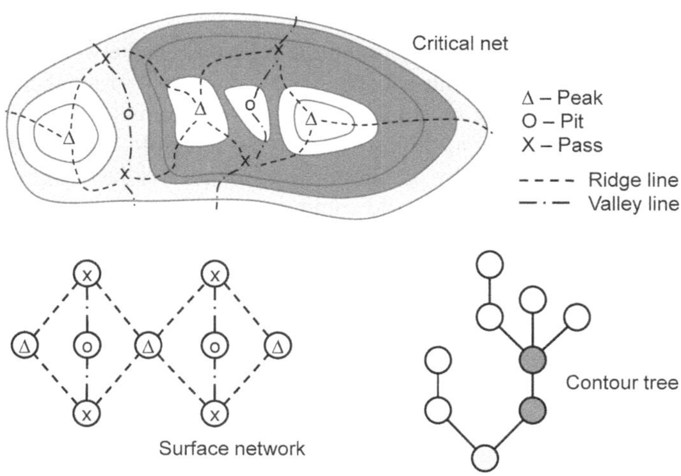

Fig. 8.2 Topological structures of a terrain: critical net, surface network and contour tree

resolutions by removing points while preserving feature lines. As highlighted by Jenny et al. (2011), the emphasis was on performance rather than cartographic generalisation. They are therefore more relevant to model generalisation.

8.3 Object-Oriented Classification of Landforms

Landform recognition has received more attention in the last decade and methods have been developed for the characterisation of specific landforms (Feng and Bittner 2010; Straumann and Purves 2011) and for the representation of landforms at different levels of detail. Levels can be defined by fixed resolutions or scales of observation from a raster DTM (Chaudhry and Mackaness 2008) or based on relationships between contours (Guilbert 2013). In both cases, landforms are bounded by contours and are identified at a resolution given by the vertical interval. The objective of these methods is to enable the representation of the relief at various levels of detail and its storage in a single database.

Chaudhry and Mackaness (2008) are interested in detecting hills and ranges from a raster DTM. Contours at a given vertical interval are first computed and then summits within contours are computed. The prominence of a summit is defined by the height difference between the summit and the key contour that is the lowest contour containing this summit and no other higher summit (Fig. 8.3, left). The terrain is then classified into morphometric features using Wood's (1996) approach. Each morphometric feature which is neither a plane nor a pass is converted into a "morphologically variable polygon". The spatial extent of a summit is defined by the contour that best overlaps with the morphologically variable polygon containing the summit (Fig. 8.3, right). Overlap value is defined by the area intersection between the contour polygon and the morphologically variable polygon divided by the contour polygon area.

Once all extents are computed, partonomic relationships between summits are defined. If the extent of a summit is contained by the extent of another summit, a parent–child relationship can be defined between the summits. Based on the definition of the summit extent, a summit can only be the child of a higher summit. Authors can then set a hierarchy of summits and identify isolated mountains or a hill and a parent summit with its child summits as a range.

Guilbert's (2013) work focuses on contour maps and provides a hierarchical structure, the feature tree, which makes explicit the relationships between features. A feature is defined by a region bounded by one or several contours and can be classified as a prominence (boundary contours are lower than other contours inside the feature) or a depression (boundary contours are higher than other contours). The contour map (Fig. 8.4a) is processed first by building the inter-contour region graph (Fig. 8.4b). The structure has the advantage that contours can be either open or closed and a feature, such as a channel stretching across the map, can be delineated by several contours. Features are extracted recursively in a bottom-up

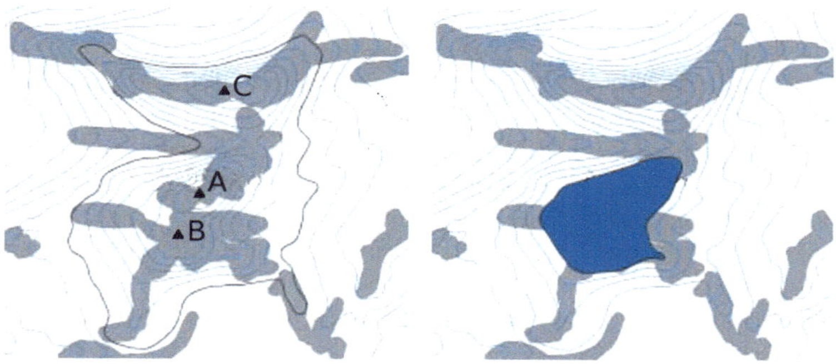

Fig. 8.3 *Left* summit *A* with key contour and morphologically variable polygons. *Right* extent of the summit in *blue* (Chaudhry and Mackaness 2008)

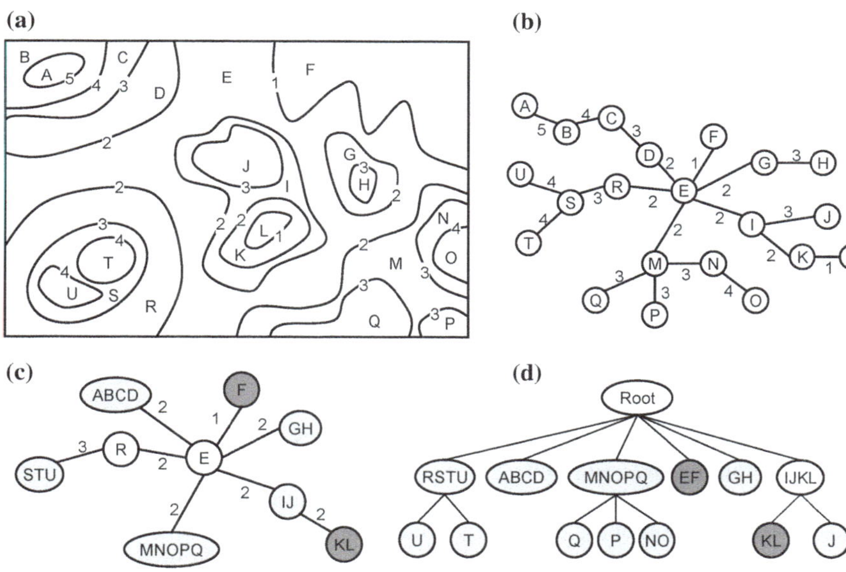

Fig. 8.4 Contour map (**a**) and its corresponding feature tree (**d**). Prominences in light *grey*, depressions in *dark grey* and unclassified features in *white* (Guilbert 2013)

approach by collapsing edges of the region graph. Each round of the process goes through three steps.

In the first step, pairs of adjacent regions which have no more than two neighbours and have the same slope direction are merged by collapsing their connecting edge, e.g. regions K and L and regions A, B, C and D of Fig. 8.4b, are respectively merged into regions KL and ABCD. In the second step, new leaves obtained are copied to the feature tree. In the first round, they form the leaves of the feature tree. In the following rounds, newly extracted features are added on top of existing ones.

In the third step, leaves are aggregated to their neighbouring regions. Candidate regions for aggregation are regions *r* from the graph for which all adjacent regions but one are leaves. The region which is not a leaf is the one connecting *r* to the rest of the graph by the edge which is the base of the region as it encloses the subset formed by *r* and its leaves. Candidate regions are classified according to their edge elevation. If edges connecting *r* to its leaves are at the same elevation different from the base, *r* contains a pass connecting the leaves (region S with leaves T and U). Other candidate regions where at least one leaf edge is at the same elevation as the base (region I with leaves J and KL) are aggregated if there is no pass left. They correspond to channels or ridges connecting different parts of the map. Regions connecting to the smallest features are aggregated first so that more prominent features are closer to the root. The process stops when the whole map is partitioned into features. Finally, spurious features may be removed. For example, F is classified as a depression in the first round and is later aggregated with E to form another depression EF. F becomes redundant as it is a part of EF carrying the same meaning and is therefore removed.

The depth of the feature tree does not depend on the scale but on the terrain morphology. The data structure extends previous works on topological structures presented above by building an explicit hierarchy of features. An example on a contour map is shown in Fig. 8.5. The method allows the identification of the channel which crosses the map. Such feature cannot be characterised with a contour tree as the feature is delineated by two contours.

Guilbert (2013) provides a richer topological structure as a summit can belong to several features delineated by different contours. Chaudhry and Mackaness (2008) associate a summit to only one hill in their hierarchy but the summit extent and summit relationships are related to the terrain morphometry. For example, in Fig. 8.3, summit C is not part of summit A. Using a feature tree, only contours are considered and A would be the summit of two features, one containing only A and one containing all three summits.

Both methods provide an object-oriented description of landforms and can be used to enrich a topographic database. Summits in the first case and features in the second can be stored and queried in a database. Geometric and semantic attributes such as the feature or summit name and height can also be added. Both methods can therefore be considered as landform generalisation as described by Weibel (1992) and allow the automatic selection and application of heuristic operators. They can be applied to either the DLM or the DCM however classifying a DLM at too high a resolution will lead to excessive decomposition of the map. Therefore they are more appropriate for producing a DCM either from an existing DCM or from a DLM which has already been simplified. Methods are limited to the description of prominences and depressions. Further knowledge could be gained through terrain analysis about the features in order to provide a more detailed classification of landforms however it would necessitate a formal description of landforms which is adapted to the map application. Such description can be achieved through an ontology but its definition is still an open problem (Smith and Mark 2003).

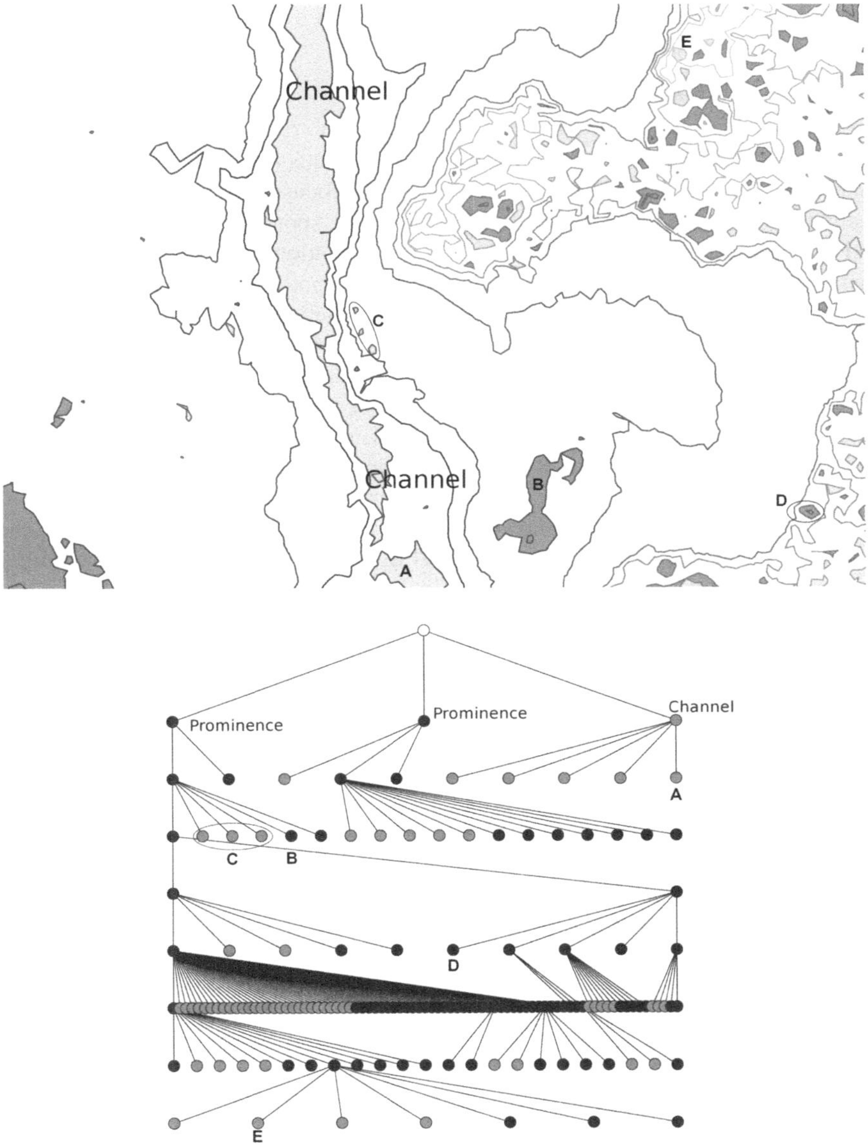

Fig. 8.5 *Top* Contour map with feature leaves in *light grey* (depressions) and *dark grey* (prominences). *Below* feature tree with depressions (*light grey*) and prominences (*black*). Below the root, the map is partitioned into three features: one channel in the middle and two prominences on each side. Features labelled on the map are highlighted in the feature tree

8.4 Generalisation Methods

8.4.1 Spot Height Selection

Automated spot height selection for topographic maps has received less attention than other design tasks and research has mostly focused on the classification of feature points through selective filtering. However, spot heights are not limited to VIP or feature points describing landforms. As mentioned in Sect. 8.2.2, filtering methods are mostly relevant for scale reduction. Spot heights on a map are also selected according to user needs and their distribution over the map. Palomar-Vázquez and Pardo-Pascual (2008) present a method where spot heights are selected according to their significance. Palomar-Vázquez and Pardo-Pascual (2008) applied their method to the production of a recreational topographic map. The selection criteria relate to proximity to hiking trails, transit points and places of touristic interest. Furthermore, morphometric points are extracted from the TIN and classified according to their type (peak, pass, depression). The importance of a peak also depends on its prominence which is modelled in terms of its height, the centrality of the peak and the mean slope (Fig. 8.6). A peak with a high centrality or steep slope marks an abrupt change in terrain and should therefore be given more significance.

Once classification is done, each peak is assigned a weight according to its type. Palomar-Vázquez and Pardo-Pascual (2008) give a higher weight to points of interest as they are the most relevant to hikers. Finally, spot heights are selected with due consideration to their distribution. This is controlled by partitioning the map into half planes forming a binary tree with approximately the same amount of spot heights in each block. Points are eliminated according to their concentration, which is defined by the length of the minimal spanning tree connecting all the points within a block. Starting from the block with the highest concentration, the spot height with the lowest weight is removed from each block until the selection percentage is reached. The principle of the method can be applied to topographic and thematic maps however constraints are specific to each type of map and spot height density is related to map scale. Non morphological constraints need to be translated into a weighting that reflects their importance and which must be assessed by a cartographer.

8.4.2 Contour Line Generalisation

Contour generalisation is performed either when moving from one scale to a smaller scale or within a given scale to improve the quality of the representation. In the first case, simplification is done either by filtering the grid or TIN DTM (whether already available or generated from the contours) and extracting contours

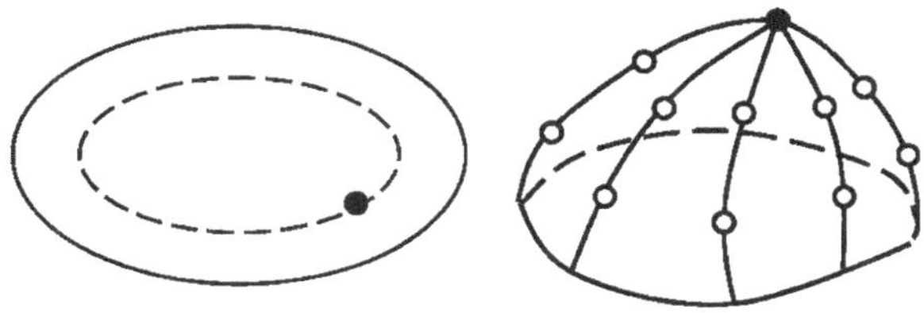

Fig. 8.6 *Left* the centrality is defined by the ratio between the area of the contour offset passing through the peak (*dashed line*) and the area of the contour (*plain line*). *Right* mean slope defined by the average of the slopes that connect the peak to the points of the *curves*

from the simplified representation or by directly simplifying the contours to the destination scale. In the second case, specific operators providing local corrections on contours are performed to fulfil cartographic constraints.

8.4.2.1 Contour Simplification

Traditional line simplification methods can apply to individual contours however they do not always maintain topological integrity. Recently, a line simplification method that preserves topological relationships and can be applied to any kind of line (contours or road networks) was presented by Dyken et al. (2009).

Specific contour line simplification methods are presented by Gökgöz (2005) and Matuk et al. (2006). Gökgöz (2005) first computes an error band around each contour. Characteristic points of the contours are then computed using a deviation angle defined by the angle between consecutive segments, which is more robust than the line curvature, and ordered according to their importance, the higher the angle the more characteristic the point. Each simplified contour is built iteratively by adding characteristic points and smoothing the line through cubic interpolation until the whole line lies within the error band. Results presented by Gökgöz (2005) show that the method provides the same amount of simplification as the (Li and Openshaw 1993) algorithm but points are not distributed regularly along the contours as no point is kept along straight lines. Gökgöz (2005) solution is more computationally expensive since it requires a priori computation of the error bands and each simplified contour is smoothed by cubic interpolation.

Matuk et al. (2006) build the skeleton of regions bounded by contours. The skeleton is formed by points which have at least two of their nearest neighbours on the boundary and is equal to the set of Voronoi edges built from the contours that do not intersect the contour edges (Fig. 8.7). A potential residual function is associated with each point of the skeleton and is defined by the distance along the boundary connecting its two nearest neighbours. Simplification is performed by pruning the skeleton (edges whose potential is smaller than a given threshold are removed) and reconstructing contours from the pruned skeleton. The method presents an original approach where the pruning threshold is set according to the

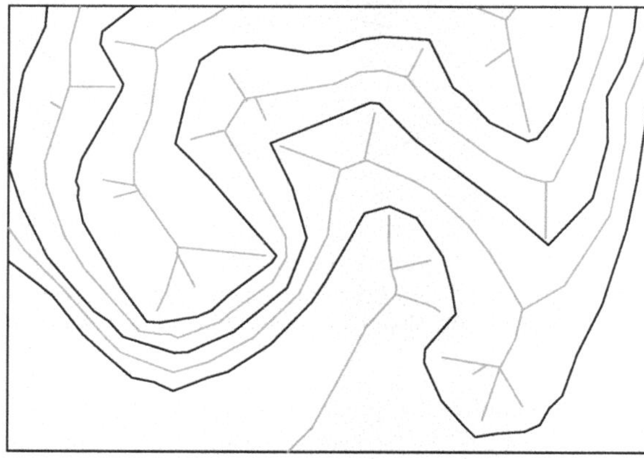

Fig. 8.7 Contour lines in *black* and skeleton in *grey* (Matuk et al. 2006)

scale factor however it may create visual artefacts if the scale factor is too big. Furthermore the algorithm does not guarantee the absence of intersections between contours. The method should therefore be applied iteratively with smaller thresholds.

In the context of nautical charts, these methods are not appropriate because the depth portrayed on the chart cannot be greater than the real depth (that is referred to as a safety constraint). Peters (2012) presents a method for extracting isobaths at a smaller scale where soundings below the plane formed by their surroundings are pushed 'upward' to smooth out the surface. Isobaths can also be aggregated and smoothed by interpolating between new soundings. The method defines a higher surface from which new isobaths are extracted, guaranteeing the safety constraint. The approach is very efficient in extracting isobaths at a smaller chart scale and is applicable to high resolution DTMs.

TIN or grid based simplification methods have the major advantage compared to line simplification methods of being more robust and applicable to large scale change since contours extracted from the simplified terrain are always topologically correct. They mostly apply to model generalisation. Line simplification methods are more appropriate for cartographic generalisation from a source DCM to a target DCM or for updating surfaces as they can directly integrate cartographic constraints such as the distance between the contours. In both cases, results yielded by these methods may still require further processing to provide a final map. Visual conflicts may remain and some terrain landforms may be removed or emphasised according to the purpose of the map.

8.4.2.2 Cartographic Generalisation Operations on Contours

In order to improve the quality of a representation, algorithms have been developed for selective contour removal (Mackaness and Steven 2006), smoothing (Irigoyen et al. 2009; Lopes et al. 2011) and displacement (Guilbert and Saux

2008). These methods apply to a set of contours which have already been rescaled to improve their legibility or to improve the aesthetics of the map by performing local corrections.

Mackaness and Steven (2006) developed an algorithm that detected and removed segments of intermediate contours in steep areas of terrain. The method is illustrated on a 1:50,000 map with index contours at 50 m interval and intermediate contours at 10 m interval. The method first computes the gradient from the DTM and defines regions where the gradient is greater than a 30° threshold. The gradient threshold depends on the scale of the map, the vertical interval and the legibility distance. One or all contour segments crossing these regions are removed according to the gradient value. The choice of contours to be removed is based upon rules set by mapping agencies.

Guilbert and Saux (2008) propose a model combining contour smoothing and displacement. The method is applied to cartographic generalisation of depth contours (isobaths) on nautical charts at fixed scale. Isobaths are not modelled by polygonal lines but by cubic B-spline curves (Saux 2003). The benefit is that curves are modelled by a mathematical expression so that derivative and curvature computations are more robust and the curve has a smooth representation. The limitation is that the quality of the approximation depends on the quality of the sampling and an ill-conditioned problem can lead to a non-reliable approximation.

For navigation safety reasons, generalisation can be done only by pushing isobaths towards areas of greater depth. Shallow isobaths are generalised first so that their displacements are propagated to deeper isobaths. Prior to its generalisation, the 'isobath admissible area' is computed. This is the area that the curve should stay within or to where it should be moved to correct conflicts with isobaths at the same and lower depths (Fig. 8.8). Deformation is performed by minimising an energy coefficient associated with each isobath. Two energy terms are defined: an internal energy related to the smoothness of the curve and an external energy related to the curve position which is non-zero if the curve is not within its admissible area. Convergence to an admissible solution is guaranteed by fixing critical points characterising shape features that one wishes to preserve and by removing bottlenecks or self-intersections that may occur during the process.

The method is applied to a set of isobaths in a semi-automatic way as all conflicts cannot be corrected by displacement (removal and aggregation may also

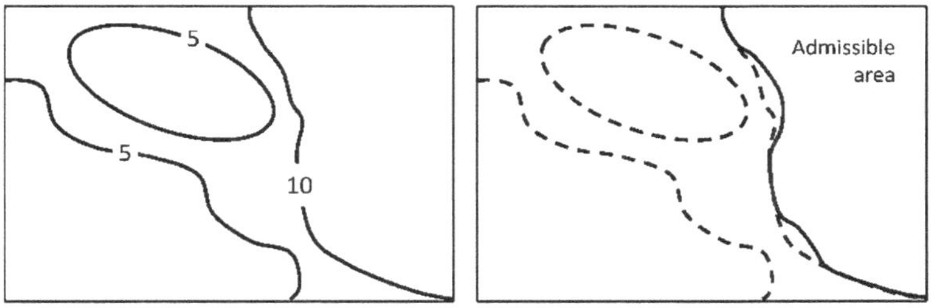

Fig. 8.8 *Left* Isobaths with depth. *Right* Admissible area for the 10 m isobath

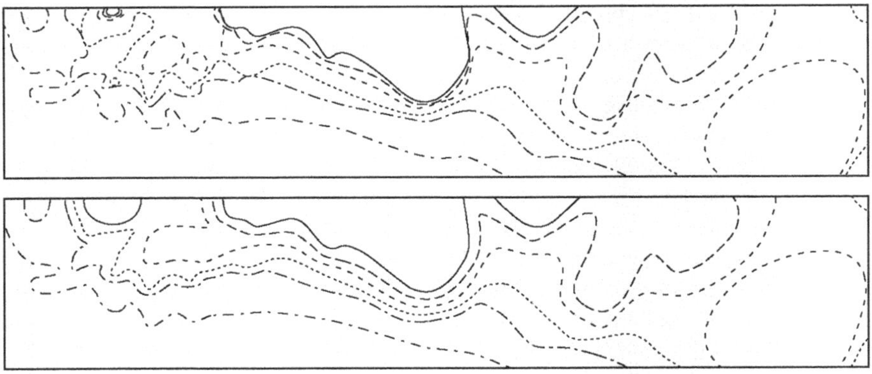

Fig. 8.9 Generalisation by smoothing and displacement. In the *centre* are examples of where isobath segments could be removed to avoid too large a displacement

be required). Propagating displacement from lower to greater depth can result in big deformations, and the artificial smoothing of steep slopes (Fig. 8.9). Finally, the method, based on iterative deformation, is quite computationally demanding.

Overall, cartographic generalisation operations are designed to correct one specific type of conflict or to perform one type of operation. These methods are still required to output the terrain representation according to mapping agency requirements that global DTM based approaches cannot consider. Mackaness and Steven (2006) correct local conflicts by removing contours. Correction remains contained in a small area and there are no side effects. This is different in the context of smoothing and displacement (Guilbert and Saux 2008) which may lead to side effects that propagate to other contours. Corrections may be applied in sequence however it is still up to the user to decide which operator to apply. For example the same conflict may be corrected either by removing a contour or by displacing it. Choosing an operation depends on the type of conflict and the terrain morphology and requires the application of a strategy that can be automated. An approach is presented in Sect. 8.6 where features formed by groups of isobaths are selected according to the morphology. Furthermore, methods presented in these sections apply to only one type of element (contours or spot heights) while conflicts can also occur between both or with other map elements. Work in this direction is discussed in Sect. 8.4.4.

8.4.3 DTM Generalisation for Relief Shading and Hypsometric Colouring

The computation of shaded relief from a DTM was pioneered by Yoeli (1966), and various computational models were documented by Horn (1982). Enhancements to relief shading methods have been proposed that seek to improve the depiction of terrain structures (Brassel 1974; Jenny 2001; Kennelly and Stewart 2006; Loisios et al. 2007; Kennelly 2008; Podobnikar 2012). Before relief shading methods can be applied to a DTM for display at medium or small scales, the DTM data often requires generalisation, because digital shaded relief at small scales is often excessively detailed when computed from high-resolution terrain models, making it difficult or impossible to perceive major landforms.

Leonowicz et al. (2010a, b) proposed a method for generalising DTM data that was developed for relief shading. First, the DTM is simplified with a strong low-pass filter, removing details such as mountain ridges or smaller valleys. In the next step, important details are detected in the original DTM using curvature operators. The detected high-frequency details are then amplified and added to the smoothed grid generated in the first step. This procedure is carried out separately for mountainous and flatter areas using separate sets of parameters. The resulting two terrain models are then combined with a slope mask, and, finally, a shaded relief image is computed.

Similarly, hypsometric colouring requires DTM generalisation as it applies to small scale maps. The process is therefore also based on terrain filtering and the identification of the main elements of relief that will be emphasised. Much recent work in this domain has been done by Leonowicz and Jenny (2011) (see Sect. 8.5).

8.4.4 Modelling the Relation Between Field and Object Type Data

A map is more accurate if the terrain generalisation is undertaken with respect to other objects on the map. Relationships between map objects can be expressed as constraints that must be maintained during the generalisation process (Harrie and Weibel 2007). With a terrain model described by a field function, the difficulty is to integrate these constraints in the generalisation process. Filtering methods rely on a mathematical representation to simplify the relief. Therefore, these constraints should be expressed as mathematical functions and integrated into the filtering process. However, these filtering methods usually only perform simplification and are not adapted for local deformation operations such as displacement or enlargement of a protrusion on the surface. Furthermore, maintaining a relationship may require modifying the terrain and the object at the same time. Therefore, both should be considered at the same time in the same model.

Most research done in this domain is concerned with maintaining topological relationships between rivers, contours and spot heights. Contours are considered as individual objects and so constraints can be defined directly between a contour and a river. Chen et al. (2007) provide a classification of conflicts between rivers and contours. Lopes et al. (2011) also take into account the relation between rivers and spot heights when deforming contours: displacement is controlled so that topological relationships between contours and spot heights are preserved and contours still cross rivers at an inflection point. Baella et al. (2007) apply Palomar-Vázquez and Pardo-Pascual (2008) method to detect spot heights for topographic maps and add a weight to selected spot heights close to points of interest (for example near roads, and inhabited zones).

However, these methods are limited to generalising one element with consideration of position constraints imposed by other objects. A more pertinent approach is to generalise the terrain and objects at the same time so that operators can be applied either to one or the other according to constraints. Such a model was proposed by Gaffuri (2007b) and Gaffuri et al. (2008). In their approach, each geometrical element of the terrain is considered as an object under constraints and conflicts are solved by using a multi-agent system approach. The model is reviewed in detail in Sect. 8.7.

8.5 Case Study I: Hypsometric Colouring

Bernhard Jenny

Hypsometric tinting is mainly used for small-scale maps (Table 8.1). Imhof (1982) recommends hypsometric tints for maps at scales of 1:500,000 and smaller. With the advent of chromolithography (the first colour printing technology) cartographers started producing maps with a variety of hypsometric colour schemes and, by the mid-twentieth century, hypsometric tints became the de facto standard for physical reference maps at small scales. For an overview of the historical development and contemporary application of hypsometric colour schemes, the reader is referred to Patterson and Jenny (2011).

A raster image with hypsometric colours can be easily derived from a DTM. A simple linear mapping of the elevation range to a colour range is sufficient to determine a colour for each cell in the DTM. Colour can be arranged in discrete steps, or can be interpolated to create continuous tone hypsometric tints. Imhof (1982) provides guidance on the vertical distribution of colour, suggesting a geometric progression, with small vertical steps between neighbouring colours for low elevations, and large vertical steps for higher elevations. Hypsometric colours are often combined with shaded relief to accentuate the third dimension of the terrain, except for extremely small scales when shaded relief is unable to effectively show terrain (Imhof 1982).

The case study discussed here generalises a DTM, which is then used to derive hypsometric colour. The method seeks to accentuate important landforms, such as major valleys and ridgelines, and remove distracting small terrain details (Leonowicz et al. 2009). The DTM is filtered with lower and upper quartile filters. These quartile filters assign to each raster cell the 25 or 75 percentile of its neighbouring values. The lower quartile filter is applied along valleys, and the upper quartile filter is used in the remainder of the DTM. Valley regions are identified based on a drainage network derived from the DTM.

When developing this method, one of the goals was to take design principles developed for manual cartography into account that had the ambition of increasing the readability of hypsometric colours, as documented by Horn (1945), Pannekoek (1962), and Imhof (1982). Manually generalised contour lines were used as a reference for evaluating this approach. These contour lines were generalised by an experienced cartographer (Emeritus Professor Ernst Spiess, ETH Zurich) in which he added hypsometric tints to a map of the Swiss World Atlas (Spiess 2008). The target map scale was 1:15,000,000. The cartographer used contour lines derived from the GTOPO30 elevation model with a 30-arc second resolution as a base for retracing the generalised contours with a digital pen tool.

When generalising terrain for hypsometric colouring, the main landforms should be accentuated, while secondary features should be eliminated. When removing elements, Horn (1945) recommends treating each landform as an entity with the ambition of either removing or retaining the entire entity. For example, if a side valley of a major valley is not important, it should not be shortened, but removed entirely.

The generalisation method applied in this study uses a series of operations performed on the GTOPO30 digital elevation model. This is illustrated on Fig. 8.10 (after Leonowicz and Jenny 2011). The flow diagram in Fig. 8.11 illustrates the sequences of the procedure.

1. The initial DTM (Fig. 8.10a) is filtered with an upper-quartile filter, which assigns to each raster cell the 75 percentile of its neighbouring values. The upper-quartile filter preserves elevated areas (ridgelines) and aggregates isolated small hills and mountain peaks. This step generates the first intermediate DTM (Fig. 8.10b).
2. The initial DTM is also filtered with a lower quartile filter, which assigns the 25 percentile of the neighbouring values to each raster cell. The lower-quartile filter preserves elevation along valley bottoms, and preserves valleys from being dissected into a series of unconnected depressions. This filter also retains mountain passes. This step generates the second intermediate DTM (Fig. 8.10c).
3. The D8 hydrological accumulation flow algorithm (O'Callaghan and Mark 1984) is applied to the initial DTM to compute a drainage network. The D8 (deterministic eight-node) algorithm first computes a flow direction for each cell (the direction of the steepest path). The value of accumulation flow is then calculated for each cell as the number of cells draining into that cell. A threshold is applied to the accumulation flow grid to identify the cells that are considered to be part of the drainage network (Fig. 8.10d).
4. The drainage network is simplified with the desired level of generalisation. Starting at each raster cell, an upstream path is created by following cells that have smaller accumulation values than the current cell. The algorithm follows the path with the smallest absolute difference. If the path is longer than a predefined threshold it is retained, otherwise it is discarded.
5. The rivers found in step 4 are enlarged by a series of buffer operators (Fig. 8.10e). The resulting grid is then used as a weight to combine the two intermediate DTMs created with the upper- and lower-quartile filters in steps 1 and 2. This weighting

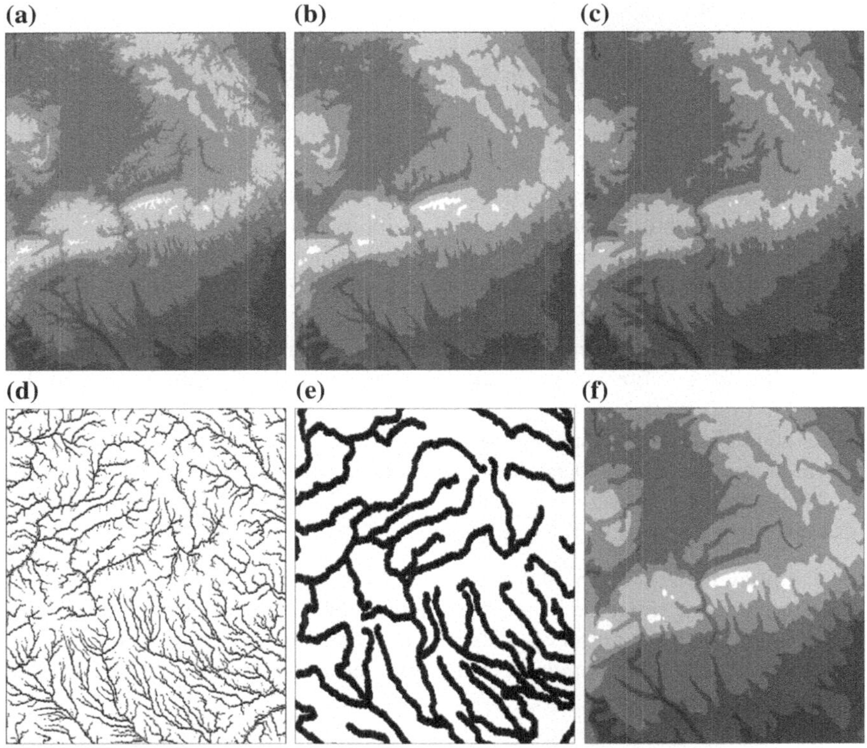

Fig. 8.10 Steps leading to a generalised terrain model for hypsometric colouring at small scales

applies the grid filtered with the lower-quartile filer to valley bottoms, and the grid filtered with the upper-quartile filter to other areas. Care is taken to create a smooth transition between valley bottoms and the surrounding areas (for details, see Leonowicz et al. 2009). The final result is shown in Fig. 8.10f.

The first map in Fig. 8.12 shows hypsometric colours derived from the un-generalised GTOPO30 DTM. The second map is the reference map—drawn manually. The third DTM is automatically derived with the method described above. The manually generalised map is slightly more generalised than the map generalised using this algorithm. The digital method successfully aggregates mountain ridges. Small valleys are removed while the bigger ones are retained, but not shortened. Though intended for hypsometric tinting at small scales, it could be adapted to the derivation of contour lines and hypsometric tints at intermediate scales, but this option has not been explored yet.

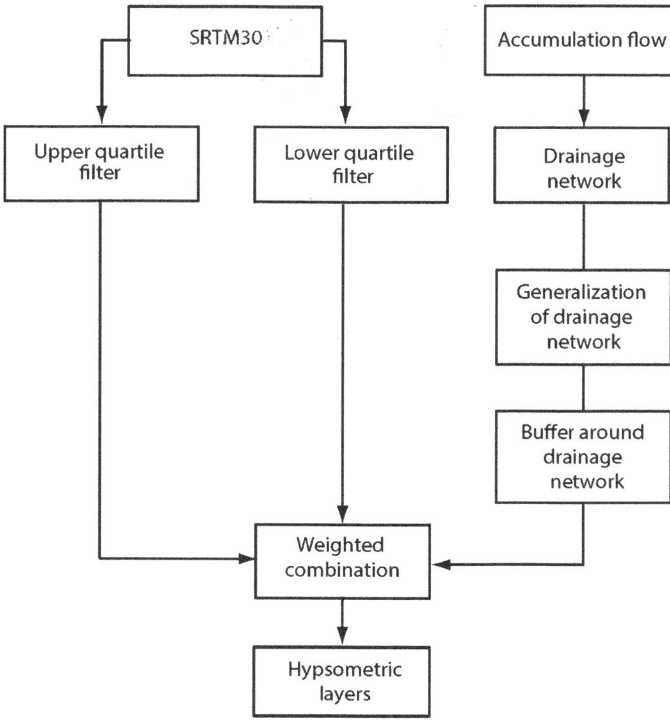

Fig. 8.11 Flow diagram for the generalisation of terrain models for hypsometric colouring at small scales

8.6 Case Study II: Isobathic Line Generalisation

Eric Guilbert

Nautical charts provide a schematic representation of the seafloor, defined by soundings and isobaths, and are used by navigators to plan their routes. As the seafloor is not visible to navigators, they have to rely on the chart to identify hazards (reefs, shoals) and fairways. As a consequence, the depth reported on the chart must never be deeper than the real depth to ensure safety of navigation and submarine features are selected according to their relevance for navigation (Fig. 8.13). Indeed, nautical charts provide a more schematic representation of landforms when compared to topographic maps. As reported in NOAA (1997, pp. 4–11), "[cartographers] do, deliberately and knowingly, and on behalf of the navigator, include all lesser depths within a contour even if it means that [their] catch includes many deep ones as well".

Fig. 8.12 A comparison of ungeneralised, and manually and digitally generalised hypsometric colouring. 1:15,000,000, southeast France

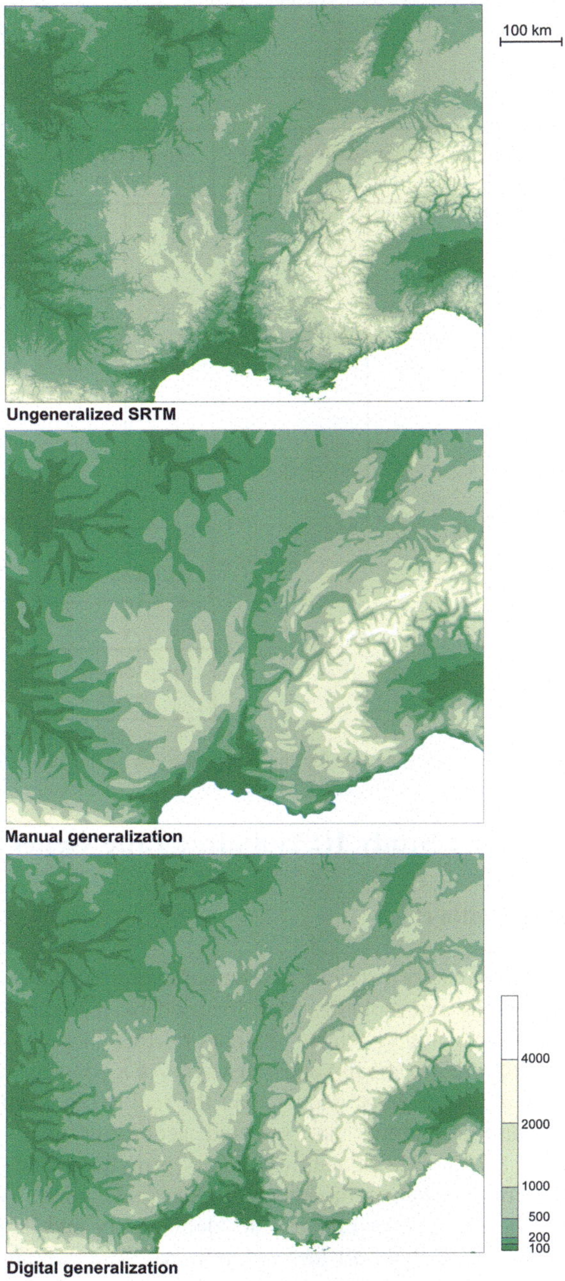

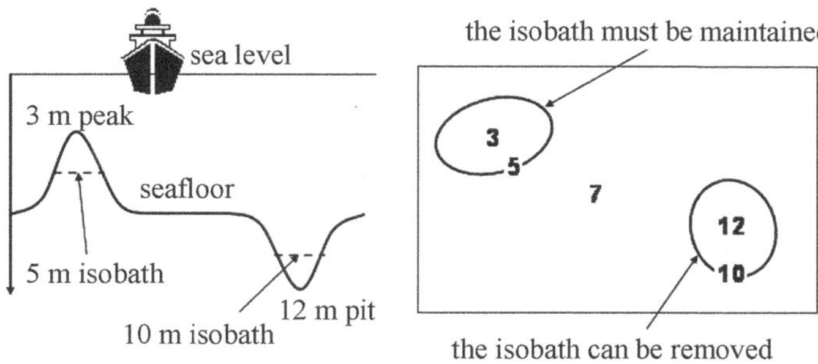

Fig. 8.13 Isobaths are generalised according to the type of feature they characterise

Isobaths can be extracted from the DTM using DTM based methods (Peters 2012) however emphasising features to produce the DCM is done using heuristic methods: isobaths are enlarged, displaced or removed according to the landforms they characterise. In order to mimic the manual process done by cartographers, the seafloor relief portrayed on the chart is perceived as a set of discrete submarine features, which need to be generalised according to their significance from the navigator's point of view. Constraints can be classified into (Guilbert and Zhang 2012):

- The legibility constraint: generalised contours must be legible by observing a minimum size or distance between them;
- The position and shape constraints: position and shape of isobaths are preserved as much as possible;
- The structural and topological constraints: spatial relationships as well as distribution and mean distance between isobaths are preserved;
- The functional constraint: a reported depth cannot be greater than the real depth and navigation routes are preserved.

The first three constraints (legibility, position and shape) apply to individual contours or locally to groups of isobaths. The objective of structural and topological constraints is to maintain morphological details by preserving groups of isobaths corresponding to submarine features. Constraints are expressed not only at the local level but also at more global levels on larger features.

8.6.1 A Feature Driven Approach to Isobath Generalisation

In this research, the initial set of isobaths was provided by the French hydrographic service. Isobath extraction from the bathymetric database was done first by simplifying the original set of soundings by sounding selection (interpolation, displacement or modification of soundings were not considered by cartographers because such soundings cannot be reported on the chart) and then by extracting isobaths by interpolation. The objective was to select the isobaths according to the features they characterise. Automating the generalisation process requires the identification of features formed by groups of isobaths and the definition of a strategy that applies various operators. Characterising the features utilised the approach of Zhang and Guilbert (2011) and Guilbert (2013). Topological relationships are stored in two data structures: the contour tree connecting the isobaths and the feature tree where each feature is composed of a boundary isobath and all the isobaths within the boundary. These are classified as either a peak or a pit.

Automating the process requires the definition of a generalisation strategy so that operators are selected to satisfy generalisation constraints. Guilbert and Zhang (2012) proposed a multi-agent system (MAS) to select features formed by groups of isobaths on the chart. Features and isobaths are respectively modelled as meso-agents and micro-agents at two different levels (Ruas and Duchêne 2007). At the micro level, operations and constraints relate to a single isobath (minimum area, isobath smoothness) or to adjacent isobaths (distance between isobaths). The terrain morphology is defined at the meso level. Features hold information related to the seafloor morphology which is used to evaluate whether the morphology is preserved and which operation can be performed with respect to the safety constraint.

Feature agents are able to communicate with their environment (that is, with other features and inner isobaths), in order to evaluate their state and decide upon further actions. Isobaths, on the other hand, evaluate their environment by estimating constraints (area, distance to neighbouring isobaths) and act based only on information received from a feature. The whole generalisation process is therefore driven by the features and each feature agent goes through a series of steps. These are summarised in Fig. 8.14.

The feature first evaluates if generalisation must be performed by communicating with the contour agent forming its boundary. The feature passes information about the neighbouring features and the direction of greater depth. The contour checks if any area or distance conflict has occured and returns the result to the feature. The feature then evaluates its situation with regard to the different generalisation constraints that apply to features:

- A feature on the chart must be large enough to contain a sounding marking at its deepest or shallowest point;
- A pit cannot be enlarged or aggregated with another feature;
- A pit that is too small or not relevant is removed;
- A peak cannot be removed;
- A peak that is too small is enlarged or aggregated to an adjacent peak;
- A minimum distance must be observed between adjacent features.

If some constraint is violated, a list of plans is set by the feature (Table 8.2). Each plan consists of one or several generalisation operations which are of two kinds: continuous and discrete operations. Continuous operations consist of deforming the boundary isobath in order to modify the extent of the feature. They are performed by applying a ‘snake model’ where an internal energy term expresses

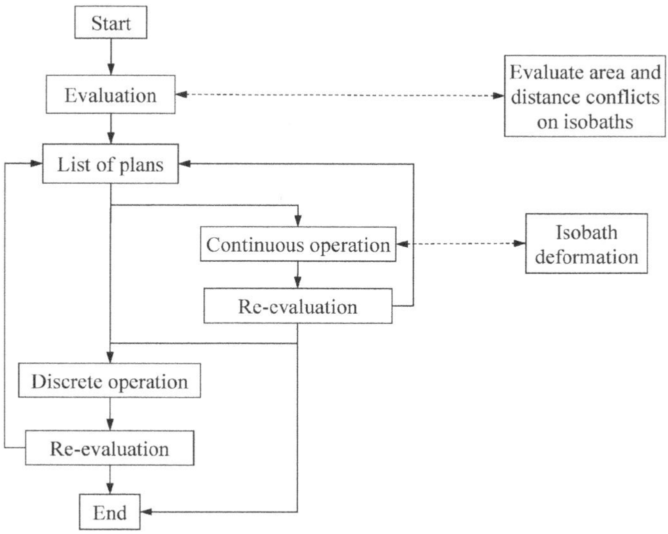

Fig. 8.14 Flowchart of the feature generalisation process (Guilbert and Zhang 2012)

Table 8.2 The list of actions (after Zhang and Guilbert 2011)

Feature type	Conflict	Plan
Peak	Small peak	Enlargement
	Close peaks	Aggregation
	Close and small peaks	Enlargement
		Aggregation
		Enlargement and aggregation
Pit	Small pit or close pits	Omission

the shape preservation constraint and an external energy term models other constraints (distance, safety, area). The snake model is detailed in Guilbert and Zhang (2012). Such operations do not modify the structure of the terrain and so the contour tree and feature tree are not affected. The safety constraint is guaranteed by imposing that the force applied to a point in the snake model is oriented towards the greater depth. Discrete operations, on the other hand, may remove isobaths and features and so may update both topological structures. It should be noted that aggregation is seen as a two-step operation including a continuous deformation, where features are deformed until their boundaries overlap, followed by a discrete transformation where the new boundary contour is created and the feature tree is updated. In this way, the deformation is performed smoothly and distance constraints with other neighbouring isobaths are also taken into account.

When processing a plan, the topological and safety constraints are always maintained as any operation that violates these constraints would be rejected. Once a feature has reviewed all its plans, the best plan is selected by checking which one best preserves the terrain morphology: feature areas are compared and the plan

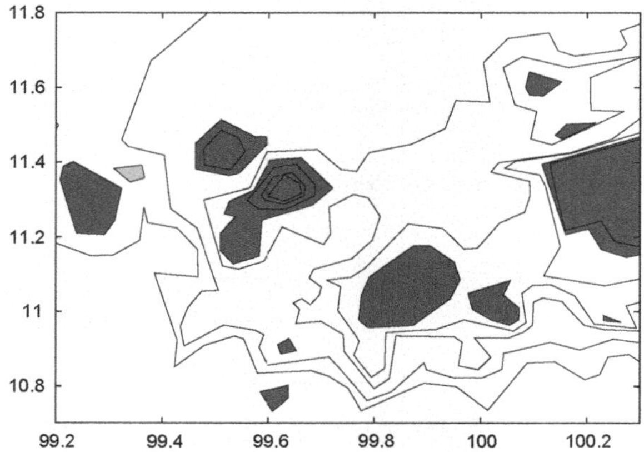

Fig. 8.15 Partial view of the original map (units in cm) with feature tree leaves. *Dark grey* peaks, *light grey* pits

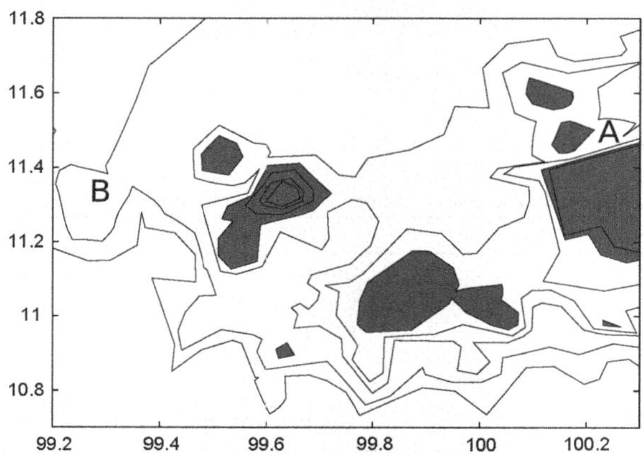

Fig. 8.16 The map after processing

with the smallest variation of area is selected. Aesthetic and shape constraints are not considered and consequently the boundary of aggregated features presents sharp angles at the place where isobaths are merged.

Results for the generalisation of isobaths of Fig. 8.15 are presented in Fig. 8.16. Figure 8.17 presents the feature trees before and after processing. The process was performed automatically from the building of the feature tree to the application of generalisation operators. The MAS approach has the advantage that the user does not impose an order to the operations and the process keeps going until no further operation can be performed. At mark A, the grey feature was enlarged after the larger peak was aggregated, providing space for enlargement. Similarly, at mark B, the peaks were aggregated after the pit was removed. Some small features were not enlarged or aggregated because no valid solution was found.

This work provides a basic strategy for automatic generalisation however it has to be noted that only feature selection was considered and that legibility conflicts between isobaths were not corrected—no smoothing and no displacement were

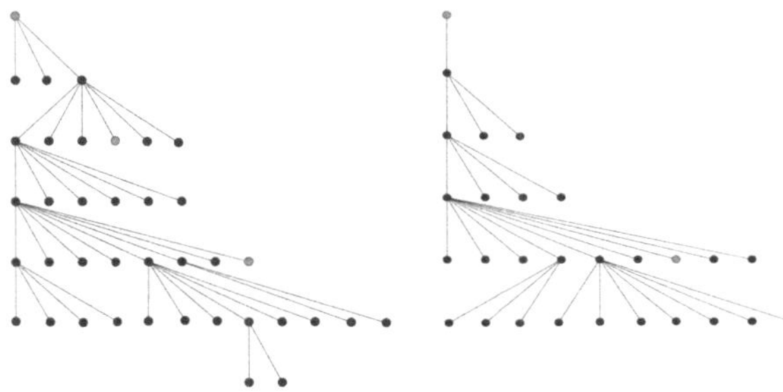

Fig. 8.17 Feature trees before and after generalisation

performed, although legibility distance was considered during the deformations. As a consequence, the result is not acceptable as it is and the model needs to be extended by giving more autonomy to isobath agents in evaluating and correcting local conflicts.

8.7 Case Study III: Preserving Relations with Other Objects During Generalisation

Julien Gaffuri

The first generalisation models were mainly focussed on the generalisation of individual objects or object groups belonging to the same data layers. This chapter presents the GAEL generalisation model (Gaffuri 2007b; Gaffuri et al. 2008) dealing with the co-generalisation of two layer types: objects and fields. Fields, also called coverages, are a common method in GIS and cartography for representing phenomena defined at each point in geographic space. Relief is an example of such a field: it exists everywhere and other objects, such as buildings, roads and rivers lie upon it. As a consequence, many relations exist between these objects and the relief that it would be important to preserve. For example, river objects should flow down the relief field and remain in their valleys. This section presents how the GAEL model handles such object-field relations throughout the generalisation process and allows the co-generalisation of objects and fields.

The principle of the GAEL model is to explicitly represent relations between objects and fields and to include constraints on these relations' preservation in the generalisation process. The fields are deformed by the objects, and the objects are constrained by the fields (Fig. 8.18) in order to preserve the relations that they share.

The following sections present in more detail these mechanisms using the river-relief outflow relation as an example.

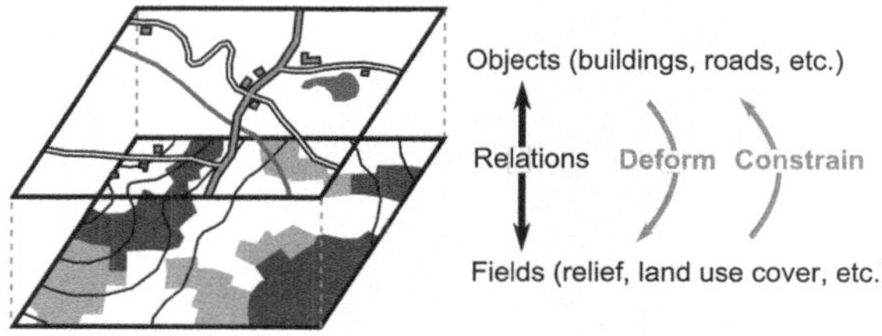

Fig. 8.18 Field-object relations in the GAEL model

8.7.1 Object-Field Relations and Their Constraints

Specific spatial analysis methods can be used to make explicit the relations between objects and fields. In the case of the river-relief outflow relation, an indicator is defined that assesses how the river flows down the relief: a river is considered to be flowing 'downwards' if each segment composing its geometry is directed toward the relief slope. With this indicator, rivers that do not flow properly on the relief (or even sometimes appear to flow 'up') are detected. A qualitative satisfaction function representing how the outflow relation is satisfied is then defined from this indicator.

In order to consider field-object relations in the generalisation process, constraints on these relations are defined. One constraint is defined for the field and another one for the object. The purpose of each constraint is, of course, to force the relation satisfaction to be as high as possible. The modelling of these relations and their associated constraints uses the same modelling pattern as the CartACom model (Gaffuri et al. 2008; Duchêne et al. 2012): a relation object is shared between both objects involved in the relation. Its role is to assess the satisfaction state of the relation between both objects. The two objects bear one constraint each, which models how each object sees the relation and how it should be transformed to improve the relation satisfaction state. Object-field constraints are included in generalisation processes whose purpose is to balance all generalisation constraints. Any constraint-based generalisation process may be used. For our experiments, we used the agent generalisation model of Ruas and Duchêne (2007). The following section describes the algorithms used to transform objects and fields in order to satisfy their common object-field constraints.

8.7.2 Algorithms for Object-Field Relation Preservation

The GAEL model includes a generic deformation algorithm whose principles are:

1. To decompose the objects into small components such as points, segments and triangles.
2. To define constraints on these components depending on the deformation requirements. Some of these constraints may be preservation constraints (to

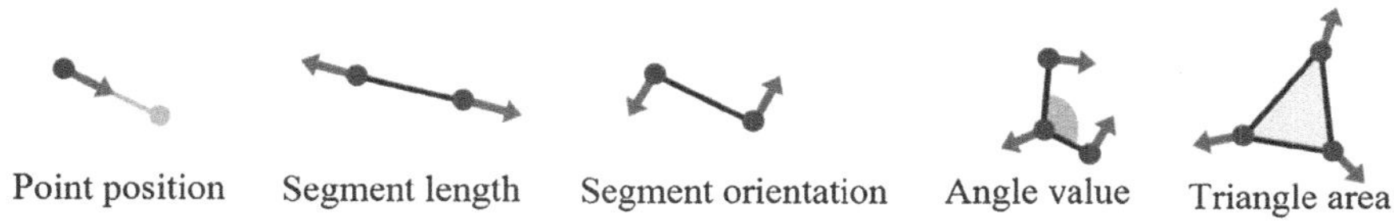

Fig. 8.19 Constraints connected with the deformation algorithm

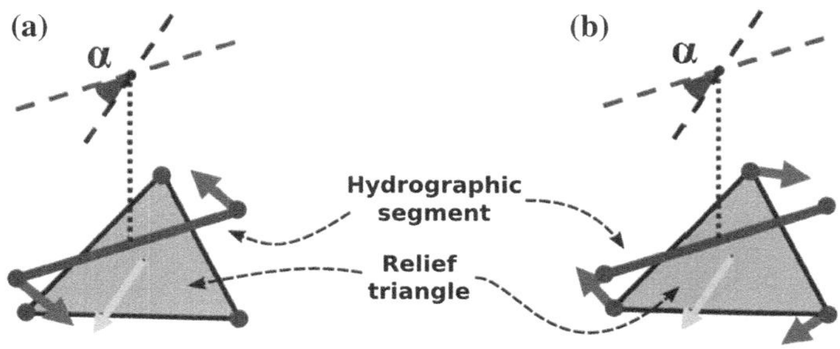

Fig. 8.20 Outflow constraints for hydrographical segments (**a**) and relief *triangles* (**b**)

force the object to keep its initial shape) or deformation constraints (to force the object to have its shape changed). Figure 8.19 shows example of such constraints.

3. To balance the preservation and deformation constraints by moving the points. This balance is found using an agent optimisation method. The advantage of this method is to perform deformations locally, only around the location where the deformation is required.

In the example of the river-relief outflow relation, the relief is represented as a TIN constrained by the contour line geometries. The following preservation constraints are used:

1. Triangle area preservation constraints;
2. Contour segments length and orientation preservation constraints;
3. Point position preservation constraint.

Both the relief and the hydrographic network are modelled as deformable features. In order that the outflow constraint is satisfied, both constraints of Fig. 8.20 are used. Their purpose is to have the angle α between the hydrographical segments (in dark grey) and the triangle slope (in light grey) as small as possible. River segments are constrained to rotate toward the slope direction, like a compass needle (Fig. 8.20a), while relief triangles are also constrained to rotate toward the flow direction of the rivers above them (Fig. 8.20b).

Using both deformation algorithms, the relief is deformed by the hydrographical network and the hydrographical network is deformed by the relief in order to have their common outflow relation preserved. Figure 8.21 shows the result of the

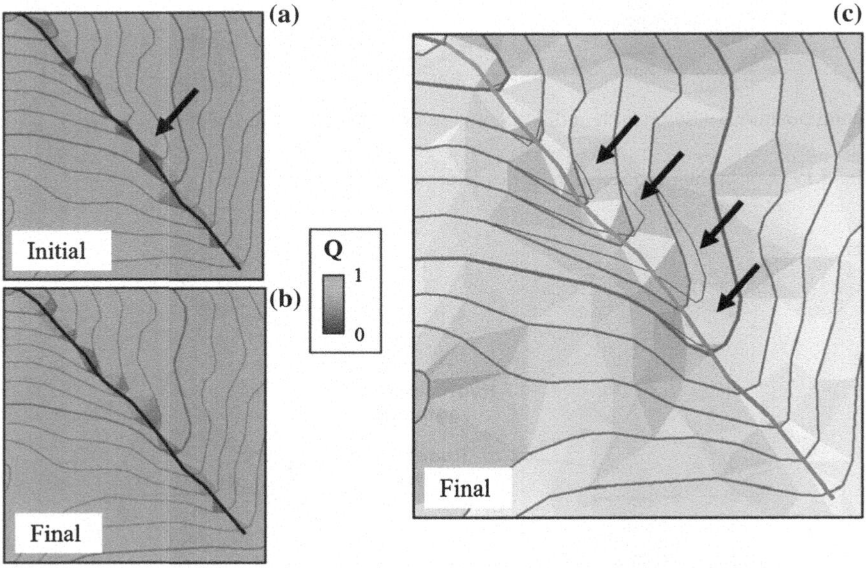

Fig. 8.21 Relief deformation. Initial state (**a**): Some *triangles* in *dark grey* are not well oriented according to the river over them. Final state (**b**): The relief has been deformed according to the outflow *triangle* constraint. The valley has 'shifted' so it is aligned with the river (**c**)

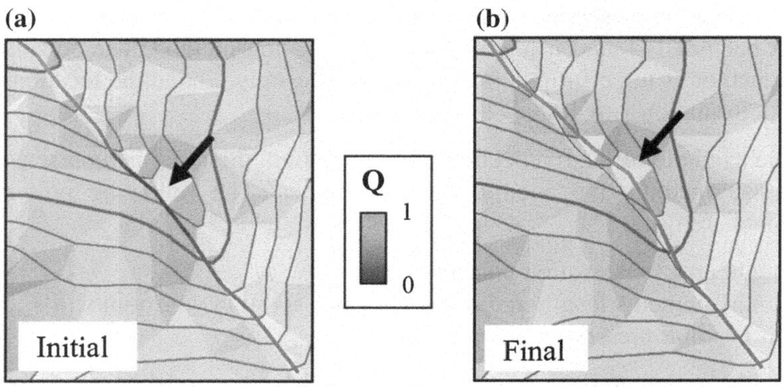

Fig. 8.22 River deformation. *Initial state*: Some segments in *dark grey* do not flow correctly with the relief *triangles* under them. *Final state*: The river has been deformed according to the outflow segment constraint. The river now falls in its valley

relief deformation for a 'fixed' river. Figure 8.22 shows the result of a river deformation over a fixed relief. In both cases, the outflow relation between them has been preserved.

Further details on this outflow relation example are given in Gaffuri (2007a). The same approach can be applied to other kinds of object-field relations (Fig. 8.18). It requires:

- Spatial analysis methods to measure the relation satisfaction;
- Field decomposition and constraints to perform the field deformation;
- Object deformation or displacement algorithms.

Gaffuri (2008) proposes such elements for other relief relations (with buildings objects for example) and with a land cover field. The GAEL model is now part of the production environment of the 1:25,000 base map of France.

8.8 Conclusions

This chapter has reviewed recent research in terrain generalisation. Following Weibel's (1992) classification of generalisation methods, current thinking in terrain generalisation is built around ideas of selective filtering and heuristic methods. Filtering methods provide a simplified representation of the terrain represented by a field function but can less easily take into account non morphometric constraints. In heuristic methods, features composing the terrain are seen as objects and are generalised by applying individual operators that allow us to model constraints related to the purpose of the map and the relation with other objects portrayed on the map. Therefore, a first step in the generalisation process is to extract terrain information. Although much work has focused on the classification of point and line features for filtering methods, new approaches presented in Sect. 8.3 were developed to characterise landforms as objects defined with their own spatial and non-spatial attributes and on which heuristic operators could be applied.

Section 8.4 described new advances in different representation techniques. The focus has been on utilising DTMs that are stored in a grid, TIN or contour form. It can be seen that terrain generalisation for cartographic purposes does not solely focus on simplification and on performance or compression aspects but, as illustrated in the different examples, also on the information retained on the map according to the quality of the visual information (Sect. 8.5), its purpose (Sect. 8.6), and on the integration of terrain with other map elements in the generalisation process (Sect. 8.7).

Although terrain representation is a major aspect of cartographic generalisation, this review also shows that much work still remains. Classification of landforms as individual objects is still limited to morphometric classification and to basic features such as hills and valleys. As mentioned by Smith and Mark (2003), such classification is complex and the definition of an ontology of landforms is still an open problem. Another research area is in the logical definition of constraints and operators that apply to these landforms. Work presented in Sect. 8.6 is limited to a small number of constraints and operators. As discussed in Chap. 3, ontologies that formalise the generalisation process for terrain generalisation need to be developed in order to extract more knowledge from the terrain model and design efficient implementation strategies. The agent model presented in Sect. 8.7 continues to show promise in facilitating the implementation of a generalisation strategy that allows terrain data to be integrated with other layers of the map. However, the method comes with a high computational cost and its application to several types of layer greatly increases the complexity of the problem due to the large number of constraints that have to be considered.

In the context of research into continuous generalisation and on-demand mapping, the work outlined in Sect. 8.2.2 does not yet incorporate user requirements analysis. One reason for this is that representation of field data requires a large amount of data to be processed, as well as semantic knowledge and data enrichment prior to the generalisation process taking place. Modelling user's requirements in this context is difficult to model and remains another open problem.

References

Baella B, Palomar-Vázquez J, Pardo-Pascual JE, Pla M (2007) Spot heights generalization: deriving the relief of the topographic database of Catalonia at 1:25,000 from the master database. In: Proceedings of the 11th ICA workshop on generalisation and multiple representation, Moscow, Russia, 2007

Beucher B (1994) Watershed, hierarchical segmentation and waterfall algorithm. In: Proceedings mathematical morphology and its applications to image processing, pp 69–76

Brassel K (1974) A model for automatic hill-shading. Cartography Geogr Inf Sci 1(1):15–27

Chaudhry O, Mackaness W (2008) Creating mountains out of mole hills: automatic identification of hills and ranges using morphometric analysis. Trans GIS 12(5):567–589

Chen J, Liu W, Li Z, Zhao R, Cheng T (2007) Detection of spatial conflicts between rivers and contours in digital map updating. Int J Geogr Inf Sci 21(10):1093–1114

Chen Y, Wilson JP, Zhu Q, Zhou Q (2012) Comparison of drainage-constrained methods for DEM generalization. Comput Geosci 48:41–49

Danovaro E, De Floriani L, Magillo P, Mesmoudi MM, Puppo E (2003) Morphology-driven simplification and multiresolution modeling of terrains. In: Hoel E, Rigaux P (eds) The 11th international symposium on advances in geographic information systems. ACM press, pp 63–70

Danovaro E, De Floriani L, Magillo P, Vitali M (2010) Multiresolution morse triangulations. In: Proceedings of the 14th ACM symposium on solid and physical modeling, pp 183–188

Deng Y (2007) New trends in digital terrain analysis: landform definition, representation, and classification. Prog Phys Geogr 31(4):405–419

Duchêne C, Ruas A, Cambier C (2012) The CartACom model: transforming cartographic features into communicating agents for cartographic generalisation. Int J Geogr Inf Sci 26(9):1533–1562

Dyken C, Dæhlen M, Sevaldrud T (2009) Simultaneous curve simplification. J Geogr Syst 11:273–289

Fei L, He J (2009) A three-dimensional Douglas–Peucker algorithm and its application to automated generalization of DEMs. Int J Geogr Inf Sci 23(6):703–718

Feng CC, Bittner T (2010) Ontology-based qualitative feature analysis: Bays as a case study. Trans GIS 14(4):547–568

Fisher P, Wood J, Cheng T (2004) Where is Helvellyn? Fuzziness of multi-scale landscape morphology. Trans Inst Brit Geogr 29(4):106–128

Frank A (1996) The prevalence of objects with sharp boundaries in GIS. In: Burrough A, Frank AU (eds) Geographic objects with indeterminate boundaries. Taylor and Francis, London, pp 29–40

Gaffuri J (2007a) Outflow preservation of the hydrographic network on the relief in map generalisation. In: International cartographic conference, International Cartographic Association, Moscow

Gaffuri J (2007b) Field deformation in an agent-based generalisation model: the GAEL model. In: GI-days 2007—young researchers forum 30, pp 1–24

Gaffuri J (2008) Généralisation automatique pour la prise en compte de thèmes champs: le modèle GAEL. PhD thesis, Université Paris-Est

Gaffuri J, Duchêne C, Ruas A (2008) Object-field relationships modelling in an agent-based generalisation model. In: Proceedings of the 12th ICA workshop on generalisation and multiple representation, jointly organised with EuroSDR commission on data specifications and the Dutch program RGI, Montpellier, France, 2008

Gökgöz (2005) Generalization of contours using deviation angles and error bands. Cartographic J 42(2):145–156

Guilbert (2013) Multi-level representation of terrain features on a contour map. Geoinformatica 17(2):301–324

Guilbert E, Saux E (2008) Cartographic generalisation of lines based on a B-spline snake model. Int J Geogr Inf Sci 22(8):847–870

Guilbert E, Zhang X (2012) Generalisation of submarine features on nautical charts. In: ISPRS annals of photogrammetry, remote sensing and spatial information sciences vol I-2, pp 13–18

Harrie L, Weibel R (2007) Modelling the overall process of generalisation. In: Generalisation of geographic information: cartographic modelling and applications. Elsevier, Oxford

Horn BKP (1982) Hill shading and the reflectance map. Geo-Processing 2:65–144

Horn W (1945) Das Generalisieren von Höhenlinien für geographische Karten (Generalisation of contour lines for geographic maps). Petermanns Geogr Mitt 91(1–3):38–46

Imhof (1982) Cartographic Relief Presentation. Walter de Gruyter & Co, Berlin

Irigoyen J, Martin M, Rodriguez J (2009) A smoothing algorithm for contour lines by means of triangulation. Cartographic J 46(3):262–267

Jenny B (2001) An interactive approach to analytical relief shading. Cartographica 38(1–2):67–75

Jenny B, Jenny H, Hurni L (2011) Terrain generalization with multi-scale pyramids constrained by curvature. Cartography Geogr Inf Sci 38(1):110–116

João E (1998) Causes and consequences of map generalisation. Taylor and Francis, London

Jordan G (2007) Adaptive smoothing of valleys in DEMs using TIN interpolation from ridgeline elevations: an application to morphotectonic aspect analysis. Comput Geosci 33:573–585

Kennelly PJ (2008) Terrain maps displaying hill-shading with curvature. Geomorphology 102(3):567–577

Kennelly PJ, Stewart AJ (2006) A uniform sky illumination model to enhance shading of terrain and urban areas. Cartography Geogr Inf Sci 33(1):21–36

Kweon IS, Kanade T (1994) Extracting topographic terrain features from elevation maps. CVGIP: Image Underst 59(2):171–182

Leonowicz AM, Jenny B (2011) Generalizing digital elevation models for small scale hypsometric tinting. In: Proceedings of the 25th international cartographic conference ICC, Paris, 2011

Leonowicz AM, Jenny B, Hurni L (2009) Automatic generation of hypsometric layers for small-scale maps. Comput Geosci 35:2074–2083

Leonowicz AM, Jenny B, Hurni L (2010a) Automated reduction of visual complexity in small-scale relief shading. Cartographica: Int J Geogr Inf Geovisualization 45(1):64–74

Leonowicz AM, Jenny B, Hurni L (2010b) Terrain sculptor: generalizing terrain models for relief shading. Cartographic Perspect 67:51–60

Li Z, Openshaw S (1993) A natural principle for objective generalization of digital map data. Cartographic Geogr Inf Sci 20(1):19–29
Loisios D, Tzelepis N, Nakos B (2007) A methodology for creating analytical hill-shading by combining different lighting directions. In: Proceedings of 23rd international cartographic conference, Moscow, 2007
Lopes J, Catalão J, Ruas A (2011) Contour line generalization by means of artificial intelligence techniques. In: Proceedings of the 25th international cartographic conference, Paris, 2011
Mackaness W, Steven M (2006) An algorithm for localised contour removal over steep terrain. Cartographic J 43(2):144–156
Matuk K, Gold C, Li Z (2006) Skeleton based contour line generalization. In: Riedl A, Kainz W, Elmes GA (eds) Progress in spatial data handling. Springer, Berlin, pp 643–658
NOAA (1997) Nautical chart user's manual. National Oceanic and Atmospheric Administration, U.S. Department of Commerce
O'Callaghan JF, Mark DM (1984) The extraction of drainage networks from digital elevation data, computer vision. Graphics Image Proc 28(3):323–344
Palomar-Vázquez J, Pardo-Pascual J (2008) Automated spot heights generalisation in trail maps. Int J Geogr Inf Sci 22(1):91–110
Pannekoek AJ (1962) Generalization of coastlines and contours. Int Yearb Cartography 2:55–75
Patterson T, Jenny B (2011) The development and rationale of cross-blended hypsometric tints. Cartographic Perspect 69:31–45
Peters R (2012) A Voronoi- and surface-based approach for the automatic generation of depth-contours for hydrographic charts. Master's thesis, TU Delft
Podobnikar T (2012) Multidirectional visibility index for analytical shading enhancement. Cartographic J 49(3):195–207
Rana S (ed) (2004) Topological data structures for surfaces. An introduction to geographical information science. Wiley, New York
Ruas A, Duchêne C (2007) A prototype generalisation system based on the multi-agent system paradigm. In: Generalisation of geographic information: cartographic modelling and applications. Elsevier, Oxford
Saux E (2003) B-spline functions and wavelets for cartographic line generalization. Cartography Geogr Inf Sci 30:33–50
Smith B, Mark DM (2003) Do mountains exist? Towards an ontology of landforms. Environ Plann B: Plann Des 30(3):411–427
Spiess E (eds) (2008) Schweizer Weltatlas (Swiss World Atlas). Konferenz der Kantonalen Erziehungsdirektoren, Zürich
Straumann RK, Purves RS (2011) Computation and elicitation of valleyness. Spat Cogn Comput 11(2):178–204
Takahashi S (2004) Algorithms for extracting surface topology from digital elevation models. In: Rana S (eds) Topological data structures for surfaces. An introduction to geographical information science. Wiley, New York, pp 31–51
Weibel R (1992) Models and experiments for adaptive computer-assisted terrain generalization. Cartography Geogr Inf Syst 19(3):133–153
Wood J (1996) The geomorphological characterisation of digital elevation models. Unpublished PhD thesis, University of Leicester
Yoeli P (1966) Analytical hill shading and density. Surveying Mapp 26(2):253–259
Zhang X, Guilbert E (2011) A multi-agent system approach for feature-driven generalization of isobathymetric line. In: Ruas A (eds) Advances in cartography and GIScience. Lecture notes in geoinformation and cartography. Springer, Berlin

Chapter 9
Evaluation in Generalisation

Jantien Stoter, Xiang Zhang, Hanna Stigmar and Lars Harrie

Abstract This chapter presents the context, the issues and the research associated with the evaluation of map generalisation output as well as of map readability. Two main approaches of evaluation are described, i.e. visual and quantitative evaluation. Visual evaluation is subjective, qualitative, and time-consuming, while it is argued that quantitative evaluation is only appropriate for assessing specific aspects. Since automated evaluation is becoming very important in the field of automated generalisation, this chapter further explores the topic of automated evaluation. The previous frameworks for automated generalisation are reviewed and the three main components of automated evaluation are explained. Related to automated evaluation of generalisation output are formulas to automatically evaluate map readability. These are also discussed. This chapter ends with three case studies. The first Case study identifies and evaluates generalised building patterns. It demonstrates the three-step approach of data enrichment, data matching and constraint evaluation. The second Case study deals with formulas to automatically evaluate map readability and the third Case study carries out a comprehensive evaluation demonstrating the main aspects described in this chapter. Both visual and quantitative evaluation are applied of which the last one includes the three main components of automated evaluation. The chapter closes with conclusions and highlights research issues in evaluation.

J. Stoter (✉)
Delft University of Technology and Kadaster, Delft, The Netherlands
e-mail: j.e.stoter@tudelft.nl

X. Zhang
Wuhan University, Wuhan, China
e-mail: xiang.zhang@whu.edu.cn

H. Stigmar · L. Harrie
GIS Centre, Lund University, Lund, Sweden
e-mail: hanna.stigmar@gis.lu.se

L. Harrie
e-mail: lars.harrie@nateko.lu.se

D. Burghardt et al. (eds.), *Abstracting Geographic Information in a Data Rich World*, Lecture Notes in Geoinformation and Cartography, DOI: 10.1007/978-3-319-00203-3_9,

9.1 Introduction

Evaluation in map generalisation is the process of examining and checking whether the desired characteristics of the resulting data are satisfactory for a given task. Quality evaluation in map generalisation differs from measuring the absolute loss in data quality, which is not *per se* a problem in map generalisation. It is unavoidable that map generalisation will change spatial data in terms of data quality (Müller 1991). However, these changes may be very helpful given a specific map purpose and scale (a fitness for use concept). Apart from measuring the quality of generalisation output, evaluation also provides users with necessary information to understand the ability and limitations of a system or data in use.

It is acknowledged that evaluation is one indispensable component of a holistic generalisation process in both manual and (semi-)automated generalisation. This is because assessment of generalisation results can help to tune parameters and to select the best algorithms or sequence of algorithms for the generalisation process (Mackaness 1991; Weibel 1991, 1995; Ruas and Plazanet 1996; João 1998; Weibel and Dutton 1998; Galanda 2003).

Evaluation of a generalised map includes two parts. The first part considers the representation of features and the second part concerns the readability of the map. Therefore, there is an overlap between the generalisation evaluation methods and the method to define map readability analytically, e.g. by using map readability formulas (Rosenholtz et al. 2005; Stigmar 2010).

This chapter presents the context, the issues and the research that has been carried out in the domain of evaluation in map generalisation and map readability and extends the work of Zhang (2012). First the purposes of evaluation in generalisation are explained (Sect. 9.2). Then the two main approaches of evaluation are described in Sect. 9.3. The quantitative approach gets further attention in Sects. 9.4 and 9.5 describing frameworks for automated evaluation. Related to the automated evaluation are map readability formulas, which are discussed in Sect. 9.6. Sections 9.7–9.9 present three case studies on the evaluation of generalised maps. The chapter closes with conclusions and further research.

9.2 The Purposes of Evaluation

Evaluation of generalised maps is beneficial for different user groups. The evaluation helps cartographers in detecting errors and it thus facilitates the interaction between human and computer. For system designers, evaluation returns useful feedback, which improves the development of automated generalisation. For data users, evaluation results provide knowledge on the content of the data in use.

Earlier research on quality evaluation in map generalisation (Weibel 1995), suggested that evaluation can be made at three different stages in the generalisation process: *a priori*, *posteriori* and *ad-hoc* evaluation. Likewise, Mackaness and

Ruas (2007) summarised three scenarios where the evaluation may be applied: evaluation before, during and after generalisation. These three scenarios are motivated by different purposes:

Evaluation for tuning: acts as a support for setting parameters and tuning generalisation systems prior to processing.

Evaluation for controlling: acts as an intermediate step to the generalisation process. It can be used for (1) identifying conflicts and errors and (2) improving the process by triggering algorithm(s) or modifying the sequence of operations.

Evaluation for assessing: evaluation of generalisation output after the process is complete.

This chapter specifically focuses on the evaluation carried out during and after the generalisation process. These two evaluations are detailed in the subsections below.

9.2.1 Evaluation for Controlling the Generalisation Process

Earlier theoretical analysis on evaluation for controlling a holistic generalisation process was made by McMaster and Shea (1992), who suggested using 'carto-metric evaluation' to answer the question of when to generalise. They further suggested to use a series of 'spatial and holistic measures' (e.g. measurement of density, distribution, shape) to detect undesired situations (conflicts) and to select operations to resolve those identified conflicts. Later, Ruas and Plazanet (1996) and Weibel and Dutton (1998) proposed similar generalisation models which treat automated evaluation as an integral part of the generalisation process, based on the idea of cartographic constraints (i.e. map specifications, see also Sect. 9.4). In their proposed models, the evaluation is used for structure recognition, conflict detection and quality assessment (AGENT 1998). Brazile (2000) designed an architecture for automated generalisation incorporating quality assessment for controlling the whole process. At the moment, an implementation for those proposed conceptual models is missing, due to the complexity of the process and technological limitations.

With the development of constraint-based approaches, more and more advanced techniques are being adapted from other domains in order to control generalisation. Multi-Agent System (MAS) for example is a technique to implement constraint-based generalisation, in which evaluation can be performed before and after each step of the generalisation in order to reach the optimal solution among a set of constraints. The research of Galanda and Weibel (2002) and Galanda (2003) focused on the generalisation of categorical (land use) data using MAS techniques. The evaluation mainly focused on the calculation of some basic metric and topological constraints, such as maintaining minimum distance and avoiding self-intersection. Based on measured values and their discrepancy to the goal values, the violation (or satisfaction) of a constraint can be calculated and a list of generalisation plans can be automatically proposed. On the other hand, optimisation techniques were also widely used to implement constraint-based generalisation, (for example Harrie 1999, 2001; Harrie and Sarjakoski 2002; Sester 2005).

In summary, quality evaluation is an indispensable component to a successful automated generalisation system. Evaluation as part of a generalisation system has the advantage of being able to direct the system to its final goal (optimal solution). But it has two disadvantages. First, evaluation criteria (i.e. constraints) may well be the same as the constraints used to steer the generalisation process. In such a case, generalised data usually satisfy the evaluation criteria. Consequently the evaluation as such may give over-optimistic results (Harrie and Weibel 2007), because the data may violate other important constraints that are not defined. Another disadvantage is that the comparison between different generalisation outputs (i.e. evaluation for grading) performed on different systems is not possible for systems with self-evaluation capabilities (Ruas 2001).

9.2.2 Evaluation for Assessing Generalisation Output

Evaluation for controlling the generalisation process is not a good option for assessing the overall quality or for making comparison between different solutions. To achieve the goal of assessing the final output of generalisation, an independent evaluation is necessary. According to Bard and Ruas (2004), evaluation for assessing can be further divided into three types:

Evaluation for editing: focused on detecting cartographic errors and inconsistencies. It is performed at the final stage of the generalisation process and is used to aid subsequent manual or interactive editing, where necessary.

Evaluation for grading: attempts to find an aggregated value that reflects the overall quality of the generalised data and compares different generalisation solutions in order to determine the optimal solution or to identify poor generalisation solutions given specific generalisation tasks.

Descriptive evaluation: provides general information of the modifications performed on the generalised data. Such information can be used to improve the quality description (e.g. metadata) of the final products (e.g. what has been removed, or emphasised? How much has the data changed?)

Little research relates directly to the evaluation of generalised data at the end of the process. The first contribution to this topic was that of Ehrliholzer (1995). The important ideas emerging from this study were: (1) the use of three levels of abstraction (i.e. element, group and map level) for assessment, (2) the need to standardise the scale of various criteria (e.g. a value from, say, 1 to 5) for comparison purposes, and (3) the combination of criteria to provide a summary of evaluation results. The study remains theoretical and only a few generalisation algorithms were evaluated. Other experiments such as the ones developed as part of the AGENT project (Ruas 2001; AGENT 1999) helped the community to develop various evaluation techniques.

Up to the present time, the most comprehensive study on the automated evaluation of generalised data was undertaken by Bard (2004a, b). One of the objectives of his study was to develop a synoptic evaluation framework in which the relative quality of generalisation outputs can be graded (using evaluation grading categories described above). The author identified a set of reference functions for automated evaluation and the interaction between object types, object characters, reference functions and threshold values of these functions according to a scale transition. The author evaluated map requirements on buildings to show how these functions work in the evaluation process. A more recent project exploring evaluation of generalisation output was the EuroSDR project (Stoter et al. 2009b; Stoter 2010), —described in Sect. 9.9.

Also the work of Touya (2012) is relevant. He examined the use of social welfare theory to aggregate single satisfaction values to global legibility assessment at the map level.

9.3 Visual and Quantitative Evaluation on Map Generalisation

Two approaches are commonly used as a basis for evaluation: visual evaluation and quantitative evaluation.

9.3.1 Visual Evaluation

Traditionally, cartographers undertake a visual evaluation of the quality of generalisation output. The judgement on the quality depends largely on knowledge and experiences of the specific cartographer. The visual evaluation approach persists in (semi-) automated generalisation. That is, the evaluation of maps from automated solutions still resorts to visual comparison with 'optimal' solutions (often a paper map), in order to validate the automatic solutions or algorithms and to identify cartographic conflicts.

Usually, visual evaluation is carried out by asking cartographic experts to grade the quality of solutions based on certain criteria (Weibel 1995; Mackaness and Ruas 2007). The expert judgement can be made on a global criterion such as 'maintenance of overall character of the original map', or the criteria can include a very detailed checklist. Grading can be used (such as an index of 1–5 representing the quality of generalisation from 'poor' to 'excellent'). The overall quality can then be determined based on the qualitative grading of different cartographers. This process is also called expert evaluation. The difficulty with expert evaluation is that the criteria may be too general and different experts may have different opinions as to what makes a 'good generalisation'. Thus, a standard is needed for such an evaluation. One approach is to produce a comprehensive and common form of checklist and a questionnaire developed in close collaboration with different experts (Weibel 1995).

The visual evaluation is the simplest form of assessment to implement, but the most time consuming (Bard 2004a). Furthermore, visual assessment is subjective and comments obtained from experts might be very general or based on rather vague criteria (João 1998; Weibel and Dutton 1999; Brazile 2000; Ruas 2001; Kazemi et al. 2005; Mackaness and Ruas 2007).

9.3.2 Quantitative Evaluation

In addition to the visual assessment approach, many researchers have been looking for quantitative methods for evaluating the quality of generalisation. The idea is to identify one or more quality measurements to determine the fulfilment of the product with respect to map requirements. Some early studies concentrated on the quantification of content change, while others focused on the measuring of specific primitive geometries before and after generalisation.

The *radical law* is one of the earliest attempts to formalise the change of map content over scale transitions (Töpfer and Pillewizer 1966). The radical law formulates the number of map objects (symbols) that should be selected according to different scale transitions. As a result, the Law can be used as a criterion for controlling the number of items to be selected. However, the Law cannot address the question of where and which objects should be selected, and consequently it is not capable of controlling semantic and structural aspects of the generalisation process.

Important work was done to study extensions to the Radical Law, for example to focus computations on line length (Buttenfield et al. 2010, 2011). Brewer et al. (2013) 'inverted' the Radical Law to determine optimal target scales for a given number of feature labels. Both are examples of 'Generalisation for Controlling', discussed in Sect. 9.2. These studies have demonstrated how the Radical Law can be extended in order to work within automated environments.

There was once a trend in utilising information theory (entropy) to model the changing levels of map content that arise from the generalisation process (e.g., Neumann 1994; Bjørke 1996, 1997, 2003; Li and Huang 2002). On the other hand, using information theory to evaluate the map content could be problematic. Mackaness and Ruas (2007) for instance argued that it is extremely difficult to formalise the change of map content caused by generalisation in terms of information entropy, since information content varies depending on the map purpose, application of map symbols and map reader's interpretation skills.

Earlier attempts at the evaluation of primitive objects focused on linear geometries (such as roads, rivers, contours, railways). McMaster (1983, 1987) proposed measures to quantify the positional change in the vertices of lines before and after generalisation, based on the concept of 'vector displacement'. The main focus of this assessment is on geometric precision. A later work by Buttenfield (1991) expanded the description of linear features to five measures (length, width band, segmentation, and error variance), which makes up the 'structural signature' of the lines. The aim of Buttenfield (1991) was to formulate the changes of the 'structural signature' over scale transitions. These two contributions have inspired other attempts for describing the characteristics of linear features, including geometric and structural aspects (Plazanet et al. 1998; Wang and Müller 1998; Ai et al. 2000; Van der Poorten and Jones 2002). In addition, some studies have focused on evaluating the change of different types of linear objects in terms of line length and 'vector displacement' e.g. Barber et al. (1995), João (1998). Finally, several studies suggested a methodology for evaluating the quality of generalised lines based on the description of the structure and shape of the lines (Skopelity and Tsoulos 2000; Skopelity and Lysandros 2001).

Work evaluating polygonal objects is a more recent topic of research. Generally two feature types are treated separately: buildings and land use. These two types are differentiated by their different properties in shape and topology. The building features are anthropogenic constructs with regular shape and are usually spatially disjoint; whereas land use features have irregular shape and are adjacent to each other. Their evaluation therefore requires different approaches. Regnauld (1998) focused on the generalisation of buildings. He argued that there is a need to evaluate whether the generalisation results correspond to the following three aspects: (1) characteristics of buildings (e.g. size, ratio between the number of buildings at the target and initial scale), (2) spatial distribution of buildings (e.g. alignments and patterns), and (3) density of buildings. Other measures for evaluation of polygonal generalisation have been proposed by Peter (2001). Dettori and Puppo (1996) proposed the extension of quality evaluation of polygonal objects using topological qualities such as connectivity, and inclusion.

However, evaluation of primitive geometries is not enough, since multiple objects are involved when assessing the quality of a generalised map (Weibel and Dutton 1998). Relationships between objects (either within a thematic class or between different classes) should be formalised and taken into account by the evaluation process.

Visual evaluation is subjective, qualitative, and time-consuming, while quantitative evaluation has tended only to examine certain characteristics and is therefore only appropriate for assessing specific requirements. It is difficult to adapt and integrate these measures into a holistic automated generalisation process. Comprehensive solutions for quantitative, automated evaluation are needed and this remains an area of further research (Burghardt et al. 2008).

This chapter seeks to provide an overview of research in this area, and what remains in need of further attention.

9.4 Frameworks of Automated Evaluation

As suggested by Weibel (1995), the process of automated evaluation involves two tasks or components:

- Formalisation of map specifications: map specifications must be formalised to describe what is a ‘good’, ‘acceptable’, or ‘poor’ generalisation for a given application and mapping scale.
- Evaluation process: applying methodologies to determine the degree to which the specifications are respected for a given generalisation output. This process requires both measure for automated evaluation as well as data matching techniques to enable the automated evaluation.

These two components were also explored in the work of Bard and Ruas (2004). Mackaness and Ruas (2007) further pointed out that a challenge of automated evaluation is to decompose the evaluation process into a set of values that can be calculated automatically and that remain meaningful to users. It is important to know which characteristics are pertinent to a specific map task and scale, and how to precisely define them (formalisation). It is also important to develop various techniques to qualify and quantify the actual changes of characteristics of data objects, which are later compared with the formalised map requirements (evaluation process). In the AGENT (1998) project, the evaluation process was further divided into three steps: (1) measurement, which is used for the computation of raw parameters (e.g. area of buildings), (2) characterisation, which aims to interpret those measured parameters (e.g. using Min, Mean, and Max values of computed areas to classify buildings), and (3) evaluation, which determines the characterisation by comparing the measured value and the goal value (e.g. if a building is too big or too small with respect to the specifications).

Bard and Ruas (2004) proposed a comprehensive framework of automated evaluation of generalised data which extended the work of AGENT (1998). The components of their framework and the interactions among the components are shown in Fig. 9.1.

In their evaluation model, **Specification** (map requirements) contains lists of characteristics, reference functions, and parameters. The establishment of map specifications is referred to as the formalisation process. In Bard and Ruas’s work, the specifications mainly contained formal requirements of buildings (e.g. size, shape, and orientation). Their focus is on the framework development and the evaluation process. In their model, the evaluation process is modelled as two classes, namely characterisation and evaluation. **Characterisation** holds the measures (algorithms) to calculate the properties of geographic objects (both initial and target objects), such as size, shape and orientation of buildings. **Evaluation** first uses the Compare() method to calculate the distance between measured values of a characteristic and the target value defined in the specifications. It then re-invokes Evaluate() to interpret the distance into a qualitative evaluation result according to the defined tolerance. For example, if the distance between the

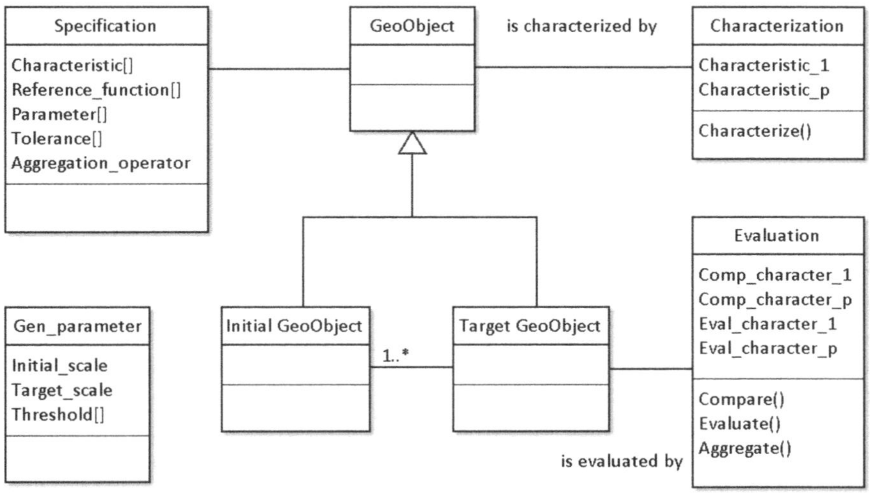

Fig. 9.1 A model to evaluate generalisation output by Bard and Ruas (2004)

measured size and target size of a building is 15 map mm^2, and the tolerance is 20 map mm^2, then the generalisation of this particular building is considered to be 'good' in terms of its size property.

It is noticeable that the evaluation process in this model involves measuring and comparing properties of objects at both the initial and target scales—that is to say, corresponding relations between target and initial objects are explicitly modelled. Burghardt et al. (2007) and Stoter et al. (2009a) also suggested that the evaluation of generalisation output should be based on the specification defined as a set of constraints, and partly on the initial dataset. As a consequence, data matching for automated evaluation of generalisation output is needed.

An improved framework for automated evaluation of generalised building patterns has been proposed in Zhang (2012). This framework is detailed further in the Case study in Sect. 9.7.

9.5 Components of Automated Evaluation

Starting from these frameworks, the following sections further explain the three main components of automated evaluation, namely (a) definition and formalisation of map requirements, (b) measures for automated evaluation and (c) data matching techniques. Together, these sections outline the major difficulties of the research on automated evaluation.

9.5.1 Defining and Formalising Map Requirements

The process of defining and formalising user requirements in map specifications has been described in Chap. 2. In an automated context, the challenge of this process is in specifying the map requirements in a computer understandable way (Burghardt et al. 2007; Stoter et al. 2009a). The challenge is in knowing which constraints are appropriate in order to assess the quality of generalised solution and how the constrained properties change at scale transitions. The former aims to select a set of constraints, whereas the latter aims to specify the goal values for the constraints (e.g. the minimum area of buildings after generalisation). Specifying goal values for constraints is however not always straightforward, since it depends (1) on the specifications of generalisation, (2) on the initial state of the property (Burghardt et al. 2007). Results from the EuroSDR project (see Sects. 2.5 and 9.9) show that constraints defined on two objects and on groups of objects are more difficult to interpret and evaluate by generalisation systems than those defined on individual objects. The formalisation of such constraints is something that requires further attention.

A closer look at the literature reveals that, low-level constraints such as those on one object (e.g. minimum dimensions) and between two objects (e.g. topology) are relatively easy to formalise. However, the constraint on general shape, which is also a low-level constraint, is harder to formalise. Shape is concerned with our ability to recognise the original form, silhouette or outline of objects and is different from positional accuracy (Müller et al. 1995). Shape needs to be preserved during generalisation as long as the details are still legible. In the constraint 'target shape should be similar to initial shape', shape is not a mathematically defined concept and the operation 'similar' is not well-defined either. Defining and precisely measuring shape is necessary for automated evaluation. The 'similarity' of two shapes can be formalised by specifying the scale-dependent modification to the measured values. A potential way of acquiring the scale-dependent knowledge is by studying map series at different scales (i.e. reverse engineering) as indicated in Harrie (2001). The results of the acquired change of a property can be stored in an 'evolution function' as proposed by Bard (2004b). This is a function that expresses how properties change over scale.

In addition to the shape constraint, the constraints dealing with higher-level concepts such as pattern, distribution, and network are also difficult to formalise. Most of these constraints define restrictions to the generalisation of groups of objects. These higher-level map requirements are important for defining the quality of generalisation output over large changes in scale, since the concepts describe important characteristics of the map at smaller scales and hence should be visually more discernable at these scales (Müller 1991; Ruas and Plazanet 1995; Mackaness and Edwards 2002; Mackaness and Ruas 2007).

At the conceptual level, Steiniger (2007) sought to formalise patterns using 'horizontal relationships', which describe the interactions between map objects within the same map scale. His reason for formalising patterns was to be able to recognise patterns, which is the first step in automated evaluation. The recognition of patterns is viewed as a data enrichment process by some (Zhang 2012). A complete view of the formalisation of higher-level constraints for automated evaluation should also involve modelling the change of these concepts at scale transitions. This means that how the concepts modify themselves at scale transitions needs to be formalised. However, the knowledge of these constraints is rarely expressed in a clear and computer-readable way. Therefore, knowledge acquisition techniques are still needed to extract the knowledge for automated evaluation of generalisation output from various sources.

9.5.2 Measures for Automated Evaluation

As described in Sect. 9.4, automated evaluation is performed via three major tasks: defining and formalising map requirements; developing measures which automatically quantify the properties that need to be evaluated; and, evaluating by comparing between corresponding objects or with respect to a set of map specifications.

Measures play an important role in automated evaluation: they are used to detect the constraint violations and describe the severity of a violation. For example, a target building with an area slightly larger the size threshold is more acceptable than if it is smaller than the threshold. Using sensitivity ranges (tolerances), a measure can evaluate the degree to which a metric is violated.

This section further explains the concept of measures and the way they are used in automated evaluation.

A measure is defined in the AGENT (1999) project as a procedure to compute geometric properties or relationships, which is the basis for evaluating characteristics of map objects as well as for assessing the need for generalisation (as in the case of conflict detection) and the success of generalisation (that is, evaluation of generalisation output). Measures consist of four main aspects. Consider a distance measure for example, which has an operational aspect (a distance metric), a mathematical aspect (such as *Hausdorff* distance), an algorithm (implementation of *Hausdorff* distance) and a measured value. A measure can be used to quantify a property of an object with respect to a constraint. A measure may consist of a formula or involve a complex algorithm including computation of representations and auxiliary data structures. Representations here describe the geometry of objects derived by a transformation from 2D Euclidean space to another reference-system (e.g. Fourier transform) or by a transformation between dimensions (e.g., a collapse operation). Auxiliary data structures capture the relationships among a group of objects within higher-level concepts such as alignments and networks. Examples of these data structures are Delaunay triangulation and Minimum Spanning Tree (MST) (AGENT 1999).

Mathematical measures can be exact or approximate, according to the implementation algorithm. For example one can use an approximate *Hausdorff* algorithm but also precise ones such as Euclidean distance to calculate the distance between two lines. Computing a precise *Hausdorff* distance is time-consuming. A measure may have either a precise or approximate value, e.g. numeric [e.g. distance (a, b) = 3.2], ranked distances, (as in the earliest versions of the Douglas-Peucker simplification algorithm), or boolean [e.g. disjoint (a, b) = false]. Mathematically precise measurements may not be adequate to describe higher-level concepts fully such as patterns and structures using a single value. In many cases they have to be characterised geometrically using auxiliary data structures such as MST (Regnauld 1996).

McMaster and Shea (1992) described several measures in digital cartography for determining when to apply generalisation operations: for example density, distribution, length and sinuosity, shape, distance, Gestalt, and abstraction. To support quality evaluation, Weibel (1995) proposed a classification of measures into global, geometric, topological, and software-related categories. At present, the most comprehensive classification of measures for quality evaluation is proposed by Mackaness and Ruas (2007). It has two aspects:

(1) A measure can be internal or external:
   (a) Internal: measure of objects at the same scale (within a dataset)
   (b) External: measure of objects between two scales (before and after generalisation).

(2) A measure can be Micro, Meso or Macro:
   (a) Micro: measure on individual objects or parts of an object
   (b) Meso: measures on groups of objects
   (c) Macro: measures on all objects of a feature class.

As defined in AGENT (1999), there are two classes of external measures. An implicit external measure is computed by comparing two internal measurements calculated at the initial and target scale (e.g. change rate of size by dividing the two areas). Explicit external measures compare two objects at different states (e.g. *Hausdorff* distance for describing the shape change of lines at two scales). The measures at micro- and meso-level are in line with the map requirements of one object, between two objects or on groups of objects. For instance, measures at the micro-level relate to position, orientation, size, and shape constraints on one object. Measures at the meso-level deals with topology, spatial arrangement/distribution, and density constraints on groups of objects.

Shape describes the geometric form of individual spatial objects (Wentz 1997) and should be maintained during the generalisation process. General shape is difficult to measure. There are many internal shape measures (e.g. sinuosity, complexity, elongation, compactness, concavity, shape index) AGENT (1999). The measures describe certain aspects of shape using scalar values.

While all of the individual metrics can be computed, they do not readily summarise to a general shape measure that is readily understand. The work of Skopelity and Tsoulos (2000) and Skopelity and Lysandros (2001) advanced the formalisation of shape metrics in an important way. They reduced these structure metrics to only three primary metrics using Principle Component Analysis. Similarity measures of shapes have also been studied. For example, Veltkamp and Hagedoorn (1999) summarised several techniques for shape matching (e.g. Fréchet distance, Hausdorff distance, turning function, signature function). Ai et al. (2008) proposed a shape descriptor for polygonal objects based on Fourier transform. These measures can be used to compare the similarity of shapes at different map scales.

For each of these measures we might ask: Which is the best-suited measure for the description of general shape in map generalisation? How do we combine different measures together to describe general shapes? If we can answer these questions then shape measures can be incorporated into the evaluation process (Burghardt et al. 2008).

High-level concepts or phenomena, such as road and water networks, building alignments and patterns are especially important at smaller scales. A definition of these concepts is hard to come by: What are the characteristics of these high-level concepts? Can these characteristics be measured? Do these measures really reflect the nature of the concepts in the experts' mind? How do these high-level phenomena change at scale transitions? These questions are pertinent to automated evaluation. Measuring higher-level concepts might involve complex computations of auxiliary data structures, such as Delaunay triangulation, Minimum Spanning Tree, and Voronoi diagram.

Various authors have examined how these high level concepts can be measured. Regnauld (1998) used MST in conjunction with road networks to measure the characteristics of groups of buildings. Christophe and Ruas (2002) used a straight-line scanning technique and statistical analysis to detect building alignments and measure their distribution. A density measurement based on skeletonisation of gap space was developed to evaluate the spatial distribution of the density of features (Zhang et al. 2008). For network features, 'continuity' is regarded as a measure for evaluating the quality of generalised network features (Edwardes and Mackaness 1999), as is "connectivity" (Li and Zhou 2012). Other methods for describing network features utilise graph theory (Mackaness 1995; Jiang and Claramunt 2004; Heinzle et al. 2005, 2006, 2007).

Despite these efforts, it is not yet known which one is suited best for modelling a certain characteristic with respect to a constraint. Relatively little work has been done in formalising the modification of these high-level concepts at scale transitions. Initial work on formalising and evaluating high-level concepts has been done by Zhang (2012) who studied the automated evaluation of building patterns in generalised maps based on a classification of building alignments (see Sect. 9.7).

9.5.3 Data Matching Between Initial and Target Data

One difficulty of automated evaluation is that the reference data (which is more detailed) and the data being evaluated are at different scales (Bard and Ruas 2004; Mackaness and Ruas 2007). Consequently, corresponding relations identifying objects that represent the same real-world entities (an object or a group of objects) at different scales are hard to establish (Bobzien et al. 2008). In order to measure the properties of spatial entities at different scales and to compare the measured values of both entities, the relations between entities at the two scales should be explicitly known.

In the generalisation domain, this kind of relation is referred to as a vertical relation, and obtaining this relation-information is considered as a data enrichment task (Neun 2007). The vertical relation has three different classes: 1-to-1 relation, *n*-to-1 relation and *n*-to-*m* relation (Ai and van Oosterom 2001). The 1-to-1 relation is the simplest relations that can be identified during generalisation. The automated evaluation against preservation constraints on one object (e.g. absolute position, orientation, shape) depends on the 1-to-1 relation of corresponding objects. The 1-to-1 relation usually results from generalisation operations such as simplification, exaggeration and displacement (Neun and Steiniger 2005). The n-to-1 and n-to-m relations result from applying generalisation operations such as aggregation and typification.

Mao et al. (2012) proposed the usage of Attributed Relational Graphs (ARG) as a similarity measure of n-to-1 and n-to-m relationships. In their work they used the method for comparing original and generalised city models consisting of 3D buildings. The approach used was that the original buildings were described by one ARG (graph) and the generalised by another ARG where the nodes contained visual features of the buildings (height, ground plan area, etc.) and the edges relationship between the buildings (e.g. distance between buildings). To model the similarity between the original and generalised buildings then become an ARG matching problem. A common method to perform ARG matching is the Nested Earth Mover's Distance (NEMD) method (Kim et al. 2004). This type of approach is common in pattern recognition, but more studies are required to evaluate its usefulness in cartographic generalisation.

The vertical relations of topographic data at different scales are not an explicit and standardised part of most data models. The links have to be created by means of automatic matching techniques. The primary objectives of data matching between different levels of detail are for establishing MRDB and for spatial database integration (Bernier and Bédard 2007). Various studies on matching network data such as roads can be found in the literature (Devogele et al. 1996; Zhang and Meng 2007; Mustière and Devogele 2008). It is clear that automated evaluation against the constraints of 1-to-1 will benefit from a data matching process relationships (e.g. preservation of absolute position, orientation). But it is not yet clear the extent to which data matching is required for automated evaluation and how this works for n-to-1 and n-to-m relations.

Some work has evaluated generalisation results by automated comparison to a previous accepted cartographic solution, or benchmark dataset. Stanislawski (2009) presents a Coefficient of Line Correspondence (CLC) in a gridded solution to map the distribution of how well a generalised set of lines matches benchmark lines for the same area. The method does not relate between the two sets, but rather marks features in both sets as either matching or not matching. A confidence interval is generated for the study area CLC value (Stanislawski et al. 2010). The CLC has been used to evaluate different generalisation solutions for hydrography (Stanislawski and Buttenfield 2011; Buttenfield et al. 2011). The CLC is analogous to a "similarity" measure (Tversky 1977; Willett 1998), which has been used to compare graph structures and was recently used by Li and Zhou (2012) to compare a generalised road network against a benchmark. The CLC has been used in conjunction with raster-based line density differencing to quantitatively and visually evaluate how well original and generalised hydrographic lines conform to terrain-derived drainage lines (Stanislawski et al. 2012, 2013). Raster density differencing is a relatively fast automated method to map the distribution of local differences in the context of two sets of vector data at various resolutions.

9.6 Map Readability Formulas

The evaluation of a generalised map requires several balanced perspectives. Preservation constraints are needed to protect the presence and arrangement of objects, and this must be countered by readability constraints to ensure that the map is readable. Readability constraints are linked to map readability formulas that are analytical expressions that describe how legible a map is. They can be used to evaluate any type of map. Map readability formulas are useful in several mapping applications, for example when combining geographic data from several sources to create real-time maps. In the context of map generalisation, map readability formulas are vital for defining map readability constraints.

Readability formulas are currently not as well developed for maps as they are for written text. Readability formulas were introduced in the 1920s and although criticised (Chomsky 1957) for not reflecting the full semantics of deep meaning, they are still used. Several hundred readability formulas have been published and used for predicting the difficulty of reading texts. In order to create readability formulas for maps, the cartographic variables need to be converted to measures in a similar manner as the linguistic variables have been. The following section presents a number of proposed measures for map readability. In Sect. 9.8, a Case study concerning the suitability of these measures is presented.

9.6.1 Measures for Map Readability

There are several measures for map readability. They can be categorised into the following types: the amount of information, the spatial distribution, object complexity and graphical resolution. Some of these measures can be used for other purposes such as for automated evaluation (see Sect. 9.4) and as constraints to trigger generalisation.

The amount of information measured refers to the number and size of map objects. Some measures that have been proposed for map readability are: the number of objects (e.g. Phillips and Noyes 1982; Wolfe 1994; Oliva et al. 2004), the number of objects of a particular type (Töpfer and Pillewizer 1966), the number of vertices (Woodruff et al. 1998; MacEachren 1982; Stigmar and Harrie 2011), the number of nodes, links and areas (MacEachren 1982), and the amount of occupied space (Frank and Timpf 1994; Oliva et al. 2004).

Measures of *spatial distribution* are determined by the density and distribution of map objects. This type of measure can reveal if objects on the map are too densely positioned, and if the map gives a homogeneous or heterogeneous impression. Some of the measures proposed include: the distribution of objects (MacEachren 1982), their symmetry and organisation (Oliva et al. 2004), entropy measures for objects and points (Bjørke 1996; Li and Huang 2002; Harrie and Stigmar 2009), homogeneity and the number of neighbours (Li and Huang 2002), and the density of objects (Hangouët 1998).

Measures of *object complexity* are determined by the shape and size of map objects. Examples of such measures are: sinuosity (João 1998), total angularity (McMaster 1987), and line connectivity (Mackaness and Mackechnie 1999; Fairbairn 2006; Li and Zhou 2012).

Measures of *graphical resolution* are determined by the size and colour of symbols. Examples include contrast of the visualised objects (Eley 1987; Oliva et al. 2004), contrast in brightness and hue (Rosenholtz et al. 2005; Stigmar et al. 2013) and line thickness (Spiess 1995).

A single measure cannot describe the readability of a map, however. We need information about how many objects there are, how they are distributed, and the graphical properties of their symbols. That is, we need to develop a synthesis of measures to establish readability formulas.

The measures above are concerned with the graphical properties of the map objects. Critics argue that these types of quantitative measures do not sufficiently reflect the true readability as they do not capture the whole meaning of the objects, as interpreted by the map reader. However, one can also argue that the meaning of the map objects can be very subjective, and that they therefore are the carriers of the meaning, the graphical symbols, that should be the focus of any measure. When the quantitative measures and readability formulas that best reflect the readability have been found, the next challenge is to quantify semantic aspects into measures and add to the readability formulas.

9.6.2 User Studies for Evaluation of Measures for Map Readability

User studies have assessed the usefulness of these measures. Ideally map readability formulas should be based on them. The general approach of these studies has been to create a number of map samples. A set of measures for each map sample then estimates the degree of map readability. Then user study evaluates the readability of the map samples. Finally, an evaluation compares the analytical map readability with the user studies. Based on this evaluation, conclusions can be drawn if any of the measures, or syntheses of the measures, are successful in predicting the readability of the map.

There have been a number of user studies performed on map readability measures. Phillips and Noyes (1982) tested the complexity of the topographic basis of geological maps. The aim was to enable recommendations for ways of improving 1:50,000 scale geological maps. In this test, five versions of the same map with different topographic bases were compared for map reading performance. They found that the amount of information on a map reflects the performance. The reduction of points produced the largest increase in map reading performance. The results also supported the idea that similar symbols tend to each other, both concerning symbol type (points vs. points, lines vs. lines, etc.) and colour.

Rosenholtz et al. (2005, 2007) present a feature congestion index, which is a synthesis of measures for colour, contrast, and orientation. Their measure of feature congestion is based on the local variability of certain key features. Different features are combined at each point, which results in a clutter map, and combined across an area to obtain a clutter measure (in the form of a raster map). Rosenholtz et al. (2005) tested their feature congestion index by asking 20 users to rank the clutter on 25 different paper maps. The maps were compiled at different scales and used different symbologies. A statistical evaluation showed an agreement in ranking and perceived clutter. Rosenholtz et al. (2007) found that identification of objects and the time taken for search were both depended on the clutter measure.

Stigmar (2006) studied subjective preferences for maps with a specific interest in the amount of information (here represented by the number of vertices in the objects). Twelve test participants were asked to rank a number of test maps according to preference for use as navigational aids in mobile map services. The result showed that while the amount of information was important, participants preferred the presence of "all" map objects over a smaller amount of information. As long as "all" objects were present, the maps were considered better even if they were generalised to a larger degree. In Stigmar and Harrie (2011), 17 measures related to the amount of information, spatial distribution and object complexity were evaluated. The aim was to see whether they could be used to describe map legibility. Twelve test participants were asked to discuss the legibility of a number of test maps and to rank some maps according to their perceived legibility. The results showed that some measures of the amount of information and spatial distribution corresponded well to the participants' opinions. The measures of object complexity did not show the same correspondence. In Stigmar et al. (2013) the studies were extended to include syntheses of measures (see Sect. 9.8).

Most map readability formulas are based upon the amount of information and spatial distribution of information rather than map design. It is an intriguing question as to how map design really affects map readability. For example, Raposo and Brewer (2011) made an user study of map design on top of orthoimages. Their study revealed that the information content affects map reading more than the design of the map. They found that map design alone did not significantly affect map reader's impressions, performance or certainty.

The use and design of map readability formulas is still a research issue. One question is to ask, how general the map readability formulas could be? The user studies reported above use small map samples, and it is not certain that formulas based on small sample sizes can be extended to larger maps. Another difficulty is that geographic data do vary substantially between data sets, which make it difficult to create general formulas. On the other hand, it would be very useful to have map readability formulas to trigger, control and evaluate the map generalisation process. The research on the issue has been quite limited so far; to really state the value of map readability formulas would require more user studies and probably the design of new formulas.

9.7 Case Study I: Automated Evaluation of Generalised Building Patterns

Xiang Zhang

The aim of this Case study (also reported in Zhang et al. 2013b) is to demonstrate how preservation constraints can be evaluated automatically for groups of objects (e.g. building alignments). As previously noted, preservation constraints (i.e. constraints that keep important characteristics of geographic phenomena) and constraints for groups of objects (i.e. constraints that deal with groups such as networks and alignments) are more difficult to formalise and evaluate. This is because corresponding relations between source and target objects and patterns, which are implicit in the datasets, have to be made explicit in order to support automated evaluation. In this Case study, the components of pattern recognition and data matching will be discussed.

9.7.1 Methodology

The evaluation methodology can be described as a three-step approach starting from data enrichment through to data matching and to constraint evaluation). For more details the readers are referred to Zhang (2012). Before presenting the

Table 9.1 Harmonised constraints on alignments (after Stoter et al. 2009a)

Constraint ID	Geometry type	Kind of group	Kind of objects of the initial data composing the group	Constrained property	Condition to be respected
CI	Point/ polygon	Alignments	[N/A]	Alignment	Alignment should be kept
C2	Point/ polygon	Alignments	[N/A]	Alignment orientation	Target orientation should be similar to initial orientation
C3	Polygon	Building alignment	Buildings aligned	Spatial distribution	Target distribution should be similar to initial distribution

three-step evaluation, map specifications related to the generalisation of building groups are analysed.

The specifications (i.e. constraints) used in this Case study are based on the generic constraints from the EuroSDR project on automated generalisation (Sect. 9.9). The EuroSDR constraints were harmonised based on expert surveys intended for capturing NMA specifications across different mapping organisations and universities in Europe. These constraints are regarded to be generic for many countries.

Table 9.1 lists three group-level constraints that are relevant to building features, though they are also applicable to other feature classes that contain groups of objects. These constraints describe existence (C1), orientation (C2) of alignments, and spatial distribution of buildings in alignments (C3).

It is useful to further distinguish two levels of constraints. Entity-level constraints (e.g. C1) are in effect when the 'constrained property' in Table 9.1 (e.g. alignment in C1) applies to the entities themselves. Another example of this type of constraint is: 'important buildings should be kept'. Property/relation-level constraints evaluate the property of, or relation between, entities rather than the entities *per se* (e.g. target building shape should be similar to initial shape). C2 and C3 are property-level constraints.

In general, automated evaluation of preservation constraints should be carried out by linking and comparing target datasets with source datasets (Burghardt and Schmid 2010). The above distinction implies two schemes in which the comparison can be performed. In one scheme, property-level constraints perform the comparison by checking the states of the entities that are kept in the target dataset and referring back to their initial states. In the second scheme, the evaluation of entity-level constraints starts from the source dataset and checks if initial entities are still represented in the target dataset. In this case, any deletion of entities (alignments in this Case study) will be penalised by entity-level constraints.

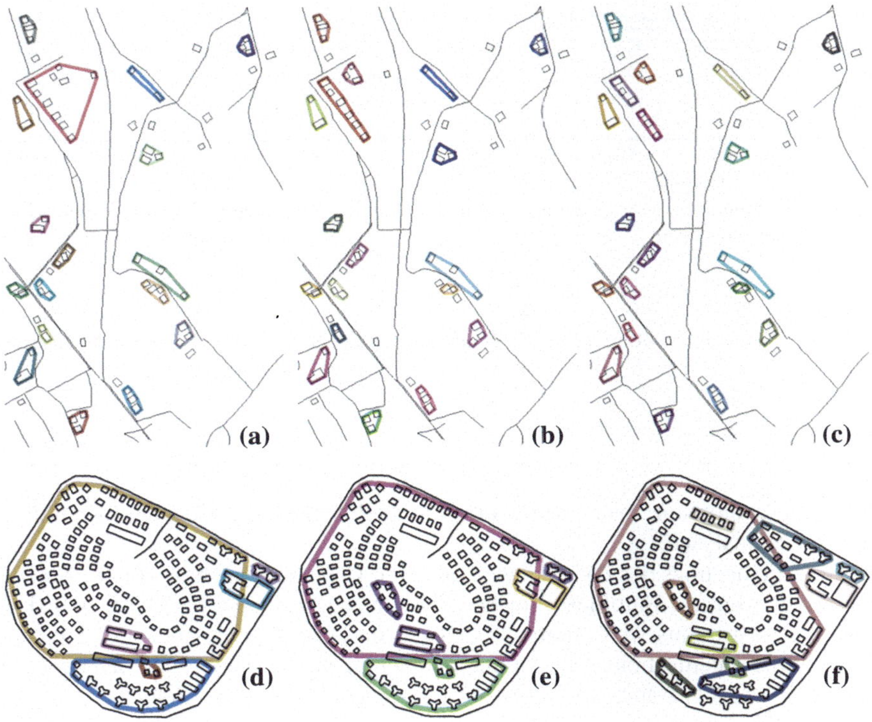

Fig. 9.2 Building clustering based on the method of Zahn (1971) with $n = 3$ (*left column*), 2 (*middle*) and 1 (*right*) (as the parameter n decreases more finer-grained groups are detected; a random colour scheme is used to delineate individual clusters; *Data*: *Top* Kadaster, NL, *Bottom* Shenzhen, China)

Building patterns discussed here are local arrangements of building groups, to which the human eye is usually attracted. These are important characteristics to retain on topographic maps. Typical forms of local building patterns include linear and nonlinear clusters. The former can be divided into collinear, curvilinear and align-along-road alignments; whereas the latter can be refined into grid-like and unstructured clusters (Zhang et al. 2013a; Anders 2006).

Building clusters are usually detected using proximity graphs, such as minimum spanning tree, relative neighbourhood graph and Delaunay triangulation (Regnauld 1996; Anders 2003). Regnauld (1996, 2001) automatically recognised building clusters for contextual building generalisation using a well-known graph-theoretic point clustering algorithm (Zahn 1971). However, recognising useful patterns is more than just identifying clusters, as is shown in Fig. 9.2. The clustering algorithm cannot explicitly recognise more refined building patterns such as alignments, and especially in densely populated areas the detected patterns are less meaningful (Fig. 9.2d–f). More details of the parameters used in the clustering can be found in Zahn (1971) and Zhang (2012).

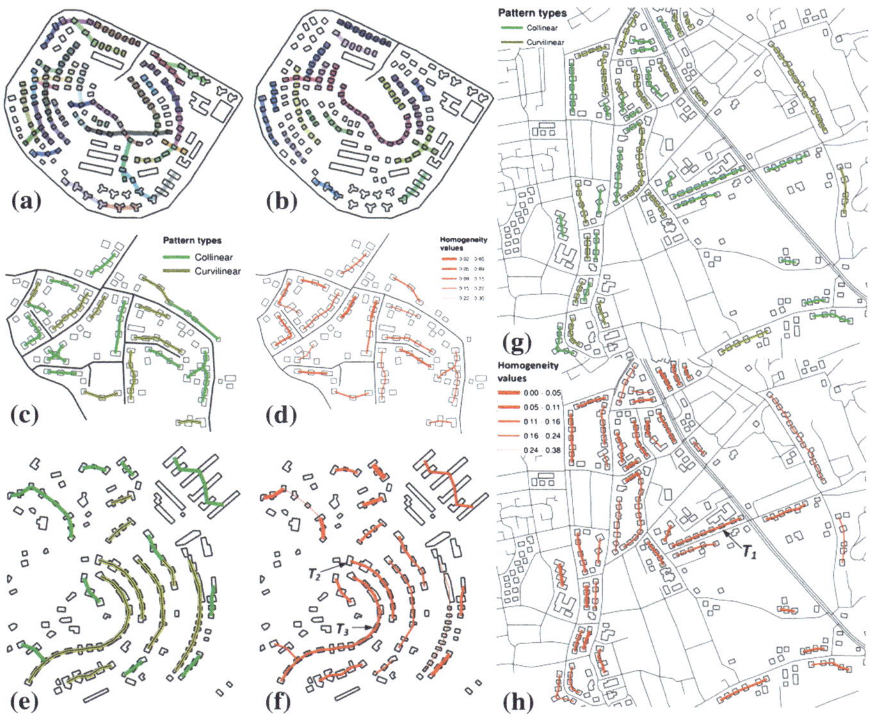

Fig. 9.3 Building alignments recognised from topographic datasets: results by a collinear algorithm (**a**) and a curvilinear algorithm (**b**); combined results of various algorithms which are visualised by pattern types (**c**), (**e**) and (**g**), and by pattern regularities (**d**), (**f**) and (**h**)

The above clustering method is based solely on a distance (proximity) relationship and ignores other important factors (e.g. size, orientation, shape, good continuity) that may impact upon the grouping process. Zhang et al. (2012) therefore proposed algorithms that recognise align-along-road and unstructured clusters. An algorithm that detects grid-like patterns was previously described by Anders (2006) for grid pattern typification. Later, Zhang et al. (2013a) proposed a recognition framework that integrates aspects of computational geometry, graph-theoretic concepts and theories of visual perception. In this framework several algorithms are used in parallel to detect collinear and curvilinear alignments, after which a mechanism was designed to combine conflicting results to improve the recognition quality. Figure 9.3 illustrates some of the recognition results.

The recognition framework described in Zhang et al. (2013a) deals explicitly with alignments and also improves the general clustering method (Figs. 9.3a–b and 9.2d–f). For the evaluation in this Case study, this recognition framework was applied with the same parameter values to both source and target datasets for matching and evaluating building alignments. According to the constraints

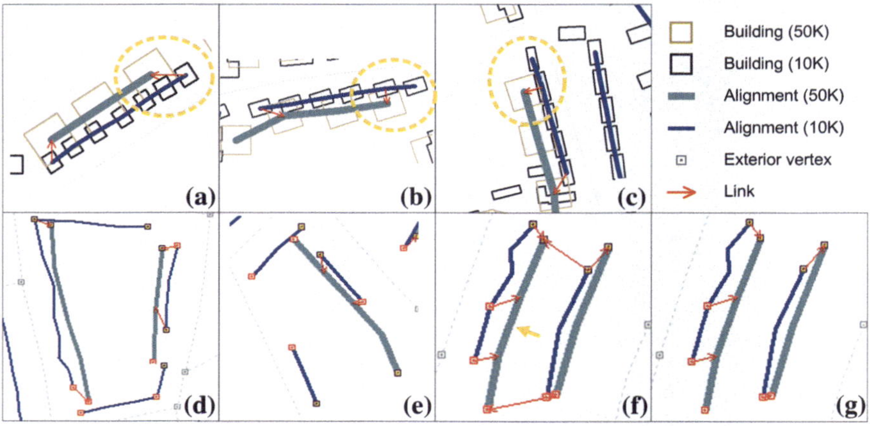

Fig. 9.4 Example results of alignment matching: (**a**)–(**c**) details of the sub-alignment matching; (**d**)–(**e**) elimination of irrelevant candidates; (**f**)–(**g**) removal of inconsistent many-to-many correspondences; reprinted by permission of the publisher (Taylor and Francis Ltd, www.tandfonline.com)

described in Table 9.1, orientation and homogeneity were measured for the detected alignments. Since building alignments are regular distributions of buildings, regularity (measured in terms of homogeneity) can be used to measure the spatial distribution of the buildings.

In order to automatically evaluate building patterns, corresponding patterns in the source and target scales must be identified. This is completed through an automated data matching process. The matching is based on the geometric representation of detected alignments (each alignment is represented by a line connecting the buildings). Because the many-to-many and partial correspondence between source and target alignments is a side-effect of generalisation, a matching process using sub-alignment elements (i.e. buildings in alignments) is designed, such that it is possible to identify which buildings in the alignments correspond to each other. This can be one building to one building, but also one building to two or more buildings, which is facilitated through the sub-alignments. The sub-alignments help to identify parts of the alignments that correspond exactly to each other. Additionally, a confidence indicator (CI) is introduced to document the reliability of the evaluation caused by the partial matching. If two alignments properly match, CI equals to 1 (Fig. 9.4a); otherwise CI decreases as the exact corresponding parts become smaller. For a detailed description of the matching algorithm, the partial correspondence and the confidence indicator see Zhang et al. (2013b). Figure 9.4 shows some of the matching results.

After data enrichment and matching processes, generalised topographic data can be automatically evaluated against the constraints (C1–C3) defined in Table 9.1. CV is used to denote the degree of constraint violation, where CV $\in$ [0, 1] with CV = 1 indicating a complete violation and CV = 0 meaning that the

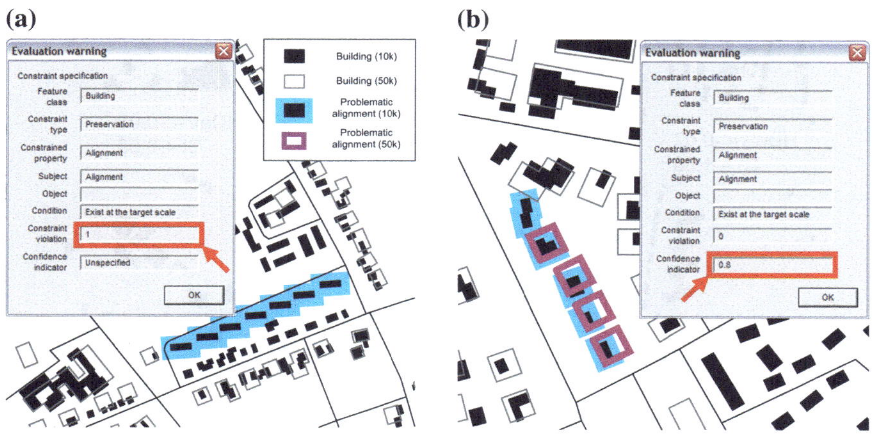

Fig. 9.5 Violation to constraint C1 (**a**) and an alert ($CI = 0.8$) of a less reliable evaluation (**b**); reprinted by permission of the publisher (Taylor and Francis Ltd, www.tandfonline.com)

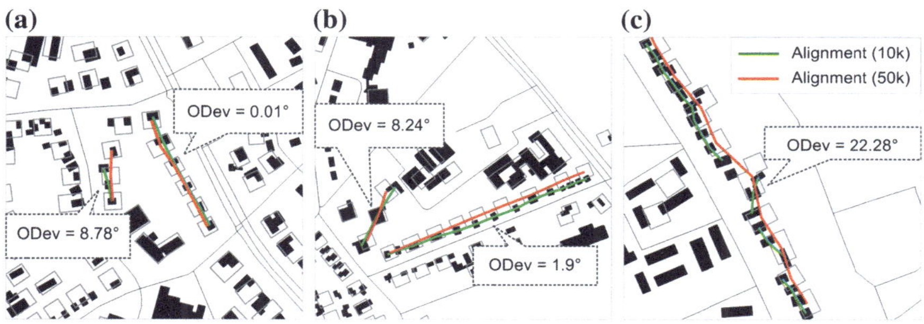

Fig. 9.6 Evaluation of C2 (ODev is defined as the deviations between alignment orientations); reprinted by permission of the publisher (Taylor and Francis Ltd, www.tandfonline.com)

constraint is not violated at all (or is satisfied). The three constraints were evaluated for individual alignments in the source and target datasets, and were aggregated to obtain an overall view of each constraint for a whole dataset. C1 is evaluated by calculating the ratio between matched target alignments and total source alignments. C2 is evaluated by calculating the deviation between orientations of corresponding alignments; the deviation is compared with a threshold to map it into a violation range [0, 1]. C3 is evaluated in a similar way by calculating the deviation between homogeneities of corresponding alignments. Individual assessments of a constraint are aggregated by weighted average. The weight is determined by each alignment: the one that is more homogeneous and contains more buildings is regarded to be more significant in the aggregation. The following figures visualise some of the evaluation results obtained in this Case study. More details can be found in Zhang et al. (2013b).

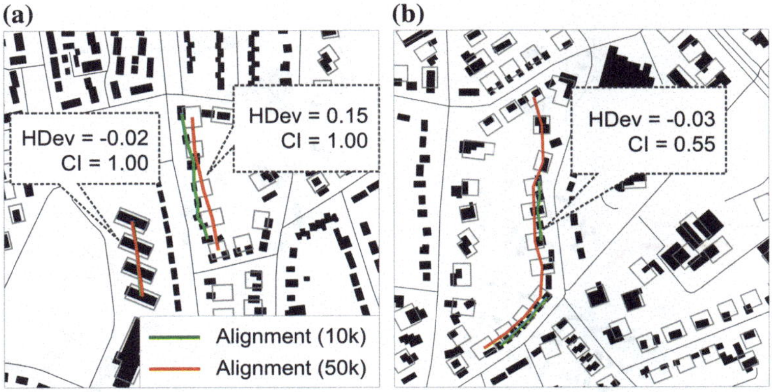

Fig. 9.7 Evaluation of C3 (HDev is defined as the deviation between homogeneities of matched alignments); reprinted by permission of the publisher (Taylor and Francis Ltd, www.tandfonline.com)

Figure 9.5a illustrates how an alignment in the source data is not kept in the target data, which constitutes a complete violation to C1 (i.e. alignment should be kept). While the source alignment is kept in the target data in Fig. 9.5b (CV = 0), an alert (CI = 0.8) indicates a less reliable evaluation caused by the partial correspondence. This can be used by cartographers to further check and improve their generalised maps (there may or may not be a problem).

Figures 9.6 and 9.7 visualise the evaluation results for C2 and C3. The confidence indicator in Fig. 9.7b shows that the result is less reliable, because a short collinear alignment is compared with a much longer and more sinuous curvilinear alignment at the target scale. Figure 9.8 visualises the evaluation of C3 for manually and automatically generalised data. It shows that more violations can be found in the automatically generalised data (alignments highlighted in green do not violate C3 while those in red and orange significantly violate C3). Further results are presented in Zhang et al. (2013b).

9.7.2 Discussion

Results in Fig. 9.8 confirm the assumptions made of the datasets. That is, the interactively generalised data has a higher quality than the automatically generated data when measured against the constraints in Table 9.1. As this Case study only concerns the evaluation of generalisation output, the process as to how the generalisation is performed is out of the scope.

This application case shows that after recognising building patterns, matching alignments between source and target datasets, the three preservation constraints on groups of objects can be evaluated automatically. The three-step evaluation

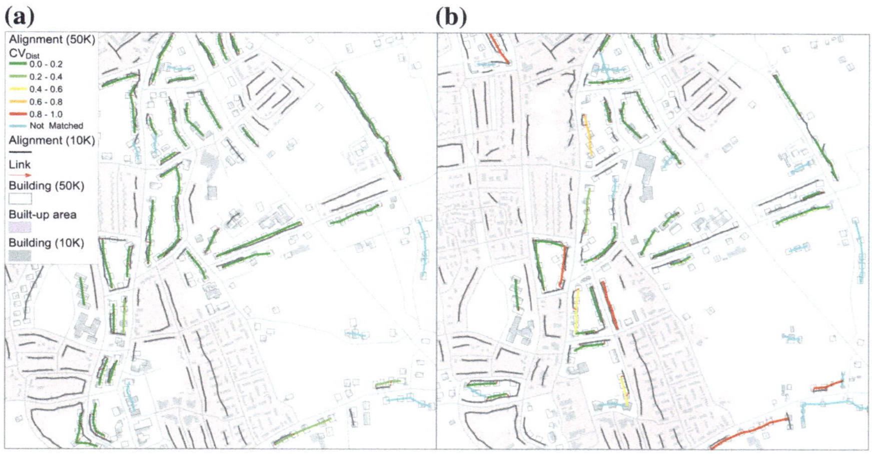

Fig. 9.8 Constraint violations are visualised for interactively (**a**) and automatically generalised data (**b**): target alignments are matched to and evaluated against C3; reprinted by permission of the publisher (Taylor and Francis Ltd, www.tandfonline.com)

approach should provide a general framework for automated evaluation of generalisation output, to compare different solutions and to summarise their data qualities. With this approach it is also possible to validate and evaluate existing and new quality measures by applying it to both interactive and automatically generalised data, which is otherwise not possible for evaluations before and during generalisation.

9.8 Case Study II: Map Readability Formulas

Hanna Stigmar and Lars Harrie

The aim of this study was to develop a map readability formula based on synthesis of measures. For details of the study see Stigmar et al. (2013). A total of 350 map samples were used in the study (Fig. 9.9). These were derived from a geographic database covering the vicinity of Helsingborg, Sweden, consisting of layers in the scale range 1:10,000–1:50,000. The map regions were chosen to present relatively homogeneous areas, and represented most typical types of areas in the map, and at the same time varying in map legibility. The maps were compiled using three levels of detail by selecting data layers with different resolutions. The maps were symbolised by two different symbologies, one traditional Swedish symbology (TS) and one "new" symbology (NS) developed by the Swedish National Land Survey.

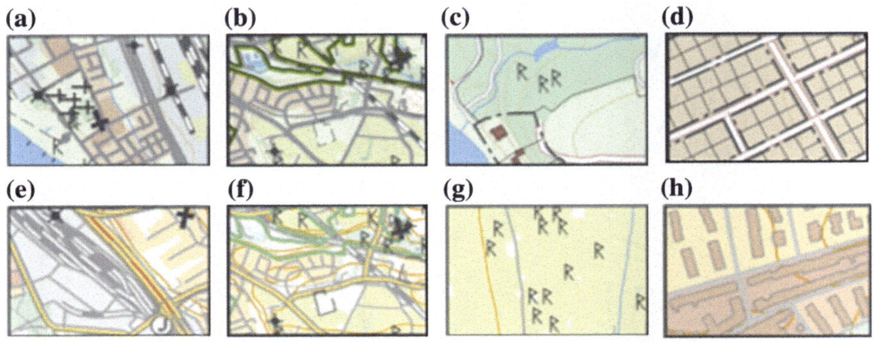

Fig. 9.9 Eight map samples used in the study. The maps are symbolised by two different symbologies, (NS and TS), at two different scales (1:50,000 and 1:10,000) and with three different levels of detail (LOD 1-3). Maps *b* and *f* only differ in symbologies

9.8.1 Methodology

The general methodology of this study was to compute several readability measures for each map sample. Then *analytical readability values* were computed using syntheses of the measures (using four different techniques). A user study was performed to obtain *perceived readability values*. Finally, in the evaluation step, the analytical and the perceived readability values were compared for each map sample.

For each map sample several readability measures were computed. Using the classification of measures given in Sect. 9.6 the following measures were used. Measures of the amount of information were: *number of objects, number of vertices, object line length, object area and number of object types*. Measures of spatial distribution were: *spatial distribution of objects, spatial distribution of vertices, degree of overlap, semantic homogeneity, number of neighbours, local density and proximity indicator*. Object complexity included five measures: *object size, line segment length, angularity, line connectivity* and *polygon shape*; while measures of graphical resolution were two: *brightness difference* and *hue difference*. For definition of the measures see Harrie and Stigmar (2009) and Stigmar et al. (2013).

The user test was designed as a web-distributed questionnaire in which participants were asked to asses the level of map legibility of the map samples as "very difficult to read", "difficult to read", "easy to read" or "very easy to read". The participants were also asked to rank some maps according to how difficult they were to read. A total of 214 participants were included in the study. Most of these were GIS professionals or GIS/geography students. Based on the user test each map sample was assigned a perceived readability value and classified as either readable or non-readable.

The evaluations were performed first for single measures and then for syntheses of measures. Four methods for syntheses were tested: manual interpretation of threshold values, multiple linear regression, support vector machines (Vapnik

1979; Tso and Mather 2009) and a supervised artificial neural network method (Biased ARTMAP, see Carpenter and Gaddam 2009).

9.8.2 Result

For the evaluation of single measures the results showed that measures of the amount of information provided the best correlation with perceived map legibility. The best correlation was given by *number of object types*, *number of vertices* and *object line length*. For the measures of spatial distribution, *proximity indicator* and *degree of overlap* showed the best correlation, while no measures of object complexity and graphical resolution alone showed a statistically significant correlation.

The first syntheses method was manual interpretation of threshold values. By manual inspection of scatter plots for various measures, we obtained a number of thresholds for non-readable maps. These are listed in Table 9.2.

A readable map was defined as a map in which all the values were below these thresholds. When all the maps were classified analytically a comparison was made with the outcome of the user test. The result revealed 85 % agreement between the classified map samples. Maps that were not correctly classified were visually investigated. Most of these maps were rather similar, and could be categorised into two groups: maps containing a high density of lines (see the example in Fig. 9.10, left), and maps with dense and overlapping point symbols in a small area (Fig. 9.10, centre). There was one exception to these categories; a map with many small, rather similar-looking area objects forming tessellations (Fig. 9.10, right).

For the three analytical syntheses methods the following measures were selected:

m_1 number of vertices
m_2 number of object types
m_3 degree of overlap of disjoint objects
m_4 brightness difference
m_5 object size.

The values of these measures were then used to classify the map samples into the classes readable or non-readable. The outcome of this classification was then compared with the user test (Table 9.3).

9.8.3 Discussion and Conclusion

The manual interpretation of threshold values gives good result. However, this method is rather limited as it is labour intensive. We would recommend using *manual interpretation of threshold values* for obtaining knowledge about the

Table 9.2 Thresholds for non-readable maps found in the manual interpretation of threshold values

Measure	Threshold
Number of object types	>17
Number of objects (for area tessellations)	>11 cm^{-2}
Object line length (for area tessellations)	>17 cm^{-1}
Object line length (for continuous fields)	>4 cm^{-1}
Object line length (for all objects)	>27 cm^{-1}
Number of vertices (for continuous fields)	>70 cm^{-2}
Number of vertices (for all objects)	>450 cm^{-2}
Proximity indicators	>80 pairs
Degree of overlap (for disjoint objects)	>3
Angularity (maximal)	>40 cm^{-1}

Table 9.3 Result of the evaluation syntheses of measures

Measures used	Multiple linear regression	Support vector machine	Biased ARTMAP
m_1	75	74	69
m_1, m_2	77	76	65
m_1, m_2, m_3	79	82	66
m_1, m_2, m_3, m_4	78	79	71
m_1, m_2, m_3, m_4, m_5	81	83	67

The columns show percentage of correctly classified map samples

Dense lines — Dense/overlapping symbols — Small, similar-looking area tesselations

Fig. 9.10 Three examples of maps that could not be identified correctly with the help of threshold values

relationship between values of the measures and the perceived legibility. For this purpose a visual method is invaluable. For the automated syntheses methods we recommend support vector machine. This method separates classes (e.g. readable and non-readable maps) using hyperplanes. This is a property that corresponds with observations from the visual interpretation of the outcome of the user study (if one or several readability measures are violated then the map is regarded as non-readable) (Stigmar et al. 2013).

From this Case study it can be concluded that several challenges need to be addressed to develop good map readability formulas that can support readability constraints in map generalisation. First, measures need to be improved. Syntactic measures that can cope with properties as in Fig. 9.10 are needed. Well-defined semantic readability measures that can be implemented through automated approaches need to be identified and developed. Additional studies should address how to synthesise these measures to classify whether a map is readable. Finally, and probably most importantly, more user preference studies and usability studies are needed to evaluate the use of these map readability formulas and their usefulness in map generalisation context.

9.9 Case Study III: The EuroSDR Project

Jantien Stoter

The EuroSDR project was introduced in Sect. 2.5. This project on the state-of-the-art of automated generalisation is a cooperative project among Universities, NMAs, and institutes across Europe (Stoter et al. 2009a; Stoter 2010). The main objective of the EuroSDR project was to study the state-of-the art of available commercial generalisation software packages to generalise a complete map. The methodology of the project consisted of (1) setting up the test cases, (2) involving commercial generalisation software vendors into the project, (3) defining map requirements, (4) setting up test and (5) designing and executing the evaluation methodology. There were four source datasets (see Chap. 2: from ICC, IGN France, OSUK and Kadaster)) and four generalisation systems: ESRI Inc. (USA), University of Hannover (Germany), 1Spatial (United Kingdom) and Axes Systems (Switzerland). The tests were performed on versions that were commercially available in June 2007. Each generalisation test was carried out by up to four testers, which resulted in around 40 generalisation outputs. The following sections describe the evaluation methodology of the project and the main lessons learned from the evaluation.

9.9.1 Evaluation Methodology

Evaluation methodology in the EuroSDR project was related to the evaluation of generalisation output after the generalisation process. It contained three basic evaluation procedures (Fig. 9.11):

- Evaluation I: automated evaluation of generalised data depends on the specification of target dataset (defined as a set of constraints), the results and partly on the initial source dataset. This type of evaluation was undertaken to answer: how many constraints are solved or violated? To what extent is the generalisation output satisfactory with respect to the map specifications?

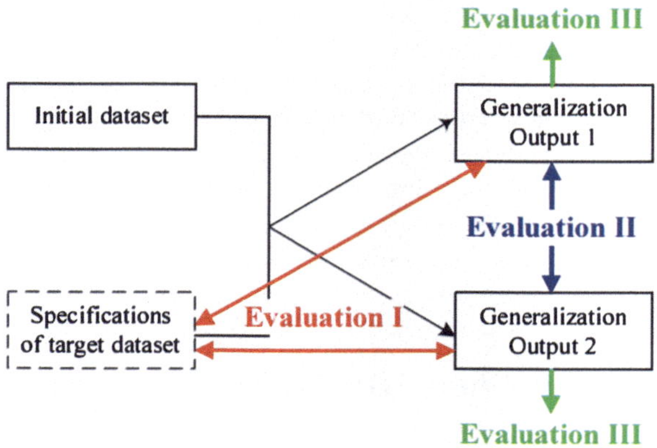

Fig. 9.11 Overview of the evaluation methodology developed in the EuroSDR project (Burghardt et al. 2007)

- Evaluation II: this type of evaluation compares the results of different outputs generalised by different software, in order to get insight into the generalisation process.
- Evaluation III: this type of evaluation requires experts to visually evaluate the final results (expert evaluation).

Within the EuroSDR project, automated evaluation was carried out only for constraints on minimum size and distance (Burghardt et al. 2008; Stoter et al. 2009a; Stoter 2010). Unfortunately, it was not possible to automatically evaluate all of constraints within the time frame of the project (Burghardt et al. 2008). However, Follow-up research was carried out to evaluate several preservation constraints on individual objects (Schmid 2008; Burghardt and Schmid 2010).

9.9.2 Lessons Learned from the EuroSDR Project

Usually, the preservation specifications in the EuroSDR project were more difficult to evaluate than the legibility specifications. This was partly due to weak formalisation of these constraints. Therefore, better understanding of preservation specifications is required in order to improve their formalisation as constraints as well as the measurement of constraint violation. This includes the concepts involved (i.e., how to mathematically describe "shape" on the basis of existing measures such as length-to-width-ratio, shape index, fractal dimension,) and of the changes allowed (how to mathematically describe acceptable modifications). Harrie (2001) obtained such information by studying existing maps at different scales. Another problem in evaluating preservation constraints is that the required correspondence with the initial data is not always easy to find, specifically when it does not concern a 1:1 relation.

The difficulty of evaluating preservation specifications was also encountered in the expert survey: it was often unclear whether a preservation constraint was assessed as "good" because the system had carefully accounted for it, or because the system had simply ignored it and at the same time had not much altered the data during the generalisation process.

This was also true for legibility constraints. For example if the system removes all the elements less than a minimum size, instead of exaggerating them, the automatic evaluation of the minimum size constraint will give a "good" result, because the constraint is not violated, but the resulting map does not represent the situation very well.

Further research is needed on how far a violation of preservation constraints is deemed tolerable. Investigations on interactively generalised data showed that cartographers also tolerate violations of legibility constraints. For both legibility and preservation constraints, a formal description of tolerated violations is therefore required. Furthermore research on the weighting between constraints and constraint violations has to be carried out to guide the generalisation process and to get an overall evaluation result for the generalised map.

The project methodology used constraints both to direct the generalisation process as well as to determine to what extent the output maps meet the specifications. The evaluation, which integrates three methods, has shown that this approach has an important limitation: the results for individual constraints are not always a good indicator of the quality of the overall solution. This has various explanations. First, some constraints may have been violated deliberately to enable good results for other constraints, e.g., by allowing (slightly) more displacement to avoid overlap. Secondly, as was observed in the automated constraint-based evaluation of interactively generalised data, one should assess not only *if* a constraint was violated but also if the violation yields an unacceptable cartographic conflict. Third, very good results for one specific constraint (e.g., minimal distance between buildings) may coincide with bad results for another constraint (e.g., building density should be kept). Fourth, a non-satisfied constraint can be due to missing functionality in a system, but can just as well be due to imprecise constraint definition. And finally, as Harrie and Weibel (2007) observed, results of constraint-based evaluation heavily depend on the defined test cases: is the constraint set complete and evenly balanced, or does it contain many constraints for very specific situations? Although the expert evaluation did evaluate generalised outputs on individual constraints taking the specific context into account, future research seek to:

- improve generalisation models and constraints to enable taking the notion of flexibility of threshold values into account;
- express constraint satisfaction in values ranging from 0 to 1, instead of in Boolean values. Boolean values may be more appropriate to identify cartographic errors. They may, however, be less appropriate for assessing the evaluation output, because they do not provide information on the degree to which a threshold is ignored;
- validate the constraint approach by considering how to aggregate "constraint-by-constraint" assessments for global indicators of map quality, specifically by

better understanding their interdependencies and impact. This also raises questions on the domain of constraint satisfaction and violation values and on their weighting and prioritising to make different constraints comparable and to enable their aggregation to global indicators. These issues have previously been addressed in the domain of constraint-based optimisation (see Ruas 1998; Bard 2004b; Mackaness and Ruas 2007);

A future test should consider selecting a representative set of constraints to better evaluate generalisation functionalities in commercial systems.

The EuroSDR study concentrated on the question of whether commercially available solutions could meet the map specifications of NMAs defined as constraints. However, during our tests several other aspects were encountered that are relevant to assessing commercial generalisation systems. For example, the testers found that in some cases topological errors were introduced during the generalisation process, and that links between generalised and ungeneralised objects, required for automated evaluation, were not created in most of the outputs. These aspects should be addressed in future tests.

The tests also highlighted the difficulty of parameterising complex algorithms. In fact, several of our evaluations showed that some vendors' solutions are better than solutions generated by the testers from NMAs and universities which shows that mastery of the software is required to obtain the best possible solutions. Software systems could help the user in finding the best parameterisation, for instance by providing tools to support interactive parameterisation (e.g. providing default parameters), or by providing tools to select similar situations, which could be generalised with the same parameterisation or tools for situation dependent, automated parameterisation. Further research should highlight parameterisation possibilities as well as user friendliness.

In addition, a future test should address aspects not amenable to constraints. The constraint approach is based on the consequences of scale changes. According to Mackaness and Ruas (2007), this bottom–up approach might work better for small-scale changes. In contrast, a top–down approach that meets the consequences of (large-) scale reduction by choosing appropriate representations for phenomena might work better over larger scale changes where changes are much more fundamental. A future test can provide more insights into the appropriateness of both approaches for automated map generalisation. Indeed, it appeared that constraints on the final result are sometimes not sufficient to fully express without ambiguity what is expected. In some cases, specifying the expected transformation can help if this transformation is always the same and if it is well known (for example building sizes should not exceed a certain limit). However, fuzzy and incomplete constraints (for example the overall landscape structure should be respected) resulted in very different interpretations and solutions among the testers. This may require a different approach in defining the requirements for automated generalisation.

The limited sizes of the four test cases precluded addressing the problems of dealing with large amounts of data (computational complexity, potential memory overflows that necessitate data partitioning, presence of numerous and various particular cases that make some algorithms fail). Future tests should define criteria as well as measuring tools to assess scalability of systems. Future tests should also quantify customisation possibilities. The most realistic way to address NMA-specific requirements may be to customise existing software. This requires facilities for writing extensions or for allowing integration with other systems.

9.10 Conclusions and Further Research

This chapter presented concepts and a review of manual and automated evaluation methodologies in the domains of map generalisation and map readability. Evaluation requires three components: (1) definition and formalisation of map requirements (2) measures for quantifying characteristics of map objects and patterns (also called data enrichment) and (3) data matching between corresponding objects.

From this chapter, we observe that the primary focus of automated evaluation research has been on evaluating constraints on one object as well as between two objects. This is evident in the evaluation model by Bard and Ruas (2004), where the model focuses on individual objects (Fig. 9.1). Less research attention has been paid to entities such as groups and feature classes and to characteristics of patterns and structures and on map readability issues for the whole map (e.g. expressing map readability formulas). This is mainly because constraints on groups, features, patterns, networks, and spatial distributions are hard to define (Burghardt et al. 2007; Stoter et al. 2009a). In addition, readability constraints have got more attention within automated evaluation research than preservation constraints. This is because preservation constraints are harder to evaluate. Readability constraints can be evaluated independently from the source data. In contrast, the evaluation of preservation constraints needs the generation and maintenance of complex (n-to-1 and n-to-m) relations between identical spatial entities.

Acknowledgments Many people contributed to this chapter. We would like to thank all of them, specifically the EuroSDR project team.

This chapter is written with the support of the Dutch Technology Foundation STW, which is part of the Netherlands Organisation for Scientific Research (NWO), and which is partly funded by the Ministry of Economic Affairs (project code: 11300).

References

AGENT (1998) Constraint analysis. Technical Report A2, Agent Consortium. Available via http://agent.ign.fr/deliverable/DA2.html

AGENT (1999) Selection of basic measures. Technical Report C1, Agent Consortium. Available via http://agent.ign.fr/deliverable/DC1.html

Ai T, van Oosterom P (2001) Spatial relation abstract in map generalisation process. In: Aref W (ed) Proceedings of the 9th ACM international symposium on advances in geographic information systems (ACM-GIS 2001), Atlanta
Ai T, Guo R, Liu Y (2000) A binary tree representation of bend hierarchical structure based on gestalt principles. In: Proceedings of the 9th international symposium on spatial data handling, Beijing, pp 230–243
Ai T, Shuai Y, Li J (2008) The shape cognition and query supported by Fourier transform. In: Headway in Spatial Data Handling (SDH'08), Springer, Montpellier, pp 39–54
Anders KH (2003) A hierarchical graph-clustering approach to find groups of objects. In: The 5th workshop on progress in automated map generalisation, Paris, 2003. Available via http://www.geo.unizh.ch/ICA/docs/paris2003/papers03.html
Anders KH (2006) Grid typification. In: Progress in spatial data handling, pp 633–642
Barber C, Cromley R, Andrle R (1995) Evaluating alternative line simplification strategies for multiple representations of cartographic lines. Cartography Geogr Inf Sci 22(4):276–290
Bard S (2004a) Quality assessment of cartographic generalisation. Trans GIS 8:63–81
Bard S (2004b) Méthode d'évaluation de la qualité de données géographiques généralisées: Application aux données urbaines. PhD thesis, Université de Paris
Bard S, Ruas A (2004) Why and how evaluating generalised data? In: Developments in spatial data handling (SDH' 04), Springer, Leicester, pp 327–342
Bernier E, Bédard Y (2007) A data warehouse strategy for on-demand multiscale mapping. In: Mackaness WA, Ruas A, Sarjakoski LT (eds) Generalisation of geographic information: cartographic modelling and applications. Elsevier, The Netherlands, pp 177–198
Bjørke JT (1996) Framework for entropy-based map evaluation. Cartography Geogr Inf Sci 23(2):78–95
Bjørke, JT (1997) Map generalisation: information theoretic approach to feature elimination. In: Proceedings of the 18th international cartographic conference (ICC'97), vol 1. Swedish Cartographic Society, Stockholm, 1997
Bjørke JT (2003) Generalisation of road networks for mobile map services: an information theoretic approach. In: Proceedings international cartographic conference (ICA), Durban, 2003
Bobzien M, Burghardt D, Petzold I, Neun M, Weibel R (2008) Multi-representation databases with explicitly modeled horizontal, vertical, and update relations. Cartography Geogr Inf Sci 35(1):3–16
Brazile F (2000) Semantic infrastructure and methods to support quality evaluation in cartographic generalisation. PhD thesis, Department of Geography, University of Zurich
Brewer CA, Stanislawski LV, Buttenfield BP, Sparks KA, McGilloway J, Howard MA (2013) Automated thinning of road networks and road labels for multiscale design of the national map of the United States. Cartography Geogr Inf Sci (in press)
Burghardt D, Schmid S (2010) Constraint-based evaluation of automated and manual generalised topographic maps. In: Gartner G, Ortag F (eds) Cartography in central and eastern Europe. Lecture Notes in Geoinformation and Cartography, Springer, pp 147–162
Burghardt D, Schmid S, Stoter J (2007) Investigations on cartographic constraint formalization. In: Proceedings of the 11th ICA workshop on generalisation and multiple representation, Moscow, Russia, 2007
Burghardt D, Schmid S, Duchêne C, Stoter J, Baella B, Regnauld N, Touya G (2008) Methodologies for the evaluation of generalised data derived with commercial available generalisation systems. In: Proceedings of the 12th ICA workshop on generalisation and multiple representation, jointly organised with EuroSDR commission on data specifications and the Dutch program RGI, Montpellier, France, 2008
Buttenfield BP (1991) A rule for describing line feature geometry. In: Buttenfield BP, McMaster RB (eds) Map generalisation: making rules for knowledge representation. Longman, London, pp 150–170

Buttenfield BP, Stanislawski LV, Brewer CA (2010) Multiscale representations of water: tailoring generalisation sequences to specific physiographic regimes. GIScience short paper proceedings. Available via http://aci.ign.fr/2010_Zurich/genemr2010_submission_11.pdf
Buttenfield BP, Stanislawski LV, Brewer CA (2011) Adapting generalisation tools to physiographic diversity for the United States National hydrography dataset. Cartography Geogr Inf Syst 38(3):289–301
Carpenter GA, Gaddam CS (2009) Biased ART: a neural architecture that shifts attention toward previously disregarded features following an incorrect prediction. Technical report CAS/CNS TR-2009-003
Chomsky N (1957) Syntactic structures. Mouton, Paris
Christophe S, Ruas A (2002) Detecting building alignments for generalisation purposes. In: Richardson DE, van Oosterom P (eds) Advances in spatial data handling (SDH 2002). Springer, Ottawa, pp 342–419
Dettori G, Puppo E (1996) How generalisation interacts with the topological and geometric structure of maps. In: Kraak MJ, Molenaar M (eds) Advances in GIS research II (SDH96). Taylor and Francis, Delft, pp 559–570
Devogele T, Trevisan J, Raynal L (1996) Building a multi-scale database with scale transition relationships. In: Kraak MJ, Molenaar M (eds) Advances in GIS Research II (SDH'96). Taylor and Francis, United Kingdom, pp 337–351
Edwardes A, Mackaness WA (1999) Modelling knowledge for automated generalisation of categorical maps: a constraint based approach. In: Atkinson P, Martin D (eds) GIS and geocomputation (innovations in GIS 7). Taylor and Francis, London, pp 161–173
Ehrliholzer R (1995) Quality assessment in generalisation: integrating quantitative and qualitative methods. In: Proceedings of the 17th international cartographic conference, Barcelona, 1995
Eley MG (1987) Color-layering and the performance of the topographic map user. Ergonomics 30:655–663
Fairbairn D (2006) Measuring map complexity. Cartographic J 43(3):224–238
Frank AU, Timpf S (1994) Multiple representations for cartographic objects in a multi-scale tree: an intelligent graphical zoom. Comput Graph 18(6):823–829
Galanda M (2003) Modelling constraints for polygon generalisation. In: Proceedings of the 7th ICA workshop on progress in automated map generalization, Paris, France, 2003
Galanda M, Weibel R (2002) An agent-based framework for polygonal subdivision generalisation. In: Richardson D, van Oosterom P (eds) Advances in spatial data handling (SDH 2002). Springer, Ottawa, pp 121–136
Hangouët JF (1998) Voronoi diagrams on segments: properties and tractability for generalisation purposes. Technical report for AGENT, Cogit, IGN
Harrie L (2001) An optimisation approach to cartographic generalisation. PhD thesis, Lund Institute of Technology
Harrie L, Sarjakoski T (2002) Simultaneous graphic generalisation of vector data sets. GeoInformatica 6(3):233–261
Harrie L, Weibel R (2007) Modelling the overall process of generalisation. In: Mackaness WA, Ruas A, Sarjakoski L (eds) Generalisation of geographic information: cartographic modelling and applications. Elsevier, Amsterdam, pp 67–87
Harrie L (1999) The constraint method for solving spatial conflicts in cartographic generalisation. Cartography Geogr Inf Sci 26(1):55–69
Harrie L, Stigmar H (2009) An evaluation of measures for quantifying map information. ISPRS J Photogrammetry Remote Sens 65(3):266–274. doi:10.1016/j.isprsjprs.2009.05.004
Heinzle F, Anders KH, Sester M (2005) Graph based approach for recognition of patterns and implicit information in road networks. In: Proceedings of 22nd international cartographic conference (ICC2005), La Coruna, 2005
Heinzle F, Anders KH, Sester M (2006) Pattern recognition in road networks on the example of circular road detection. In: Raubal M, Miller HJ, Frank AU, Goodchild MF (eds) Geographic, information science, vol 4197. Lecture notes in computer science, pp 153–167

Heinzle F, Anders KH, Sester M (2007) Automatic detection of pattern in road networks: methods and evaluation. In: Proceedings of joint workshop visualization and exploration of geospatial data (IAPRS), vol XXXVI—4/W45
João EM (1998) Causes and consequences of map generalisation. Taylor and Francis, London
Jiang B, Claramunt C (2004) A structural approach to the model generalisation of an urban streets network. GeoInformatica 8(2):157–171
Kazemi S, Lim S, Ge L (2005) Integration of cartographic knowledge with generalisation algorithms. In: Proceedings of IGARSS'05, vol 5, pp 3502–3505
Kim DH, Yun ID, Lee SU (2004) A new attributed relational graph matching algorithm using the nested structure of earth mover's distance. In: Proceedings of 17th international conference on pattern recognition (ICPR'04)
Li Z, Huang P (2002) Quantitative measures for spatial information of maps. Int J Geogr Inf Sci 16(7):699–709
Li Z, Zhou Q (2012) Integration of linear and areal hierarchies for continuous multi-scale representation of road networks. Int J Geogr Inf Sci 26(5):855–880
MacEachren AM (1982) Map complexity: comparison and measurement. Am Cartographer 9:31–46
Mackaness WA (1991) Integration and evaluation of map generalisation. In: Buttenfield BP, McMaster RB (eds) Map generalisation: making rules for knowledge representation. Longman, London, pp 217–226
Mackaness WA (1995) Analysis of urban road networks to support cartographic generalisation. Cartography Geogr Inf Sci 22:306–316
Mackaness WA, Mackechnie GA (1999) Automating the detection and simplification of junctions in road networks. GeoInformatica 3:185–200
Mackaness WA, Edwards G (2002) The importance of modelling pattern and structure in automated map generalisation. In: Joint workshop on multi-scale representations of spatial data, Ottawa, p 11
Mackaness WA, Ruas A (2007) Evaluation in the map generalisation process. In: Mackaness WA, Ruas A, Sarjakoski LT (eds) Generalisation of geographic information: cartographic modelling and applications., Series of international cartographic associationElsevier Science, Amsterdam, pp 89–111
Mao B, Harrie L, Ban Y (2012) Detection and typification of linear structures for dynamic visualization of 3D city models. Comput, Environ Urban Syst 36(3):233–244. ISSN 0198-9715. http://dx.doi.org/10.1016/j.compenvurbsys.2011.10.001
McMaster RB (1983) Mathematical measures for the evaluation of simplified lines on maps. PhD thesis, Universite du Kansas, Etats-Unis
McMaster RB (1987) Automated line generalisation. Cartographica 2(24):74–111
McMaster RB, Shea KS (1992) Generalisation in digital cartography. Resource publications in geography. Association of American Geographers
Muller JC (1991) Generalisation of spatial data bases. In: Maguire DJ, Goodchild MF, Rhind DW (eds) Geographical information systems: principles and applications. Longman, London, pp 457–475
Müller JC, Weibel R, Lagrange JP, Salgé F (1995) Generalisation: state of the art and issues. In: Müller JC, Lagrange JP, Weibel R (eds) GIS and generalisation: methodology and practice. Taylor and Francis, United Kingdom, pp 3–17
Mustiere S, Devogele T (2008) Matching networks with different levels of detail. Geoinformatica 12(4):435–453
Neumann J (1994) The topographical information content of a map/an attempt at a rehabilitation of information theory in cartography. Cartographica 31(1):26–34
Neun M (2007) Data enrichment for adaptive map generalisation using web services. PhD thesis, Department of Geography, University of Zurich
Neun M, Steiniger S (2005) Modelling cartographic relations for categorical maps. In: Proceedings of XXII international cartographic conference

Oliva A, Mack ML, Shrestha M, Peeper A (2004) Identifying the perceptual dimensions of visual complexity of scenes. In: Proceedings of the 26th annual meeting of the cognitive science society
Peter B (2001) Measures for the generalisation of polygonal maps with categorical data. In: 4th ICA workshop on progress in automated map generalisation, Beijing
Phillips RJ, Noyes L (1982) An investigation of visual clutter in the topographic base of a geological map. Cartographic J 19(2):122–132
Plazanet C, Bigolin NM, Ruas A (1998) Experiments with learning techniques for spatial model enrichment and line generalisation. GeoInformatica 2(4):315–333
Raposo PJ, Brewer CA (2011) Comparison of topographic map designs for overlay on orthoimage backgrounds. In: Proceedings of ICC, Paris, 2011
Regnauld N (1996) Recognition of building clusters for generalisation. In: Kraak MJ, Molenaar M (eds) Advances in GIS research II (SDH'96). Taylor and Francis, Delft, pp 4B.1–4B.14
Regnauld N (1998) Généralisation du bâti: Structure spatiale de type graphe et representation Cartographique. PhD thesis, Université de Provence, Aix-Marseille
Regnauld N (2001) Contextual building typification in automated map generalization. Algorithmica 30(2):312–333
Rosenholtz R, Li Y, Mansfield J, Jin Z (2005) Feature congestion: a measure of display clutter. In: Proceedings of SIGCHI conference on human factors in computing systems
Rosenholtz R, Li Y, Nakano L (2007) Measuring visual clutter. J Vis 7(2):1–22
Ruas A (2001) Automatic generalisation research: learning process from interactive generalisation. Technical report 39, OEEPE
Ruas A, Plazanet C (1995) Data and knowledge modelling for generalisation. In: Müller JC, Lagrange JP, Weibel R (eds) GIS and generalisation: methodology and practice. Taylor and Francis, United Kingdom, pp 73–90
Ruas A, Plazanet C (1996) Strategies for automated generalisation. In: Kraak JM, Molenaar M (eds) Advances in GIS research II (SDH'96). Taylor and Francis, United Kingdom, pp 319–336
Schmid S (2008) Automated constraint-based evaluation of cartographic generalization solutions. Master's thesis, Department of Geography, University of Zurich
Sester M (2005) Optimization approaches for generalisation and data abstraction. Int J Geogr Inf Sci 19(8):871–897
Skopelity A, Lysandros T (2001) A methodology for the assessment of generalization quality. In: Proceedings of the 5th ICA workshop on progress in automated map generalization. Beijing, China, 2001
Skopelity A, Tsoulos L (2000) Developing a model for quality assessment of linear features. In: Proceedings of the 4th international symposium on spatial accuracy assessment in national resources and environmental sciences, 2000
Spiess E (1995) The need for generalisation in a GIS environment. In: Müller JC, Lagrange J-P, Weibel R (eds) GIS and generalisation, Gisdata 1. Taylor and Francis, United Kingdom, pp 31–46
Stanislawski LV, Buttenfield BP, Brewer CA, Anderson-Tarver C (2013) Integration metrics for cartographic generalisation: assessment of 1:1,000,000-Scale hydrography and Terrain. In: 26th international cartographic conference, Dresden, 25–30 Aug 2013
Stanislawski LV, Buttenfield BP and Samaranayake VA (2010) Generalisation of hydrographic features and automated metric assessment through bootstrapping. In: Proceedings of the 13th ICA workshop on generalisation and multiple representation, Zürich, Switzerland, 2010
Stanislawski LV, Buttenfield BP (2011) Hydrographic generalisation tailored to dry mountainous regions. Cartography Geogr Inf Syst 38(2):117–125
Stanislawski LV, Doumbouya AT, Miller-Corbett CD, Buttenfield BP, Arundel ST (2012) Scaling stream densities for hydrologic generalisation. In: 7th international conference on geographic information science, Columbus

Stanislawski LV (2009) Feature pruning by upstream drainage area to support automated generalisation of the United States national hydrography dataset. Comput, Environ Urban Syst 33:325–333

Steiniger S (2007) Enabling pattern-aware automated map generalisation. PhD thesis, Department of Geography, University of Zurich

Stigmar H, Harrie L (2011) Evaluation of analytical methods to describe map readability. Cartographic J 48(1):41–53

Stigmar H, Harrie L, Djordjevic M (2013) Analytical estimation of map legibility. Submitted

Stoter JE (2010) State-of-the-art of automated generalisation in commercial software. Number 58. Official publication—EuroSDR 2010. Available via http://www.eurosdr.net/projects/generalisation/eurosdr_gen_final_report_mar2010.pdf

Stoter JE, Burghardt D, Duchêne C, Baella B, Bakker N, Blok C, Pla M, Regnauld N, Touya G, Schmid S (2009a) Methodology for evaluating automated map generalisation in commercial software. Comput, Environ Urban Syst 33(5):311–324

Stoter JE, van Smaalen J, Bakker N, Hardy P (2009b) Specifying map requirements for automated generalisation of topographic data. Cartographic J 46(3):214–227

Topfer F, Pillewizer W (1966) The principles of selection. Cartographic J 3(1):10–16

Touya G (2012) Social welfare to assess the global legibility of a generalized map. In: Proceedings of GIScience 2012, Springer, Berlin

Tso B, Mather PM (2009) Classification methods for remotely sensed data, 2nd edn. CRC Press, United States

Tversky A (1977) Features of similarity. Psychol Rev 84(4):327–352

Van der Poorten P, Jones C (2002) Characterisation and generalisation of cartographic lines using Delaunay triangulation. Int J Geogr Inf Sci 16(8):773–794

Vapnik (1979) Estimation of dependences based on empirical data. Nauka, Moscow (in Russian)

Veltkamp RC, Hagedoorn M (1999) State-of-the-art in shape matching. Technical report, Department of Information and Computing Sciences, Utrecht University

Wang Z, Müller JC (1998) Line generalisation based on analysis of shape characteristics. Cartography and geographic information science 25(1):3–15

Weibel R (1991) Amplified intelligence and rule-based systems. In: Buttenfield BP, McMaster RB (eds) Map generalisation: making rules for knowledge representation. Longman, London, pp 172–186

Weibel R (1995) Three essential building blocks for automated generalisation. In: GIS and generalisation: methodology and practice, vol 1 of Gisdata. Taylor and Francis, London, pp 56–69

Weibel R, Dutton G (1998) Constraint-based automated map generalisation. In: Poiker TK, Chrisman N (eds) Proceedings 8th international symposium on spatial data handling, IGU: Geographic Information Science Study Group, Vancouver, pp 214–224

Weibel R, Dutton G (1999) Generalising spatial data and dealing with multiple representations. In: Longley PA, Goodchild MF, Maquire DJ, Rhind DW (eds) Geographical information systems: principles, techniques, management and applications, geoinformation international, 2nd edn. Cambridge, pp 125–155

Wentz E (1997) Shape analysis in GIS. In: Proceedings of AUTO-CARTO 13

Willett P (1998) Chemical similarity searching. J Chem Inf Comput Sci 38(6):983–996

Wolfe JM (1994) Guided search 2.0: a revised model of visual search. Psychon Bull Rev 1(2):202–238

Woodruff A, Landay J, Stonebraker M (1998) Constant information density in zoomable interfaces. In: Proceedings of advanced visual interfaces'98

Zahn CT (1971) Graph-theoretical methods for detecting and describing gestalt clusters. IEEE Trans Comput C-20(1):68–86

Zhang M, Meng L (2007) An iterative road-matching approach for the integration of postal data. Comput Environ Urban Syst 31:597–615

Zhang X (2012) Automated evaluation of generalised topographic maps. PhD dissertation, ITC, University of Twente. Available via http://www.itc.nl/library/papers_2012/phd/zhang.pdf

Zhang X, Tinghua A, Stoter J (2012) Characterization and detection of building patterns in cartographic data: Two algorithms. In: Advances in spatial data handling and GIS. Lecture notes in geoinformation and cartography, Springer, pp 93–107
Zhang X, Tinghua A, Stoter J, Kraak MJ, Molenaar M (2013a) Building pattern recognition in topographic data: examples on collinear and curvilinear alignments. GeoInformatica 17:1–33
Zhang X, Stoter J, Tinghua A, Kraak MJ, Molenaar M (2013b) Automated evaluation of building alignments in generalised maps. Int J Geogr Inf Sci. doi:10.1080/13658816.2012.758264
Zhang X, Tinghua A, Stoter J (2008) The evaluation of spatial distribution density in map generalisation. Int Arch Photogrammetry, Remote Sens Spat Inf Sci 37:181–188

Chapter 10
Generalisation in the Context of Schematised Maps

William Mackaness and Andreas Reimer

Abstract In the last decade schematised maps have garnered substantial research interest from disciplines such as cartography, computational geometry and spatial cognition. More often than not, the approaches have been following their own specific goals, leaving the question of what they have in common relatively open. Most research has had metro maps and their automated creation as its focus. In this chapter we seek a more systematic treatment of what constitutes schematised maps. This chapter organises and differentiates the understanding of what schematisation is and how it relates to generalisation. Three cases studies variously explore and illustrate developments in the automatic generation of schematised maps.

10.1 The Nature of Schematised Maps

Conventionally the focus of application of map generalisation techniques has been on topographic mapping (Stoter et al. 2009). Typically a broad collection of generalisation operators are variously applied to control content (model generalisation) and its display (cartographic generalisation) with the ambition of creating a map relevant to a wide audience. The topographic map typically contains a broad range of entities—both natural and anthropogenic and throughout the generalisation process the desire is to conserve the positional accuracy of the features represented on the map (their relative distances, shapes and areas). This enables the map reader to readily move between the objects as represented on the map and the features experienced in the landscape. In such contexts, we might say that 'x and y are

W. Mackaness (✉)
School of GeoSciences, The University of Edinburgh, Edinburgh, UK
e-mail: william.mackaness@ed.ac.uk

A. Reimer
Institute of Geography, University of Heidelberg, Heidelberg, Germany
e-mail: andreas.reimer@geog.uni-heidelberg.de

D. Burghardt et al. (eds.), *Abstracting Geographic Information in a Data Rich World*, Lecture Notes in Geoinformation and Cartography, DOI: 10.1007/978-3-319-00203-3_10,

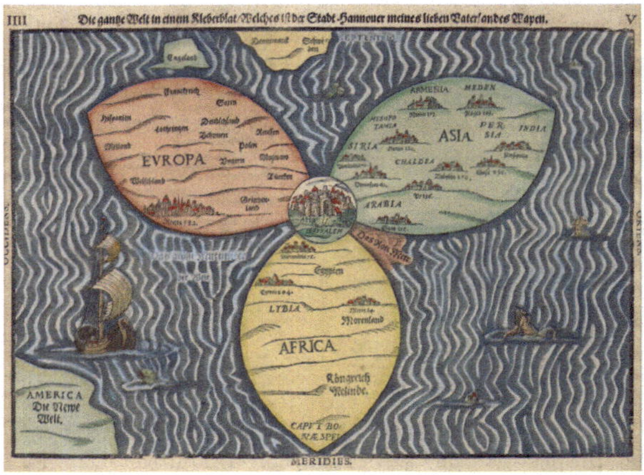

Fig. 10.1 Bünting's 'cloverleaf' world from 1581, showing the relationship between Jerusalem and its surrounding geography. http://upload.wikimedia.org/wikipedia/commons/8/87/1581_Bunting_clover_leaf_map.jpg

sacrosanct', in that we wish to preserve as far as is possible, the correct geographical location of objects both in absolute and thus relative terms. But there are other types of maps that are highly thematic in nature, for which the ambition of abstraction is just as important, if not more so. One such type is 'schematised maps'—maps that we might characterise as bereft of detail, emphasising properties such as causality, connectivity, and flow. In schematised maps, x and y are highly malleable—the movement of features are allowed if it results in simpler forms, that are more readily understood. By malleable, we mean that locational accuracy is 'sacrificed' if it results in a simpler, more consumable conceptual understanding of the relationship between the mapped entities. As such, schematic maps are narrow in their function and task, often stripped bare of their geography in the interests of instantaneous comprehension. These are maps that reflect the ultimate in generalisation—since they are abstractions of reality reduced down to their absolute essential elements. Metro maps are a frequently cited example of schematised maps (Roberts 2012), though in reality there are many other examples (Fig. 10.1).

It is relatively recently that attention has focused on the role of map generalisation techniques in the automated production of schematised maps (Reimer and Fohringer 2010). In trying to automate their design, the challenge is in (1) linking the task requirements (what are the user's needs) to a set of generalisation techniques, and (2) modelling the point at which a satisfactory design has been reached (neither too much nor too little abstraction).

As part of the generalisation process we must take care not to exceed the point at which we lose all notion of the geographic (Fig. 10.2). The limit of distortion that the user will tolerate will depend on factors such as familiarity, perceptibility and recognition (Montello 2002). Modelling constraints in highly abstracted forms is something that generalisation researchers have not had much reason to think

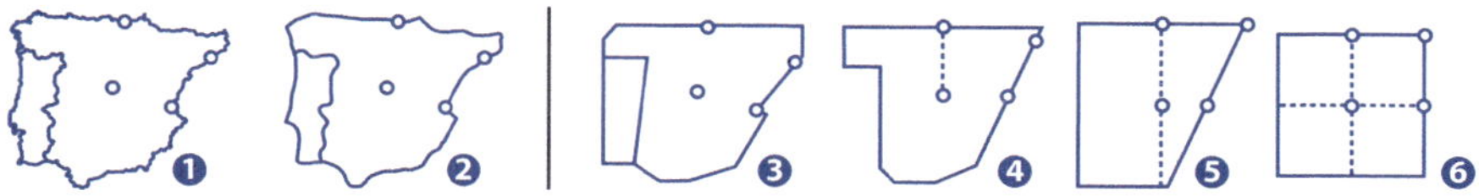

Fig. 10.2 Increasing levels of abstraction in the representation of Spain: modified from Ferras (1986)

about because of their focus on topographic maps where traditionally the degree of application of generalisation techniques has been limited by the need to preserve the underlying geography. What appears to be required is 'pattern aware' map generalisation (Steiniger 2007); by this we mean an environment in which the essential properties (metric, topological, semantic) of a geographic entity can be explicitly measured against increasing levels of abstraction—such that we can model the changing emphasis being given to certain map qualities. These significant changes arising from the process of abstraction amount to what Muller referred to as 'conceptual cusps' (Muller 1991). Typically conceptual cusps are points in the map generalisation process where the dimensionality of objects change. There is something of a paradox in that we intentionally distort and alter the geography in order to transit these conceptual cusps and that we do this precisely to gain these higher synoptic 'states' and different conceptualisations of the geography represented by the entities.

Trying to determine which is the 'best generalised solution' highlights the need to consider context and task, though trying to link the two has proved elusive (Head 1991). Some researchers have tried to link map generalisation to task and context for example in the context of mobile mapping (Nivala and Sarjakoski 2003; Ware et al. 2006), but it remains an area very much in need of research. In the meantime it is the cartographer who must decide how far to push the design of their schematised map.

10.2 A Definition of Schematisation

We define schematisation in cartography as a process that uses cartographic generalisation operators in such a way as to produce maps of a lower graphical complexity compared to maps of the same scale; the process aims to maximise task-adequacy while minimizing non-functional detail. This is in contrast to traditional cartographic generalisation which seeks to maximise functional detail with task-adequacy (in the form of legibility) as a constraint—a significant difference being that schematised maps are much less beholden to their locational constraints (Klippel et al. 2005) whereas traditional maps are *constructed from precise measurements, in which the outlines of features are shown as accurately as is possible within the limitations of scale* (ICA 1973). Topographic, thematic and schematised map can usefully be differentiated by examining the relationship between map detail and intent. Imagine a two axis graph, the y axis ordered from

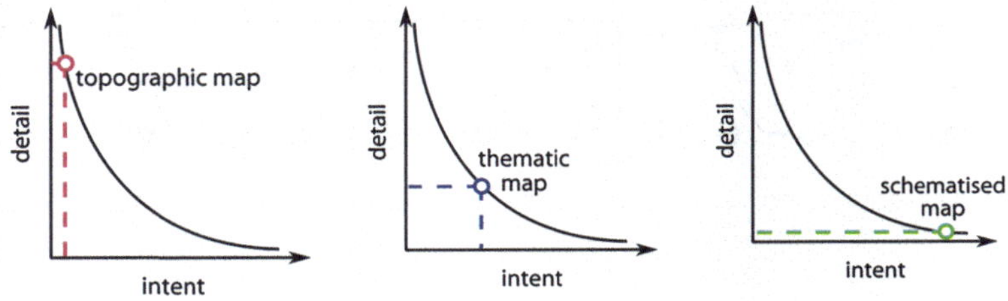

Fig. 10.3 Differentiating topographic, thematic and schematised maps

'sparse' to 'detailed', and an x axis of intent, ranging from the very general to the highly specific. For a two axis graph of geographic detail against intent, we can define a topographic map as something intended for a broad audience, containing high levels of detail. Thematic maps by definition have a specific audience in mind, and often contain topographically correct contextualising information. This is in contrast to schematised maps, that are highly specific in their ambition, and minimalist in their content (Fig. 10.3).

10.3 A Classification of Schematised Maps

We have identified seven types of schematised maps (Table 10.1); these include mental maps and sketch maps, educational, propaganda, mass media info-graphics, and schematic maps. There is some overlap between schematised maps and what Muehlenhaus (2013) calls 'persuasive maps'. While educational, propaganda and mass media schematised maps surely could be categorised as persuasive, not all persuasive maps are actually schematised in the above definition's sense.

A mental map or sketch map reflects a person's perception of that space, based more on their feelings and experiences than a desire to record locational accuracy (Gould and White 2004; Sester and Elias 2007). Educational schematised maps are synthetic thematic depictions (often at the scale of continents or countries), using a specific symbol set on schematised as well as non-schematised and emptied base maps (e.g. Jalta et al. 2006). The objective is a map of sufficient simplicity that it can be readily memorised (Gürtler 1927; Sitte 1996).

Propoganda maps are suggestive maps that are created specifically to reinforce and emphasise how spatial processes and situations mirror ideological preconceptions about the world and how it works. They too require instant comprehension and typically use simple geometries, high contrast colours and striking symbols (Scharfe 1997; Ogrissek 1983). Also included under this heading are 'geopolitical maps' (Ogrissek 1983; Bollmann and Koch 2002).

Of growing importance are mass media maps and infographics. Their nature is decidedly ephemeral with short production cycles and low exposure times. The ambition is highly schematised depictions with low cartographic complexity and a symbology of high iconicity. This is different from Web-mapping which

Table 10.1 Examples of various types of schematised map

Type	Example	Generalisation technique
Mental/sketch	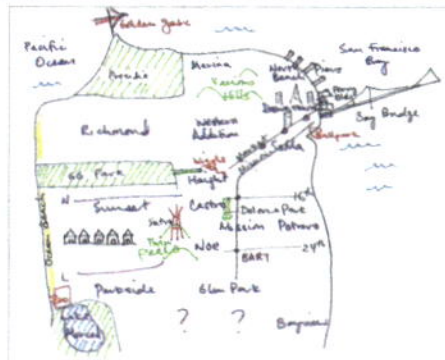	• Elimination • Smoothing
Educational	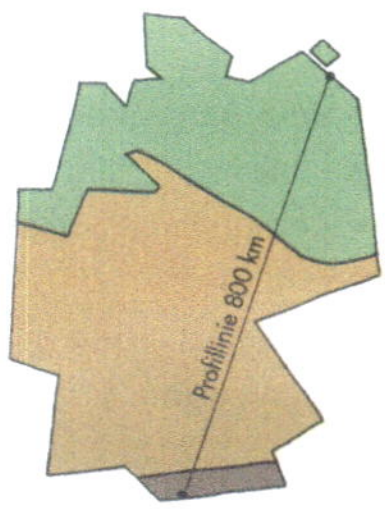	• Amalgamation • Smoothing • Geometric stylisation • Visual stylisation • Caricature - Shape elements - Arrows
Propaganda		• synthesis • geometric stylisation • iconisation • caricature - size - colour
Mass-media	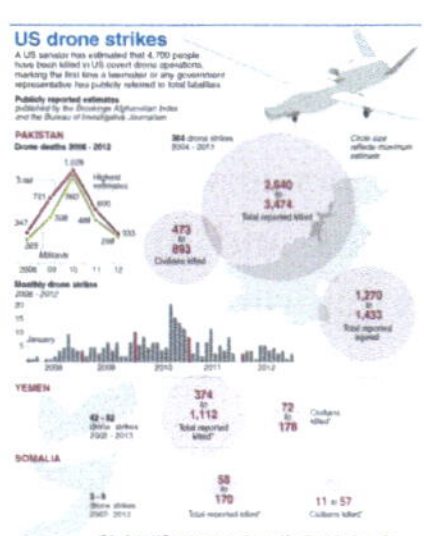	• elimination • smoothing • visual stylisation Strong interweaving with simple statistical diagrams and visual elements of high iconicity i.e. pictorial drawings
Schematic (metro)	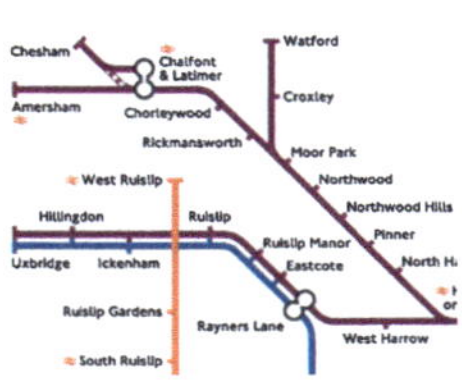	• elimination • geometric stylisation - strict angular restriction - circular arcs (experimental and Madrid metro) • smoothing (w. Bezier curves; experimental)

(continued)

Table 10.1 (continued)

Type	Example	Generalisation technique
Choreматic	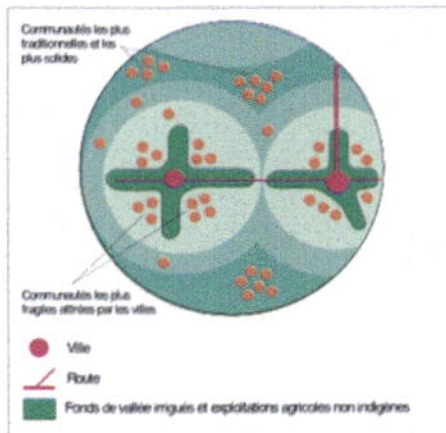	• amalgamation • synthesis • smoothing • geometric stylisation (all) • collapse (all forms) • caricature: - shape - arrows
Geodesign	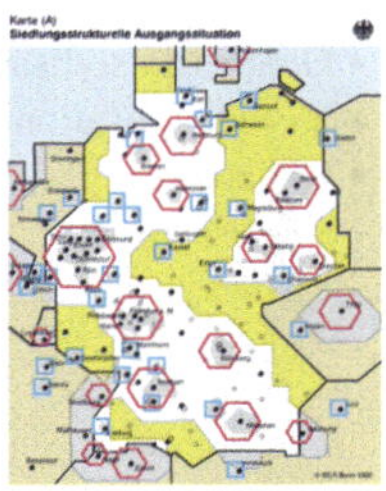	• amalgamation • synthesis • geometric stylisation • collapse: - geometrisation - iconisation

afford means of interaction and are not constrained by limited exposure time. Schematic maps such as topograms or metro maps depict (mostly in linear form) geographical entities with emphasis on the correct topology but locationally distorted in order to achieve greater clarity. The most famous example is the London Tube Map. A review of algorithmic solutions to the automated production of such maps is considered in the first case study (see also Wolff 2007).

Chorematic diagrams are all those maps and map-like diagrams that use one or more graphical embodiments of the choremes as introduced by Roger Brunet (explored in greater detail in the second case study). As a final group, we include 'geodesign maps'. These types of map have evolved in response to the needs of planners. Typically they use an intricate system of hachures and colours on top of subdued topographical information. They are perceived as being especially useful for communicating strategic political scenarios and high level policy guidelines that do not require precise locational information (Rase and Sinz 1993; Stiens 1996; Dühr 2007). In recent years, a homogenisation of EU planning cartography is in evidence in which geodesign and chorematic concepts are both toned down and retained at the same time (e.g. Schmidt-Seiwert et al. 2006; Dühr 2007).

10.4 Methods of Schematisation Production

The forms of schematisation can be readily matched with existing generalisation operators and concepts (Hake et al. 2002; Regnauld and McMaster 2007), the key difference being the intensity with which they are applied. The first step is one of synthesis—the careful selection of pertinent data (model generalisation). The second stage is a reduction in detail achieved through amalgamation, and object simplification. Typically the target contains a low number of classes, requiring strong amalgamation. Often, smaller patches of non-matching objects are reclassified and amalgamated to provide a low number of contiguous objects. Elimination is an extreme case of selection where all non-functional detail is omitted in order to reduce cognitive workload and concentrate on just a few message(s). Line and area objects typically undergo smoothing—even if the entity is angular in form. For example anthropogenic angular borders may be replaced with gentle Bezier-curves in order to emphasise the continuity of form. Additionally schematised maps are sometimes characterised by strict angular restriction and circular arcs (for example Case study I), a process which we call stylisation. Features and objects can also be collapsed (transformed from a higher differentiation schema to one of lower degree), or iconised (the process of transforming recurring shapes into glyph-like depictions). Beyond this conventional set of operators, other transformations include reduced dimensionality (e.g. polygons to lines), and representing linear and polygonal forms in symmetric shapes. Additionally, entities are often represented as caricatures of themselves—where one part of an object is emphasised over another, resulting in an uneven degree of schematisation in the interests of the message's easy recognition.

10.4.1 Governing the Generalisation Process in Schematisation

As in any generalisation process, the challenge is in knowing when to stop (i.e. how do we know that the solution is sufficient?). Essentially, in topographic mapping, the difference in geometric scale is the governing force in generalisation. The ambition in schematised maps is different; here, it is the move from higher to lower cartographic complexity that governs schematisation. Such a statement leads us to the requirement for a measure of map complexity. One way to do this has been proposed by Reimer (2010) using cartographic line frequency:

$$OLLpA = \frac{\sum_{i=1}^{n}\sum_{j=1}^{m} l_{ij}}{ab}$$

where l_{ij} is the line length of object o_{ij} of object type i, and a, b are the long and short side of the bounding box around all considered map objects. The variable i stands for the number of different object types and j for the number of objects of that type i. Structurally this is the cartographic line frequency, related to Bertin's density concept of sign per minimum visible distance, which is why we suggest the name Bertin [Bt] for this unit of measurement. This measure encapsulates various ideas, but in particular is:

- based on empirical legibility research (e.g. Harrie and Stigmar 2010)
- is congruent with information theory and Bertin's density concept
- is more expressive than other methods of measuring content (such as Töpfer 1977).

Fig. 10.4 Localised distortion to reveal detail through graph-based focus region technique (Haunert and Sering 2011)

10.5 Schematisation Metaphors in Interactive Environments

The various design ideas implicit in these highly abstracted forms have influenced the design of interactive environments, especially where space is of a premium. For example, Haunert and Sering (2011) have experimented with graph-based focus region techniques inspired by the shortcomings of fisheye lenses to reveal localised detail in the map (Fig. 10.4). The undistorted space further away from the enlarged detail acts as a contextualising frame, whilst the region immediately surrounding the detailed area 'absorbs' the distortion created by the enlargement.

A different technique for handling large numbers of linear features is 'edge bundling'. This technique groups together large numbers of linear objects that share the same pathway (Holten and Van Wijk 2009). The technique gives the appearance of a less cluttered, neater presentation of data (Fig. 10.5). Edge bundling is a technique often employed in the realm of Information Visualisation. By automatically grouping edges of a graph for a certain stretch of their length we reduce the number of line crossings. Various approaches exist, such as force-directed, a technique generally popular in graph drawing (e.g. Holten and Van Wijk 2009). Edge bundling's nearest equivalent in terms of map generalisation is aggregation. Such an approach lends itself to interactive environments, where individual bundles can be highlighted and inspected. While a plethora of techniques exist that optimise some measures, the links from measure or parameter towards task-efficiency remains unclear.

Schematised maps offer a new perspective on map generalisation, both in extending the range of techniques used, and in modelling the conditions governing

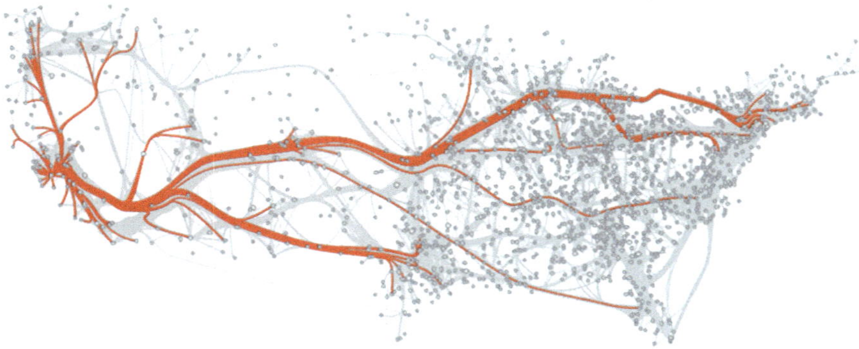

Fig. 10.5 Bundling as a means of showing predominant 'flows' (Holten and van Wijk 2009)

their degree of application. In the three case studies that now follow we explore the level of automation that has been achieved, the success of those designs, and how such maps afford rich and intuitive means of interaction.

10.6 Case Study I: Schematisation of Transportation Networks

William Mackaness

In this case study we review approaches to the automated production of schematised maps that take the form of linear abstractions of networks (Avelar 2002). We define a schematised network map as a diagrammatic representation comprising a set of distinct, but interconnected linear features, which taken together, represent a network. The ambition in design is to make simple the task of route planning and route following across a network and this typically involves making the line work simpler, re-orienting lines to avoid confusion, and making various scale adjustments to minimise cluttered areas. The most iconic form of a schematised metro map is Beck's rendering of the London Underground though this is not the earliest example of this type of thinking. For example Minard's depiction of Napoleon's Russian campaign of 1812 is an excellent example of a schematised map (Tufte 2001). But the longevity of Beck's map is testament to its utility and aesthetic appeal and it remains an inspiration to the manual and automated production of public transport maps the world over (Roberts 2012).

Table 10.2 Design criteria for schematised transportation maps

Criteria	Effect
Minimalist content: An uncluttered map by removal of contextual information *Avoidance of ambiguity*: Clutter should be minimised by localised displacement of linework	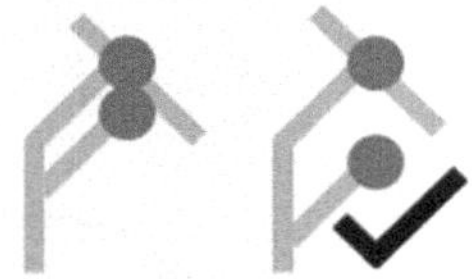
Topological preservation: Any displacement or simplification of the line work must not alter the topology of the network	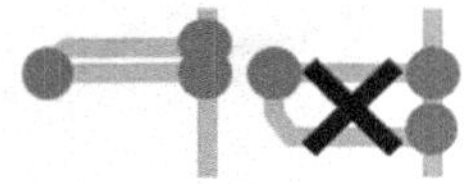
Geographical conservation: The layout should seek to minimise changes in the location of objects so as to retain the 'geography' and relative positioning of features	
Aesthetics of line work: The linework should be comprised of the fewest number of bends, Lines of minimum length, and those emanating from stations should be co-linear	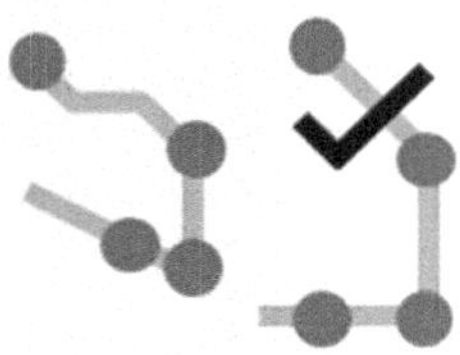
Aesthetics of station information: The lines connecting at a station should emanate evenly from the station, and conform to regular angle spacing of a minimum angular resolution	
Clarity of node information: Stations should not overlap and associated text should be unambiguous	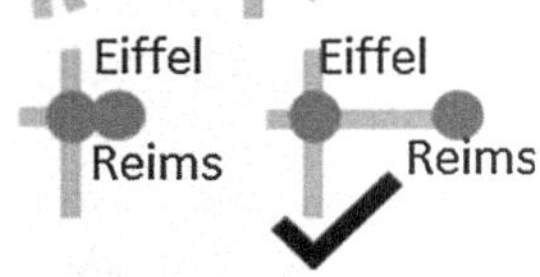

10.6.1 Design Considerations

The design of schematised network maps seeks to minimise contextual information, simplify linear form, emphasise connectivity between distinguishable parts of the network whilst preserving the topology. The ambition is to iron out unnecessary complexity in conveying connectivity between a set of connected stations. From this we can formalise a set of criteria (Table 10.2) in which lines seek to emanate at regular angles from the station, where stations are unambiguously labelled and evenly spaced, and line work has continuity of form.

10.6.2 Combining Context with Schematised Maps

The first criterion listed in Table 10.2 is a selection process in which only the salient information is retained—effectively depriving the reader of contextualising information. This gives focus to the task of route selection and following, the argument is that there is no need to retain a sense of geographical location when travelling a network. The task is often reduced to one of counting the stops, or looking out for the station name. However when the pedestrian emerges into the real world, they need to correctly geo-reference themselves. At this point there is a need for a detailed and 'true' geographic representation of the world as they 'connect back' into the real geography in which the station finds itself. This has given rise to the idea of the 'Spider map'—maps that seek to fuse the highly schematised with maps that are geographically 'correct'. They are comprised a central 'hub' that is geographically correct, together with schematised 'legs'. A level of automation has been achieved (e.g. Mourinho et al. 2011) though this remains an area in need of further development.

10.6.3 Computational Perspectives

From the discussion of these design criteria, we can see that any given solution is largely one of compromise. Though we can readily distil these criteria, automation of a process in which we collectively optimise this design has proved challenging (Avelar and Hurni 2006). This problem of 'orchestration' is common to all generalisation problems. The sorts of questions we need answers to in an automated context are: (1) how do we identify problematic regions? (2) What is the mix of generalisation techniques required (for any given region of the map)? (3) How do we know when we have arrived at a 'good' or even a 'suboptimal but acceptable solution'? What we can say, is that given the importance of preserving topological properties, it is sensible to represent the network in the form of a planar graph, in which the vertices represent stations and edges represent the routes connecting those stations (Gross and Yellen 1999; Di Battista et al. 1999).

If we imagine the geography of the network 'painted' onto a rubber sheet, the process of arriving at a schematised version of the map, is primarily one of differentially stretching and shrinking the rubber sheet until a solution is arrived at. The question, for any given vertex or node, is to decide whether any given 'stretch' (or displacement) either improves or makes worse the design, according to the criteria listed in Table 10.2. We can think of this in terms of a cost function in which we seek to minimise the loss of geography, minimise ambiguity, whilst maximising the aesthetics, clarity and thus comprehension of the map.

10.6.3.1 Hill Climbing

A number of methodologies have been proposed that model this 'compromise in design'—seeking to efficiently arrive at a solution among a large number of candidate designs. Invariably these methodologies involve an initial state for which a 'cost' is calculated (in effect the cost measures how far that initial state is from an acceptable solution). Assuming the cost is greater than a threshold of acceptance, the initial state is modified (typically displacements are applied to a randomly selected vertex), and the cost re-evaluated. Broadly speaking, if the cost is reduced via the process of displacement, then the new state is retained, otherwise it is discarded. This is how the 'hill climbing' algorithm works in which localised optimal solutions can be found. Having started from an initial state (typically a geographically correct map of the location of stations and their connecting segments), the algorithm makes iteratively small displacements in order to arrive at an improved solution (Stott and Rodgers 2004). Hill climbing is good for finding local optimums (in which the solution cannot be improved given the local neighbouring nodes and vertices), but it is not guaranteed to find a more global solution—precisely because its search space is localised.

10.6.3.2 Simulated Annealing

Another popular approach is 'simulated annealing'. Again we start with an initial state, in which candidate improvements are proposed and costed. If the cost is increased (i.e. the solution is made worse) then it is rejected. Simulated annealing seeks to address the risk that such an approach will lead to sub optimal solutions (i.e. where a localised solution is arrived at that is not a good solution overall). In order to avoid such 'locally optimal solutions' we introduce a probability function p. This enables the acceptance of worse solutions as part of the process of finding better overall solutions. The probability function p governs the likelihood of a solution being accepted or rejected, with the value of p increasing as the programme iterates over time:

$$p = e^{-\Delta E/t}$$

The value of p is dependent on two variables: ΔE (the difference in cost between the current state and proposed new states), and t—the temperature. By analogy as a metal cools, it becomes harder to forge (hence the term 'simulated annealing'). So too in this process the value of t is initially set high, and decreases in stages. When t is large, there is a greater probability that higher cost new states are retained, but as t decreases, high cost states tend to be rejected. The acceptance of higher cost solutions in the initial phases enables the solution to escape locally optimal solutions (Agrawala and Stolte 2001; Avelar and Huber 2001; Anand et al. 2007).

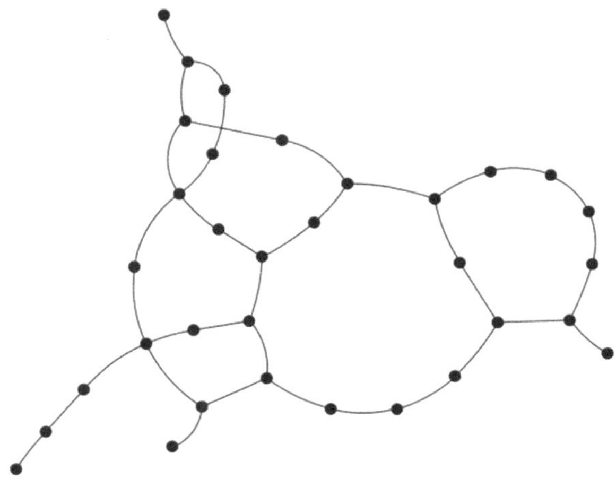

Fig. 10.6 Solution drawn in the style of Mark Lombardi (Chernobelskiy et al. 2011)

10.6.3.3 Force Directed Approach

A force directed approach is one in which optimal solutions are modelled as a set of attraction and repulsion forces (Fruchterman and Reingold 1991; Bertault 2000). An example of an attraction force, is the desire for a station to be as close to its original geographic location as possible. An example of a repulsion force is in repelling an adjacent station node in order to have clear separation between nodes. A distance decay function is used to model the influence of adjacent nodes and vertices, such that closer objects have greater 'force' than those objects that are further away. The total force action on a vertex or node is calculated by summing all the attraction and repulsion forces from which an optimal direction of movement can be assigned (Hong et al. 2006). Such an approach has been used in transportation maps (Fink et al. 2013; Finkel and Tamassia 2005), in the automated creation of cartograms, and in optimised label placement (Ebner et al. 2003).

Force directed algorithms have been extended to include the idea of an even distribution of circular arc edges that give the image a cleaner aesthetic, and better continuity in which paths are easier to distinguish, one from another. The result is a mapping that is similar to the style of the artist, Mark Lombardi, in which the edges are represented as circular arcs and the edges are equally spaced around each vertex (Duncan et al. 2012). Figure 10.6 is one such an example from the work of Chernobelskiy et al. (2011).

10.6.3.4 Mixed Integer Approaches

A fourth approach is based on ideas of 'integer programming'. Integer programming is concerned with optimisation among a large number of variables in which we wish to model binaries and logical requirements (Bertsimas and Weismantel

(a)

(b)

Fig. 10.7 **a** mixed integer approach (Nöllenburg and Wolff 2011a, b); **b** geographic counterpart of the London underground (http://www.theguardian.com/news/datablog/2013/jan/31/circle-line-tube-map-visualised)

2005). Mixed integer programming is an instance of these approaches where some of the variables are not restricted to integers. By relaxing the integer constraint and enlargening the set of possible solutions, solvers can be used and greater modelling flexibility is achieved. Nöllenburg and Wolff (2011a, b) were able to successfully generate high quality solutions using mixed-integer programming techniques, applied to both the network and the labelling of stations (Fig. 10.7a).

10.6.4 Refinements and Evaluation

Various authors have sought to improve the quality of these automated solutions. For example Fink et al. (2013) used Bézier curves to emphasise continuity of lines using a force directed approach. The ambition was similar to that of Lombardi maps in which we seek to reduce the number of inflection points in the line. Dwyer et al. (2008) refined the process of fitting station nodes to network lines, using least squares regression. By partitioning the lines into segments, they applied least squares regression to a localised set of nodes to find the best fit line. The best fit line is selected from octolinear candidates. Their approach sought to better reflect the overall geographic pattern and distribution of the network, though it did not test for changes in the topology of the network. The work of Stott et al. (2011) proposed a multicriteria optimisation approach in combination with clustering techniques in order to avoid the 'local minima' solutions associated with hill climbing algorithms. This was done by detecting local minima using clustering techniques that identified groups of stations, enabling them to be modelled as a group and thus minimising the loss of relative positioning among the group during displacement.

In combination, these approaches are generating solutions that are coming closer and closer to the hand drawn solutions with which we are so familiar (Roberts et al. 2013). Our judgement of the quality of automated solutions is, of course, influenced by that which is familiar, which begins to beg the question: 'Is there an automated design that is better than Beck'? In their work, Stott et al. (2011) compared their automated solution against geographically correct, and manually generated solutions. They found that the automated solution was preferred in the majority of cases. Nöllenburg and Wolff (2011a, b) compared their output against the automated outputs of Hong et al. (2006) and Stott et al. (2011), using three case studies. Of the three, their solution was preferred, and was deemed by experts to surpass the official version in respect of some of the design criteria.

10.6.5 Conclusions

Manually produced solutions can have a beauty and an aesthetic all of their own. For automated solutions, the challenge is in modelling the balance between displacement, line simplification and smoothing. Despite the complexity in their design, the quality of automated solutions is such that it is becoming increasingly hard to differentiate hand drawn solutions from automated ones. The advantage of automated solutions over hand drawn solutions is in their speed, and capacity to be applied to other types of 'network' data. For example schematisation algorithms have been applied to generalisation of flight paths (Hurter et al. 2010), cable plans (Lauther and Stübinger 2001), and road networks (Cabello et al. 2001; Haunert and Sering 2011; Li and Dong 2010; Casakin et al. 2000). Furthermore, all of these uncluttered forms of representation lend themselves to display on mobiles (with their typically limited screen size) (Wang and Chi 2011). More broadly, developments in automated solutions to schematised transport maps demonstrates the importance and relevance of graph drawing and optimisation algorithms to the field of map generalisation.

10.7 Case Study II: Chorematic Diagrams

Andreas Reimer

Chorematic diagrams can be understood as schematised thematic maps, which follow a certain line of geographic thinking. This school of thought is called "chorematique" and was invented by the French geographer and cartographer Roger Brunet (1980). His work has had a lasting impact on European spatial planning, school education and French geography (Reimer 2010). Their striking clarity (Fig. 10.8), coupled with their underlying materialistic-rationalistic generation philosophy make them attractive candidates for automation (Cheylan et al. 1997; Tainz and Heitmann 2005) and have inspired the work of others (Klippel 2004; Laurini et al. 2009).

Their envisaged use is in (1) overview, and interaction with, geographic data (Cheylan et al. 1997; Laurini et al. 2009; Reimer and Meulemans 2011), (2) during the consensus building phase in time-critical situations, such as disaster management (Reimer 2010), and (3) as a way of conveying spatial uncertainty—avoiding the illusion of accuracy that often arises from digital products. Because of the rich symbology and scope of themes covered, chorematic diagrams are an attractive means of generating knowledge about the schematisation process.

10.7.1 Knowledge Engineering

Chorematic diagrams are much more varied in their design as compared with other types of schematised maps. They typically employ network-forming lines and simple point symbols as well as dynamic encodings (e.g. the use of arrows). A variety of generalisation techniques are used. In the model stage, just a few entities are selected—often chorematic maps are monothematic. Entities are often simplified by collapsing the base shape of the region into a simple form, such as a

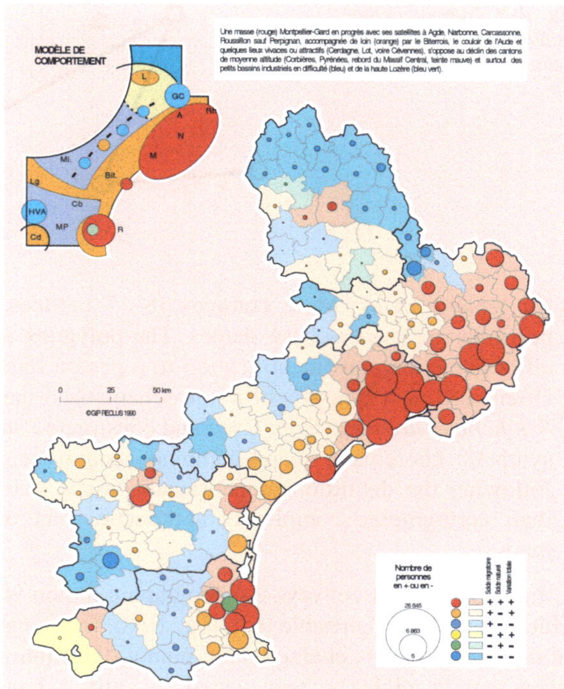

Fig. 10.8 Hand drawn chorematic diagram (*top left*) derived from the more detailed

rectangle, a circle or a hexagon (e.g. Table 10.1-geodesign). Data are re-classified, and regions amalgamated.

In general, chorematic diagrams do not show quantitative information and succinctly always rely on model generalisation techniques to deliver the reclassification of data. The most striking visual difference to non-schematised thematic maps is the absence of a proper base-map layer in many instances. The base map information is schematised (via collapse, smoothing, geometric stylisation or caricature), thereby forcing similar schematisation on the thematic layer, effectively fusing both. These symbolisation issues alone warrant a differentiation into different categories or classes of chorematic diagrams:

- *Symmetric models*: the base shape of the region of interest is collapsed into a basically undirected simple form, such as a rectangle, a circle or a hexagon (e.g. Fig. 10.2). Very sparse thematic information, often monothematic.
- *Asymmetric models*: the base shape is smoothed, geometrically stylised and possibly caricatured into a low-complexity model (5–7 vertices or control points). Very sparse thematic information, often monothematic.
- *Synthetic models (symmetric/asymmetric)*: Usually the synopsis of several preceding monothematic models, geometrically they tend to mirror the preceding models leading up to the synthetic one. They contain synthesised resultant layers that are often emphasised.

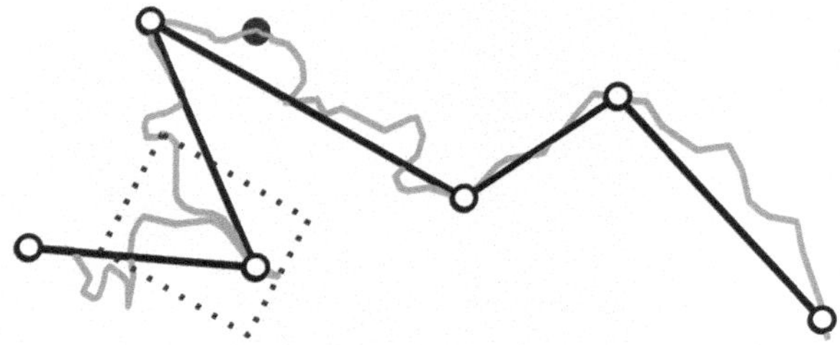

Fig. 10.9 Simplification whilst preserving characteristic points

- *Chorematic maps*: More complex (8–17 vertices or control points) irregular polygons are used as base shapes. The polygons are depicted in a manner that allows unambiguous identification to a person familiar with the original shape. Often used as a synopsis of a non-schematised thematic map.
- *Synthetic chorematic map*: Used and constructed analogous to synthetic models, synthetic chorematic maps combine geographic phenomena on a base map following the definition of chorematic maps stated above. Their content and their cartographic complexity approaches that of non-schematised thematic maps.

Information that conveys the geographic region is simplified up to the point at which it is still recognisable (e.g. Fig. 10.8). Ensuring its recognition requires us to model and retain its characteristic points (the minimum-ε Problem), particularly given that it might be represented by only a handful of points (Reimer and Meulemans 2011). Succinctly, geographic recognisability cannot always be reached without ancilliary data that models the pertinent geographic context. Research to date has concentrated on territorial outlines for countries, continents and provinces and their subdivisions. Empirical observation suggests retention of significant estuaries (defined by identifying the most land-inward part of an estuary as a characteristic point (e.g. Fig. 10.9)).

Vertices that are adjacent to administrative boundaries of at least two entities and a sea area bear special topological and conceptual relevance, albeit they might not be geometrically important within the polygon itself (for example the red dot in Fig. 10.9). This requires us to explicitly model points that serve more than one entity (i.e. having degree greater than two). In other words we cannot treat the coastal boundary in isolation of other features connected to it, including objects close to it such as towns and cities.

10.7.2 Modelling Generalisation According to Complexity

Any attempt at automated schematisation needs to be governed by the pairing of desired output complexity and generalisation operators. Generating the desired output complexity/design rule pairing from an input task can be interpreted as a function of the form:

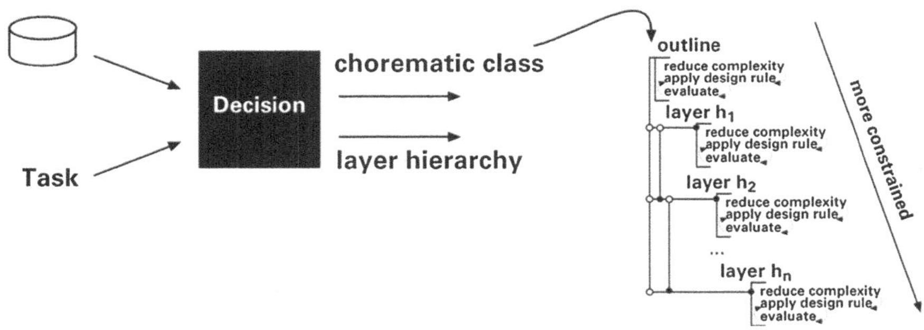

Fig. 10.10 Modelling the design of chorematic maps

$$f(t) = \sum_{i=1}^{n}(W_i s_i)$$

with $s \in S$, $W_i \in [0, 1]$ and $W_i + W_j = 1$, where W_i denotes the weight attributed to s_i, the chorematic class from the set of all classes S from an input task t. Currently there are available automated methods that model this function, so we treat it as a black box, i.e. we presuppose this decision has been made a priori. The degree to which we must apply these techniques depends largely on the complexity of the underlying data and the number of objects we wish to visualise. The classification and complexity measures that result from investigations into chorematic diagrams have the advantage of allowing us to transform the black-box problem into a matching problem.

Let T be the set of tasks and S be the set of chorematic diagram classes. Then the matching $m \colon T \times S \rightarrow \mathbb{B}$ is a Boolean function that decides whether a chorematic class in S matches the given task in T. The set $M_S(t) \colon T \rightarrow \mathcal{P}(S)$ gives for any task t in T the subset of S that matches t. That is,

$$M_S(t) = \{s | s \in S \wedge m(t, s)\}$$

Such a matching from task to the minimally allowable chorematic diagram classes expressed as $\sup\{f(s) \colon s \in S\}$ seems more tractable than a possibly AI-complete black-box function of the general case. Instead of the complete function $f(t)$, we only need to know the minimum required information for a given task to select from the known and well-described set of chorematic diagram classes.

Assuming we are given a task-derived chorematic diagram class and a layer hierarchy derived from datasets associated with the task, we can envisage the compilation stages as: the reduction in complexity of each layer, the application of various design rules, and their evaluation (Fig. 10.10).

Depending on the outcome of the evaluation, the preceding step needs to be repeated or adjusted. The layers are hierarchically inserted into the first layer (always the territorial outline) via a “mapping” utilising anchor points that preserve desirable topological relations (Saalfeld 2001). This hierarchical model mimics a process often used in manual chorematic diagram creation, in which layers are created and fused to the schematised basemap with different design principles applied to different layers.

The step of complexity reduction is crucial to constraining the design. As with schematisation in general the most common forms of complexity reduction are similar to strong reduction of detail (e.g. amalgamation, elimination and synthesis). The respective upper bound is the summed complexity expressed in OLLpA. Existing chorematic diagrams have OLLpA values in the 0.2–0.4 Bt range, depending on type (which compares with non-schematised thematic or topographic maps which typically have a value of 1 and above or transport network schematic maps with values below 0.1 Bt).

Such a strategy allows a divide and conquer approach to these diverse and varied schematic components. Compositionally we might have a polygonal outline showing a high edge parallelism with a subdivision internally smoothed with cubic Bezier curves and a simplified geometrically stylised network on top. Once a design rule is correctly identified and a measure suitable for evaluation has been found the step can be transformed into an optimisation problem. To date, this has been achieved for:

- smoothed polygonal outlines using simulated annealing and edge parallelism as evaluation measure (Reimer and Meulemans 2011)
- smoothed subdivisions using cubic Bezier curves for benign input (Reimer and Volk 2012)
- smoothed outlines and subdivisions using cubic Bezier curves (van Goethem et al. 2013) (Fig. 10.11b)
- geometrically stylised outlines and subdivisions using circular arcs (van Goethem et al. 2013) (Fig. 10.11c).

Other techniques can then be applied. For example network schematisation (Case study I) can be directly applied to achieve geometric network stylisation after the target complexity and anchor points have been completed. Not all possible design rules have yet been successfully replicated however and this remains a rich area of research.

10.7.3 Future Challenges

Chorematic diagrams make use of a large number of schematisation forms. As with many map generalisation tasks, the challenge is in integration of design rules, and the identification of stopping conditions in their application. The encyclopedic as well as procedural and modelling knowledge gained via investigating chorematic diagrams is considered beneficial for similar tasks in other schematisation efforts. The biggest open problem remains one of matching the task requirement to the selection of entities, and their subsequent analysis sufficient to determine the mix and degree of application of generalisation methods. This problem is common to automation in thematic cartography, more broadly.

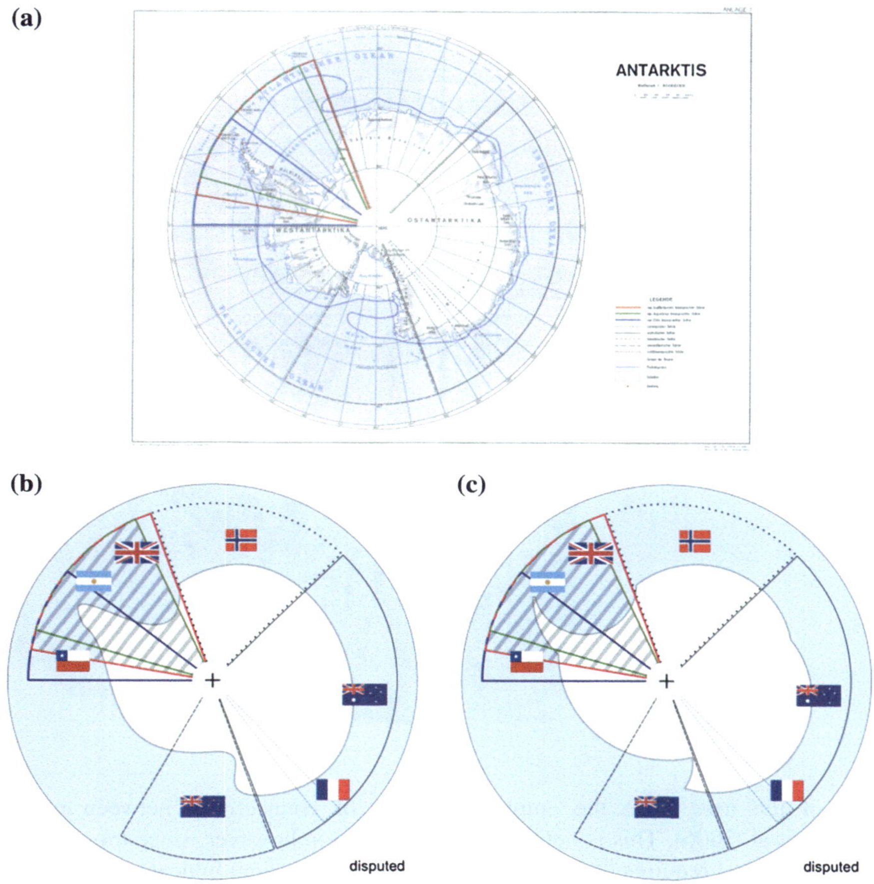

Fig. 10.11 Different ways of representing disputed land claims in Antarctica (van Goethem et al. 2013)

10.8 Case Study III: Schematised Maps for Multi Modal Travel

William Mackaness

Many different solutions exist that seek to help us navigate across cities and utilise public transport. Often schematised maps are part of that solution (e.g. Case study I; Hochmair 2009; Merrick and Gudmundsson 2007). Anything deemed not relevant to the task is omitted, and the geometry simplified to reduce the cognitive

Fig. 10.12 Connecting the detailed with the more synoptic (M. C. Escher)

load of the user since the emphasis is one of connectivity between points (Anand et al. 2006). This is not the complete solution however, since travel using public transport requires the user to operate at different geographical scales. When in a station they require maps that direct them along concourses, through ticket barriers, to a particular platform, and bus. But once on the move, synoptic information that shows time or distance is all that is required to monitor progress. However, once at the destination, more detailed information is again required that enables the pedestrian to exit the bus station, and continue their journey. In summary, the task requires 'connected' information but at varying levels of detail. For Fig. 10.7a, for example, what is additionally required are more detailed elements that would enable the pedestrian to continue their journey once they have emerged from the metro. We argue that a more complete solution would be one that delivers both the 'architectural' scale (Quigley 2001), connected to the more synoptic (Hile et al. 2009). Artistically this idea is captured in the drawings of Escher (Fig. 10.12). It is not simply the casting of a fisheye lens over a particular entity, but more of a 'semantic zoom' in which space is created in order to reveal the detail of different, albeit connected, entities.

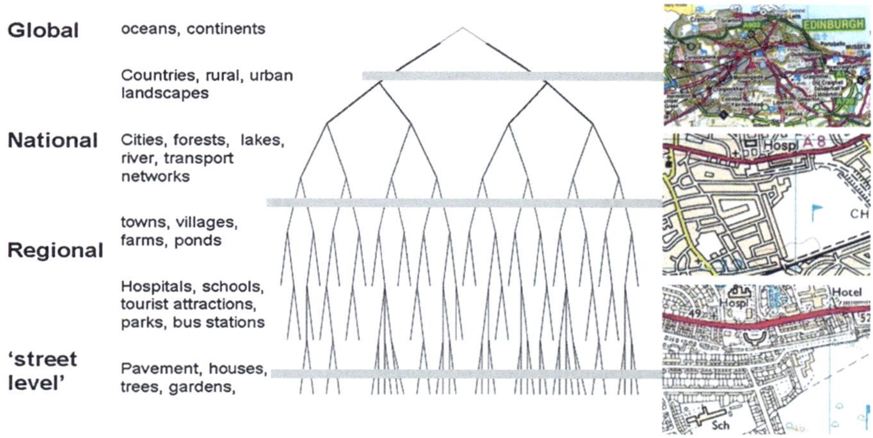

Fig. 10.13 The most granular, constituted by the most detailed

10.8.1 Hierarchical Graphs and Event Horizons

The idea that all things geographical are in some way connected, can be schematically represented as a hierarchical "compound" graph. A hierarchical compound graph consists of an 'inclusion tree' that makes explicit a set of partonomic and taxonomic relationships that exist amongst a set of entities. Such a tree reflects a multi scaled perspective, that each entity is constituted by the finer scale entities beneath it in the tree (Fig. 10.13).

Metaphorically we can use this construct as a means of selecting entities for schematic or topographic display at a particular scale, by taking a slice through the tree at a specific level. Quigley (2001) describes these slices as an 'event horizon'—the formal definition of an event horizon being a manifold through a multi hierarchical graph that represents multiple inter-related hierarchical compound graphs. Any given event horizon can be used as a means of selecting entities for a specific map task—thus creating a 'visual précis'—maps that represents an abridged subset of the graph. In Fig. 10.13 the event horizons are straight horizontal lines but Quigley argued that the event horizon could be 'deformed', so as to specifically exclude some entities, AND include other entities, which were themselves combined with entities from different scales (or levels in the tree). In this way the visual précis could encompass both a local- and a global-context (Fig. 10.14).

This event horizon gives us synoptic, global content, AND the detailed, local content. In the example of the user commuting by public transport, the detailed more local content is needed at the start and the destination points, where the pedestrian requires architectural detail in order to navigate back to the outside world.

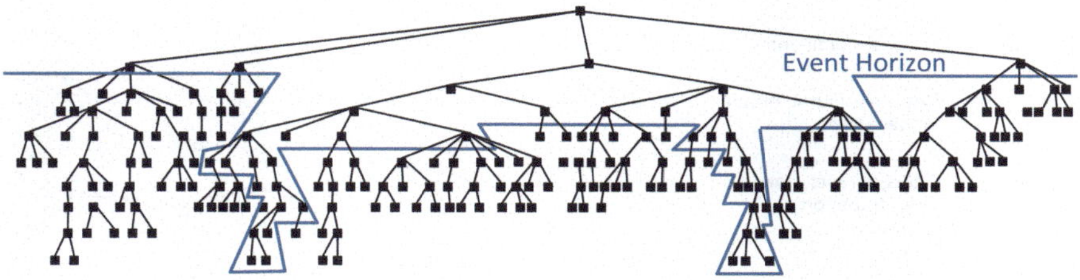

Fig. 10.14 The event horizon cuts across the hierarchical structure, essentially enabling the mixing of geographic information across different 'conceptual cusps'

Such hierarchical structures can be formulated in many ways—by spatial proximity, or partonomically—whichever fits best with our conceptual views of the world (Benslimane et al. 2003). Traditionally GIS and cartographic databases are not constituted in these partonomic forms, though a number of techniques have been developed to automatically derive such functional perspectives (Chaudhry et al. 2009).

10.8.2 Implementation

The efficacy of these ideas were explored through implementation (Mackaness et al. 2011) by bringing together various datasets combined with open source software (uDIG http://udig.refractions.net/). The data came from Ordnance Survey and Open Street Map and covered the city of Edinburgh. A journey planning algorithm was implemented on top of bus route data.

The graph database technology neo4j (http://neo4j.org/) was used to facilitate the creation of a pedestrian/public transport network. Neo4j is a flexible graph network database, supporting storage and manipulation of entities, properties and multiple relationships between entities. It was chosen because it is ideal for making explicit the relationship between different representations of concepts. In the context of journey planning, it enables us to chain together the components that make up the journey—the architectural and street level detail, with the more general and synoptic. It also enables us to model event horizons to extract those entities and their connections specific to the given task.

The link between task and both cartographic selection and symbolisation was formalised as a set of rules using drools (http://www.jboss.org/drools). The rule engine governed (1) selection of entities together with the associated contextualising information; (2) the display of text associated with any change in the task; (3) varying levels of detail within differing buffered regions; (4) colour and symbolisation. Figure 10.15 shows how the connected components enabled the

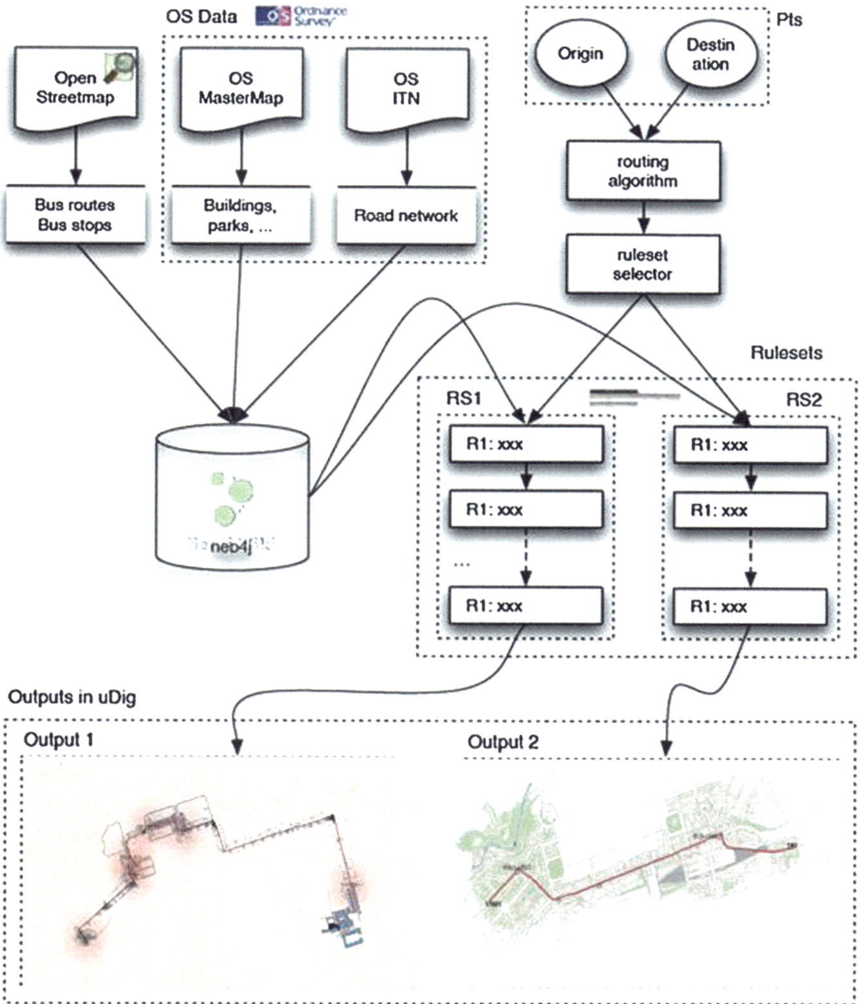

Fig. 10.15 Combining neo4j, route planner, and drools to select and symbolise differing map content

generation of different output. Figure 10.16 shows two example outputs generated using rules governing the selection of navigational cues along the route, and the level of detail associated with places of interchange between different modes of transport.

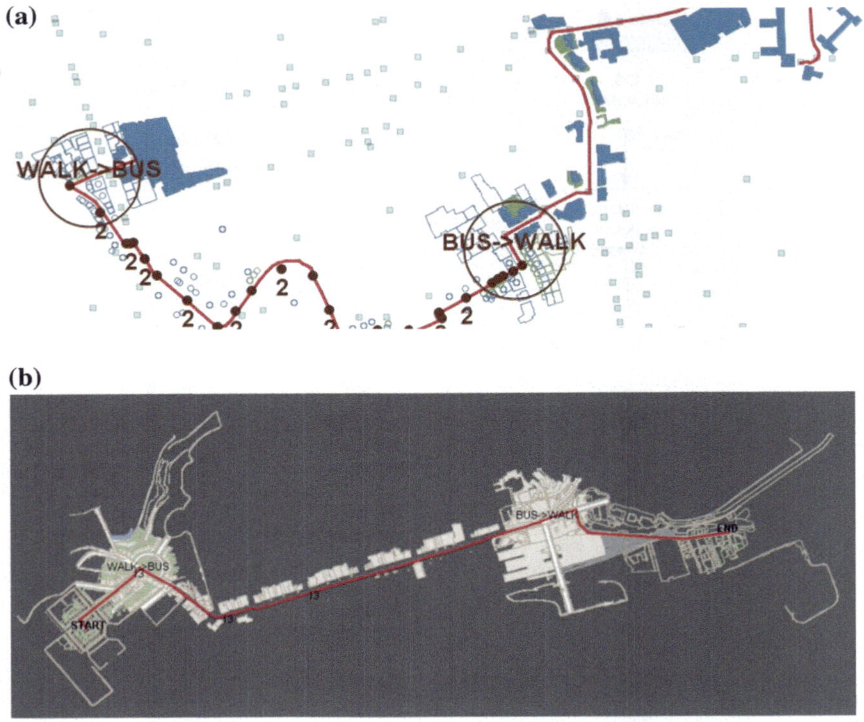

Fig. 10.16 **a** Example output, for part of Edinburgh, showing navigational cues associated with the walking task and more schematised information along the bus route; **b** alternate output from the model providing greater levels of detail at points of interchange, and a few nearby navigational cues along the route of the bus

10.8.3 Conclusions

These ideas and their implementation have focused on structures to support the selection of map content according to sub tasks within a multi modal journey. The complimentary nature of visualising geographic information in topographic and schematic form (e.g. Reimer and Meulemans 2011) has been demonstrated. The ambition of future work is to (1) model and populate multi hierarchical graphs, such that event horizons can be used to select content according to differing tasks; (2) to further develop existing visualisation techniques to support automated topographic and schematic mapping, and (3) build on existing techniques for smoothly transitioning between these complimentary but very different styles of visualisation.

10.9 Conclusions and Challenges

Schematised maps require the 'strong' application of map generalisation techniques. Different from the challenges of generalising topographic maps, their goal is one of minimum effort in comprehension, seeking to support very specific tasks or messages. They are maps that are far less constrained by the need to correctly convey locational information. In some cases their ambition is to transcend 'conceptual cusps' as they seek to represent information at much smaller scales. Such mapping points to the need for modelling complexity, and pattern aware generalisation techniques, capable of understanding the behaviour of the entity in order to determine how best to represent it. This is a topic for which many research questions remain.

References

Agrawala M, Stolte C (2001) Rendering effective route maps: improving usability through generalization. In: Fiume E (eds) Proceedings of computer graphics ACM SIGGRAPH '01, ACM Press, pp 241–249

Anand S, Ware JM, Taylor GE (2006) Generalization and schematization of large scale digital geographic datasets for Mobile GIS applications. In: Dynamic and mobile GIS: investigating change in space and time, Taylor and Francis Ltd, London

Anand S, Avelar S, Ware J, Jackson J (2007) Automated schematic map production using simulated annealing and gradient descent approaches. In: Proceedings of the 15th annual GIS research UK conference (GISRUK)

Avelar S (2002) Schematic maps on demand: design, modeling and visualization. Unpublished PhD Dissertation, Swiss Federal Institute of Technology

Avelar S, Huber R (2001) Modeling a public transport network for generation of schematic maps and location queries. In: Proceedings of the 20th international cartographic conference

Avelar S, Hurni L (2006) On the design of schematic transport maps. Cartographica 41(3):217–228

Benslimane D, Vangenot C, Roussey C, Arara A (2003) Multi representation in ontologies. In: Proceedings of the 7th East-European conference on advances in databases and information systems (ADBIS'2003)

Bertault F (2000) A force-directed algorithm that preserves edge crossing properties. Inf Process Lett 74(1–2):7–13

Bertsimas D, Weismantel R (2005) Optimization over integers. Dynamic Ideas

Bollmann J, Koch W (2002) Lexikon der Kartographie und Geomatik, vol 2. Spektrum Akademischer Verlag, Heidelberg

Brunet R (1980) La composition des mode'les dans l'analyse spatiale. L'Espace Geographique 8:253–265. Available in English as: Building models for spatial analysis. In: Two decades of l'Espace Geographique, Montpellier, 1993

Cabello S, de Berg M, van Dijk S, van Kreveld M, Strijk T (2001) Schematization of road networks. In: Proceedings of the 17th annual symposium on computational geometry, Medford

Casakin H, Barkowsky T, Klippel A, Freksa C (2000) Schematic maps as way finding aids. Lecture notes in artificial intelligence—spatial cognition II—integrating abstract theories, Empirical Studies, Formal Methods, and Practical Applications, Springer, Berlin, pp 54–71

Chaudhry OZ, Mackaness WA, Regnauld N (2009) A functional perspective on map generalisation. Comput Environ Urban Syst 33(5):349–362

Chernobelskiy R, Cunningham K, Kobourov SG, Trott L (2011) Force-directed lombardi-style graph drawing. Graph Drawing 2011:320–331

Cheylan J-P, Libourel T, Mende C (1997) Graphical modelling for geographic explanation. In: Spatial information theory: a theoretical basis for GIS. Springer, Berlin, pp 473–483

Dong W, Guo Q, Liu J (2008) Schematic road network map progressive generalization based on multiple constraints. Geo-spatial Inf Sci 11(3):215–220

Di Battista G, Eades P, Tamassia R, Tollis IG (1999) Graph drawing. Prentice Hall, Upper Saddle River

Dühr S (2007) The visual language of spatial planning: exploring cartographic representations for spatial planning in Europe. Routledge, London

Duncan CA, Eppstein D, Goodrich MT, Kobourov SG, Nöllenburg M (2012) Lombardi drawings of graphs. J Graph Algorithms Appl 16(1):85–108

Dwyer T, Hurst N, Merrick D (2008) A fast and simple heuristic for metro map path simplification. In: Advances in visual computing. Springer, Berlin Heidelberg, pp 22–30

Ebner D, Klau GW, Weiskircher R (2003) Force-based label number maximization. Technical Report TR-186-1-03-02, Institut für Computergraphik und Algorithmen, Technische Universität Wien

Ferras R (1986) Atlas Reclus: Espana/Espagne/Spain. Fayard/RECLUS

Fink M, Haverkort H, Nöllenburg M, Maxwell R, Schuhmann M, Alexander J, Wolff A (2013) Drawing metro maps using Bézier curves. In: Didimo W, Patrignani M (eds), Conference Paper : Graph Drawing (20th International Symposium, GD 2012, Redmond WA, USA, 19–21 Sept 2012. Revised Selected Papers), Lecture Notes in Computer Science, 7704. Springer, Berlin, pp 463–474

Finkel B, Tamassia R (2005) Curvilinear graph drawing using the force-directed method graph drawing. Lect Notes Comput Sci 3383:448–453

Fruchterman TMJ, Reingold EM (1991) Graph Drawing by Force-Directed Placement. Software: Practice and Experience 21(11)

Gould P, White R (2004) Mental maps. Routledge

Gross J, Yellen J (1999) Graph theory and its applications. CRC Press

Gürtler A (1927) Das Zeichnen im erdkundlichen Unterricht. Zweites Heft—Europa (ohne Deutschland), Verlag Ernst Wunderlich, Leipzig

Hake G, Grünreich D, Meng L (2002) Kartographie Visualisierung raum-zeitlicher Informationen. Walter de Gruyter, Berlin, New York

Harrie L, Stigmar H (2010) An evaluation of measures for quantifying map information. ISPRS J Photogrammetry Remote Sens 65:266–274

Haunert J-H, Sering L (2011) Drawing road networks with focus regions. IEEE Trans Visual Comput Graphics 17(12):2555–2562

Head CG (1991) The map as natural language or semiotic system: review and comment. In: Mark DM, Frank AU (eds) Cognitive and linguistic aspects of geographic space. NATO ASI series D: Behavioural and social sciences, vol 63. Kluwer, Dodrecht, pp 237–262

Hile H, Grzeszczuk R, Liu A, Vedantham R, Košecka J, Borriello G (2009) Landmark-based pedestrian navigation with enhanced spatial reasoning. In: Proceedings of the 7th international conference on pervasive computing (Pervasive '09), Springer, Berlin Heidelberg, pp 59–76

Hochmair H (2009) The influence of map design on route choice from public transportation maps in urban areas. Cartographic J 46(3):242–256

Holten D, Van Wijk J (2009) Force-directed edge bundling for graph visualization. Eurographics/ IEEE-VGTC Symposium on Visualization 28(3)

Hong S-H, Merrick D, do Nascimento HAD (2006) Automatic visualisation of metro maps. J Visual Lang Comput 17(3):203–224

Hurter C, Serrurier M, Alonzo R, Vinot JL, Tabart G (2010) An automatic generation of schematic maps to display flight routes for air traffic controllers: structure and color optimisation. In: Proceedings of the international conference on advanced visual interfaces

ICA (1973) Multilingual dictionary of technical terms in cartography by International Cartographic Association, Commission II. Wiesbaden

Jalta J, Joly JF, Reineri R (2006) Les croquis et les schémas du Bac Géographie Terminales ES, L, S: Fichier Méthodes et exercices. Magnard, Paris

Klippel A (2004) Wayfinding choremes: conceptualizing wayfinding and route direction elements. PhD thesis, Universität Bremen

Klippel A, Richter K-F, Barkowsky T, Freksa C (2005) The cognitive reality of schematic maps. In: Methods and implementations of map-based mobile services—theories. Springer, Berlin, pp 57–74

Laurini R, Sebillo M, Vitiello G, Sol-Martinez D, Milleret-Raffort F (2009) Computer-generated visual summaries of spatial databases: chorems or not chorems? SAPIENS 2:1–8

Lauther U, Stübinger A (2001) Generating schematic cable plans using spring embedder methods. In: Proceedings of symposium graph drawing

Li Z, Dong W (2010) A stroke-based method for automated generation of schematic network maps. Int J Geogr Inf Sci 24(11):1631–1647

Mackaness WA, Quigley A, Tanasescu V (2011) Hierarchical structures in support of dynamic presentation of multi resolution geographic information for navigation in urban environments. GISRUK, 2011

Merrick D, Gudmundsson J (2007) Path simplification for metro map layout. In: Kaufmann M, Wagner D (eds) Graph drawing 2006 LNCS, vol 4372. Springer, Heidelberg, pp 258–269

Montello DR (2002) Cognitive map-design research in the twentieth century: theoretical and empirical approaches. Cartography GIS 29(3):283–304

Mourinho J, Galvao, Falcao T, Cunha J (2011) Spider maps for location-based services improvement exploring services science. In: Snene M, Ralyte J, Morin J-H (eds) 2nd International conference IESS, Springer, Berlin Heidelberg, pp 16–29

Muehlenhaus I (2013) The design and composition of persuasive maps. Cartography Geogr Inf Sci 40(5):401–414

Muller JC (1991) Generalization of spatial databases. In: Maguire DJ, Goodchild MF, Rhind D (eds) Geographical information systems, vol 1

Nivala A-M, Sarjakoski LT (2003) Need for context-aware topographic maps in mobile devices. In: Proceedings of the 9th scandinavian research conference on geographic information science ScanGISí 2003, Espoo, 2003. Available via DIALOG. http://www.scangis.org/scangis2003/papers/22.pdf

Nöllenburg M, Wolff A (2011a) Drawing and labeling high-quality metro maps by mixed-integer programming. IEEE Trans Visual Comput Graphics 17(5):626–641

Nöllenburg M, Wolff A (2011b) Appendix of drawing and labeling high-quality metro maps by mixed-integer programming. IEEE Trans Vis Comput Graphics

Ogrissek R (ed) (1983) ABC Kartenkunde. Brockhaus, Leipzig

Quigley AJ (2001) Large scale relational information visualization, clustering, and abstraction. PhD Thesis, University of St Andrews. Available via DIALOG. www.cs.st-andrews.ac.uk/~aquigley/aquigley-thesis-mar-02.pdf

Rase WD, Sinz M (1993) Kartographische Visualisierung von Planungskonzepten. Kartographische Nachrichten 43:139–145

Regnauld N, McMaster RB (2007) A synoptic view of generalisation operators. In: Generalisation of geographic information: cartographic modelling and applications. Elsevier

Reimer A (2010) Understanding chorematic diagrams: towards a taxonomy. Cartographic J 47(4):330–350

Reimer A, Fohringer J (2010) Towards constraint formulation for chorematic schematisation tasks. In: Geographic Information on Demand: 13th Workshop of the ICA commission on Generalisation and Multiple Representation, Zurich

Reimer A, Meulemans W (2011) Parallelity in chorematic territorial outlines. In: Proceedings of the 14th ICA workshop on generalisation and multiple representation, jointly organised with ISPRS Commission II/2 Working group on Multiscale Representation of Spatial Data, Paris, France

Reimer A, Volk C (2012) An approach for using cubic Bézier curves for schematizations of categorical maps. In: 15th Workshop of the ICA commission on generalisation and multiple representation, Istanbul

Roberts MJ (2012) Underground maps unravelled: explorations in information design. Author publication, Wivenhoe, Essex

Roberts MJ, Newton EJ, Lagattolla FD, Hughes S, Hasler MC (2013) Objective versus subjective measures of Paris Metro map usability: investigating traditional octolinear versus all-curves schematic maps. Int J Hum Comput Stud 71:363–386

Saalfeld A (2001) Area-preserving Piecewise Affine Mappings. In: Proceedings of the 17th annual symposium on computational geometry SCG 01

Scharfe W (1997) The development of consciousness in cartography of mass media maps. In: Scharfe W (eds) Proceedings of international conference on mass media maps: approaches, results, social impact (Berliner Geowissenschaftliche Abhandlungen, Reihe C, Kartographie, Band 16), Freie Universität Berlin, Berlin, p 188

Schmidt-Seiwert V, Porsche L, Schön P (eds) (2006) ESPON ATLAS Mapping the structure of the European territory. Federal Office for Building and Regional Planning, Berlin

Sester M, Elias B (2007) Relevance of generalisation to the extraction and communication of wayfinding information. In: Mackaness W, Ruas A, Sarjakoski LT (eds) Generalisation of geographic information: cartographic modelling and applications. Elsevier, Amsterdam

Sitte C (1996) Möglichkeiten für den Aufbau besserer Raumvorstellungen und eines globalen Abbilds der Erde. GWUnterricht 64:44–52

Steiniger S (2007) Enabling pattern-aware automated map generalisation. PhD Thesis, The University of Zurich

Stiens G (1996) Prognostik in der Geographie. Westermann, Braunschweig

Stoter J, van Smaalen J, Bakker N, Hardy P (2009) Specifying map requirements for automated generalization of topographic data. Cartographic J 46(3):214–227

Stott J, Rodgers P (2004) Metro map layout using multicriteria optimization. In: Proceedings of the 8th international conference on information visualisation (IV'04). IEEE Comput Soc, pp 355–362

Stott J, Rodgers P, Martínez-Ovando JC, Walker SG (2011) Automatic metro map layout using multicriteria optimization. IEEE Trans Vis Comput Graphics 17(1):101–119

Tainz P, Heitmann J (2005) Scenarios of application of highly generalized geo-information. In: International symposium on generalization of information (ISGI 2005), Berlin, 14–16 Sep 2005

Töpfer F (1977) Kartographische Generalisierung, VEB Hermann Haack

Tufte ER (2001) The visual display of quantitative information, 2nd edn. Graphics Press, Cheshire

van Goethem A, Meulemans W, Reimer A, Haverkort H, Speckmann B (2013) Topologically safe curved schematization. Cartographic J 50(3):267–284

Wang YS, Chi MT (2011) Focus + context metro maps. IEEE Trans Vis Comput Graphics 17(12):2528–35. doi:10.1109/TVCG.2011.205

Ware JM, Anand S, Taylor GE, Thomas N (2006) Automated production of schematic maps for mobile applications. Trans GIS 10(1):25–42

Wolff A (2007) Drawing subway maps: a survey. Informatik—Forschung und Entwicklung 22(1):23–44

Chapter 11
Generalisation in Practice Within National Mapping Agencies

Cécile Duchêne, Blanca Baella, Cynthia A. Brewer, Dirk Burghardt, Barbara P. Buttenfield, Julien Gaffuri, Dominik Käuferle, François Lecordix, Emmanuel Maugeais, Ron Nijhuis, Maria Pla, Marc Post, Nicolas Regnauld, Lawrence V. Stanislawski, Jantien Stoter, Katalin Tóth, Sabine Urbanke, Vincent van Altena and Antje Wiedemann

C. Duchêne (✉)
Laboratoire COGIT, IGN, Saint-Mandé, France
e-mail: cecile.duchene@ign.fr

B. Baella · M. Pla
Institut Cartogràfic de Catalunya, Barcelona, Catalonia, Spain
e-mail: blanca.baella@icc.cat

M. Pla
e-mail: maria.pla@icc.cat

C. A. Brewer
Department of Geography, Pennsylvania State University, University Park, Pennsylvania, USA
e-mail: cbrewer@psu.edu

D. Burghardt
Dresden University of Technology, Dresden, Germany
e-mail: dirk.burghardt@tu-dresden.de

B. P. Buttenfield
Department of Geography, University of Colorado, Boulder, Colorado, USA
e-mail: babs@colorado.edu

J. Gaffuri · K. Tóth
Institute for Environment and Sustainability, Joint Research Centre, European Commission, Ispra, Italy
e-mail: julien.gaffuri@gmail.com

D. Käuferle
Federal Office of Topography swisstopo, Wabern, Switzerland
e-mail: dominik.kaeuferle@swisstopo.ch

D. Burghardt et al. (eds.), *Abstracting Geographic Information in a Data Rich World*, Lecture Notes in Geoinformation and Cartography, DOI: 10.1007/978-3-319-00203-3_11,

Abstract National Mapping Agencies (NMAs) are still among the main end users of research into automated generalisation, which is transferred into their production lines via various means. This chapter includes contributions from seven NMAs, illustrating how automated generalisation is used in practice within their partly or fully automated databases and maps production lines, what results are currently being obtained and what further developments are on-going or planned. A contribution by the European Joint Research Center reports on the use of multiple representation and generalisation in the context of the implementation of the European INSPIRE directive. The chapter finishes with a synthesis of recent achievements, as well as future challenges that NMAs have begun to tackle.

F. Lecordix · E. Maugeais
Developments department, IGN, Saint-Mandé, France
e-mail: francois.lecordix@ign.fr

E. Maugeais
e-mail: emmanuel.maugeais@ign.fr

R. Nijhuis · M. Post · J. Stoter · V. van Altena
Kadaster Geo informatie, Zwolle, Netherlands
e-mail: ron.nijhuis@kadaster.nl

M. Post
e-mail: marc.post@kadaster.nl

J. Stoter
e-mail: jantien.stoter@kadaster.nl

V. van Altena
e-mail: vincent.altenavan@kadaster.nl

N. Regnauld
Ordnance Survey Great Britain, Southampton, UK
e-mail: nicolas.regnauld@1spatial.com

L. V. Stanislawski
Center of Excellence for Geospatial Information Science,
United States Geological Survey, Rolla, Missouri, USA
e-mail: lstan@usgs.gov

S. Urbanke · A. Wiedemann
AdV Project, LGL, Stuttgart, Baden-Württemberg, Germany
e-mail: sabine.urbanke@lgl.bwl.de

A. Wiedemann
e-mail: antje.wiedemann@lgl.bwl.de

11.1 Introduction

11.1.1 Generalisation, from Research to Production

Research in generalisation has long been driven by the needs of National Mapping Agencies (NMAs).[1] NMAs remain the primary users of automated generalisation processes, although some generalisation functionalities are currently made available to more and more geographic information users via GIS software and the web.

Various schemes exist within NMAs to encourage research in this area, and then to transfer research results into production environments. Some NMAs include long-term research teams to which they can directly express future needs (e.g. IGN-France, USGS-USA, OSGB-Great Britain). This research is sometimes undertaken in collaboration with universities. Other NMAs do not have long-term research teams but collaborate with universities while subcontracting long term research tasks related to their needs (e.g. Kadaster-Netherlands, ICC-Catalonia). When it comes to industrialisation of research results, some NMAs have development teams that develop generalisation tools specific to their needs, partially based on previously published research results, usually on top of commercial off the shelf GIS software (e.g. ICC-Catalonia, Kadaster-NL, IGN-France, OSGB-Great Britain). Other NMAs subcontract the customisation of existing GIS software to GIS vendors, who also have "research and development" teams (e.g. Swisstopo-Switzerland, AdV-Germany).

In addition to this, events are organised where researchers, practitioners from NMAs and GIS vendors can meet and exchange ideas and information. The International Cartographic Conference, every two years, and the annual workshop of the ICA Commission on Generalisation and Multiple Representation, are good examples of such events. This enables NMAs to articulate their needs, researchers and vendors to understand needs, and vendors and developers from NMAs to keep up to date with recent research achievements. Other meetings have also been organised specifically for NMAs, so that they can share experiences and identify common needs that can then be communicated to the research and vendor communities (e.g. Stoter 2005; ICA and EuroSDR 2013). The survey by Foerster et al. (2010) also identified shared needs among NMAs.

Collaboration between researchers, NMAs and vendors has also been undertaken as projects. Some projects seek to build advanced prototypes on top of existing commercial or homemade software. Examples include European projects such as AGENT (Barrault et al. 2001), GiMoDig (Sarjakoski et al. 2005), ESDIN (Kereneur et al. 2011), and ELF (Jakobsson et al. 2013). A second kind of project aims at evaluating existing generalisation solutions through benchmarking: the OEEPE test (Ruas 2001), and the EuroSDR project (Stoter et al. 2009b, 2010).

---

[1] This acronym is a misnomer since some of them are national while others, within federal countries, can be regional. However, it is widely used in the ICA Community, therefore our choice to use it in this chapter.

Such tests not only enable evaluation of existing solutions, but also result in improvements, both in the short term (because the software providers were allowed to adjust the software within certain rules in order to complete benchmark tasks), and the long term (identifying shortfalls helps define future research agendas).

Finally, a last opportunity for NMAS to see their needs taken into account are the competitive tender processes they run when they renew their production lines. Indeed, during such tender processes, vendors are given the opportunity to tailor their software according to the requirements of the concerned NMA, and they often take this opportunity in order to maximise their chances of retaining contracts.

Thus NMAs gain new operational generalisation functionalities that meet their needs, both through regular improvements of generalisation functionalities of commercial GIS software, and through ad hoc customisation that can be made either from within the organisation, or by subcontracting to GIS vendors. A large part of those improvements and customisations rely on published research such as the ones presented in various chapters of this book.

11.1.2 Outline of the Chapter

The next seven sections of this chapter describe contributions from seven of the most active NMAs within the ICA Commission in Generalisation and Multiple Representation. Together they represent a broad selection of current users of automated generalisation in map production. They are ICC, Catalonia (11.2), IGN, France (11.3), Swisstopo, Switzerland (11.4), OSGB, Great Britain (11.5), USGS, USA (11.6), Kadaster, Netherlands (11.7), and AdV, a consortium group of several regional mapping agencies in Germany (11.8). Each section gives insights into the current use of generalisation and multi-representation (the main strategy of use and technical details for specific production lines), results achieved so far and future plans of the NMA. Section 11.9 reports on pieces of work conducted at the European Joint Research Center (JRC) to support the implementation of the INSPIRE European directive in European NMAs, regarding multi-scale modelling on the one hand, and the setup of generalisation services on the other. Finally, Sect. 11.10 provides a synthesis of recent achievements, current trends, on-going work and future challenges, based both on the contributions from NMAs included in this chapter and on information gathered by direct interactions with NMAs during a symposium held in March 2013 (ICA and EuroSDR 2013).

11.1.2.1 Notice to Reader

Please be aware that the facts reported in this chapter are up to date in 2013, but might evolve quite quickly since the developments in generalisation are currently particularly active in several NMAs, as a consequence of the changing context in the geoinformation world in general (see Sect. 11.10.2 and Chap. 12 of this book), and the maturity reached by research in generalisation linked to the traditional needs of NMAs (topographic data and maps production). In addition, please note that some concepts and ideas referred to by the contributions from NMAs (Sects. 11.2–11.8), such as the distinction between DLM and DCM or the star vs. ladder derivation schemes, are further elaborated in Sect. 11.10.2.

11.2 Deriving Products Through Generalisation at the Institut Cartogràfic de Catalunya

Maria Pla and Blanca Baella

Since the foundation of the Institut Cartogràfic de Catalunya (ICC) in 1982, one of the main activities was the production of georeferenced data, mainly topographic and thematic data and orthophoto images. The derivation of products at smaller scales using generalisation techniques started in 1996, after the first digital coverage at a scale of 1:5,000 of Catalonia was completed. Although initially the generalisation techniques were applied only for the derivation of vector topographic maps, some years later they were also implemented in the derivation of vector databases, raster orthoimages and map names. Generalisation workflows of topographic data and map names are semiautomatic processes that use commercial software enriched with software developed at the ICC. In the case of orthoimages the workflow is completely automatic using applications developed at the ICC.

In the last few years, high demand for derived products for visualisation via the internet and over mobile devices has introduced new requirements, both in terms of the data and in the processes used, demanding more intelligent data models and on-the-fly generalisation. The current ICC on-going work is mainly focused on these two topics.

11.2.1 Generalisation of Topographic Data

11.2.1.1 Current Production

Topographic databases at different resolutions are produced and maintained ranging from 1:1,000 scale to 1:25,000 scale. The most detailed information covers the urban areas at 1:1,000 (20 cm of accuracy), while data at 1:5,000 (1 m accuracy) and 1:25,000 (2.5 m accuracy) covers the whole Catalonia. These databases are compiled in 2.5D using stereo plotting on top of digital photogrammetric systems.

Smaller scales such as 1:50,000 and 1:250,000, also covering the whole territory, are collected by digitizing in 2D on top of orthophoto images.

After a first generation of spaghetti data collected using CAD systems, more complex models were designed and implemented that enable new data exploitation using GIS systems. The topographic data models were designed to preserve the semantic coherence between scales, but without explicit relationships between the representations of the same geographical object in the different databases. This was because of the lack of commercial tools to maintain them and because of the different update cycles of each product. The generalisation techniques were applied to the Topographic Database at 1:5,000 (BT-5M) to derive the Topographic Map at 1:10,000 (MT-10M) and the Topographic Database at 1:25,000 (BT-25M), and to the Topographic Map at 1:50,000 (MT-50M) to derive the Topographic Map at 1:100,000 (MT-100M).

The BT-5M is compiled using photogrammetric systems according to a 2.5D data model and stored in DGN files from MicroStation. During the stereo plotting process all the features required to generate a digital terrain model (DTM) and a digital surface model (DSM) are compiled together with the topographic objects. The updating cycle is 5 years over the entire territory and more frequently over the most dynamic areas, located mainly near the coast. The first compilation of the BT-5M started in 1985 and was completed in 1995. At that time, the underlying data model was based on 2.5D "spaghetti" vectors. It supported the generation of DTMs, but it was not designed for GIS. The next version of the model addressed this shortcoming. Moreover, it supported automatic generalisation and 2.5D topographic objects.

The MT-10M is just a map, not a database. It contains 2D data obtained by semiautomatic generalisation from the BT-5M. Automatic processes include elimination of some objects, class aggregation, building simplification, altimetric point selection and selection and scaling of map names. Manual editing is applied to refine the automatic results, but because the scale change between the original and the target scale is quite small, it can be performed in few hours, around 20 h per sheet (3,200 ha). The updating cycle is the same as for the original database and the updating workflow generalises again the updated BT-5M without keeping any object of the old MT-10M, because most of the generalisation operations are automatic and the manual editing is quite fast.

The need for an updated base data at 1:25,000, the existence of a production program for the BT-5M, and the experiences at the ICC in implementing generalisation workflows, allowed us to start producing the first version of the BT-25M using generalisation processes in 2003 (Baella and Pla 2003). Compared with previous ICC generalisation experiences, the workflow entailed two challenges: to obtain a topographic database, not only a map, and to generalise 2.5D data instead of 2D data. The main difference between generalisation for obtaining a map or for creating a database comes from having to preserve the topological structure of the

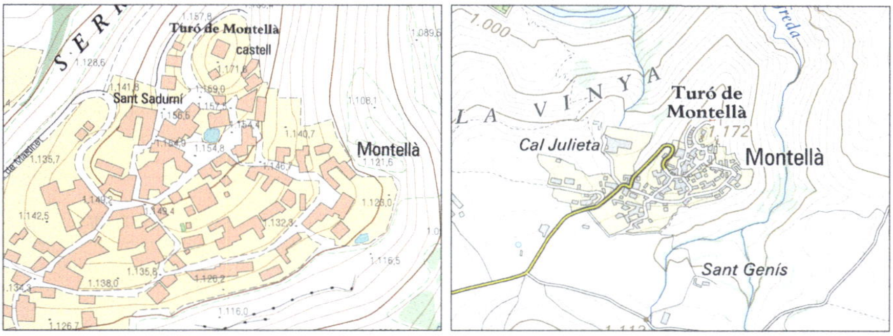

Fig. 11.1 At the *left* the original BT-5M data, at the *right* the resulting BT-25M after generalisation

data and their attributes. The 2.5D characteristic of the generalised data required new software development and a careful editing process (Baella et al. 2007; Palomar-Vázquez et al. 2011).

The BT-25M is obtained by generalisation of the BT-5M (Fig. 11.1), is updated using stereoplotting of recent photogrammetric flights and it is completed with data extracted from thematic databases and, occasionally, with data collected in the field. Automatic generalisation includes the same operations used for the MT-10M, plus the simplification of linear elements and the generalisation of the elevation of the objects. Manual editing is used for generalisation operations that cannot be automated such as typification, exaggeration, collapse, aggregation or refinement of the automatic results. This requires an average of 150 h per sheet (12,800 ha). The database is updated independently of the original database, because the updating cycle planned for the BT-25M is approximately 2 years, much shorter than the updating cycle of the BT-5M.

The MT-100M is obtained by applying 2D generalisation to the MT-50M. Automatic processes include selection, class aggregation, line simplification and selection and scaling of map names. Manual editing is applied to generalise buildings and to refine the results.

The generalisation workflows described above (summarised in Fig. 11.2) combine commercial software and ICC developments. For some specific aspects, the ICC collaborates with external research groups. The commercial software CHANGE, developed and distributed by the University of Hannover, is used for the building generalisation up to 1:25,000 scale. The ICC developments are mainly focused on line and map names generalisation, and in the tools for supporting interactive generalisation, for example collapse, exaggeration or conflict detection and resolution. Tools for terrain generalisation, as spot height selection, have been developed in collaboration with the Department of Cartographic Engineering, Geodesy and Photogrammetry of the Universitat Politècnica de València.

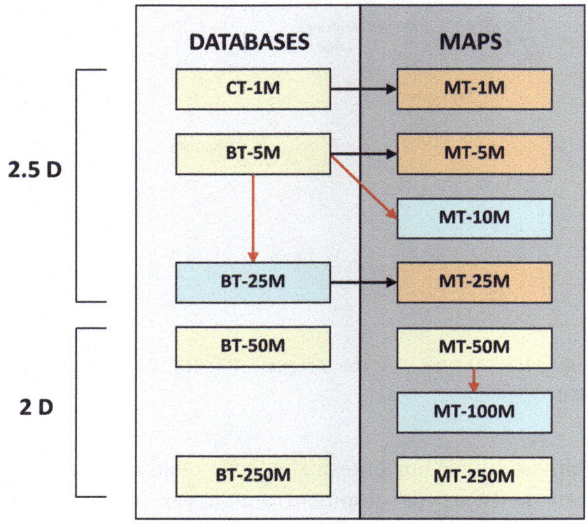

Fig. 11.2 General schema representing the existing ICC topographic databases and maps. Products compiled from primary sources are indicated in *yellow* and products derived applying generalisation in *blue*. Products in orange are derived by automatic symbolisation without any editing process

11.2.2 Generalisation of Orthophotos

The ICC has a long tradition in orthophoto production using software developed internally. Since its foundation a large collection of orthophoto products at different scales has been produced using aerial or satellite images, ranging from 10 cm to 25 m pixel ground size. At the beginning, in the 1980s, each orthophoto product used its own original image source, but in the year 2000, the production of the orthophoto of Catalonia at 50 cm pixel ground size was changed in order to ensure not only the geometrical continuity but also the radiometric continuity over the entire territory. This offered the possibility of using generalisation processes to derive orthophoto products with a larger pixel ground size.

The generalisation processes applied on the orthophotos implies, as in the case of the topographic data, a scale change, which is implemented through a reduction in the number of pixels of the image and through the change of the ground resolution size of the pixel. Basically this process has two key aspects: the determination of the new pixel position, which is a geometric operation, and the calculation of the radiometry of the new product, applying a convolution using a Gaussian bi-dimensional function for eliminating the higher frequencies of the radiometric values of the resulting image and ensuring a good radiometric result.

These generalisation processes applied to the orthophotos have optimised the production of image products at the ICC allowing a higher rate of productivity. Nowadays, from the base product, the orthophoto of Catalonia of 25 cm pixel ground size, the ICC is deriving the orthophotos of 50 cm and 2.5 m (Fig. 11.3) covering all Catalonia on an annual basis.

Fig. 11.3 At the *left* the original orthophoto at 50 cm data, at the *right* the resulting at 2.5 m after generalisation

11.2.3 Generalisation of Map Names

From its foundation, the ICC has compiled and maintained a very detailed collection of map names. The first data compilation was made through a field campaign, using orthophotos as a base reference document, started in the year 1984 and finished in 1991. The information was compiled and georeferenced on top of the orthophotos manually in the field. Following this, the alphanumerical information associated with the map names was translated to a digital text file. The collection of map names was then used in the publication of several ICC products, for example the Topographic Map of Catalonia at 1:5,000 or the orthophotomap at the same scale. Since this first campaign, the ICC has a continuous updating process using all the sources available, from new smaller field campaigns to external data incorporation. The current data model of the ICC Map Names collection is based on a set of graphical data and a set of alphanumerical information linked through an identifier. But the map names are not linked with the elements stored in the topographical database. The new topographic data model designed for MRDB purposes will include this link.

Due to the widespread use of map names in cartographic products, from the beginning there was a need for map names generalisation tools. These tools were developed at the ICC, at the beginning of the 1990s, and since then they have been continuously improved (Baella and Pla 2001).

Two types of generalisation tools were developed for map names: the automatic tools and the interactive ones devoted to helping the cartographers in the improvement and refinement of the automated results. The automatic tools include the map name selection and the new graphical placement of the map names, following the style and the scale of the new product. The interactive tools include (1) conflict detection of the generalised name placement with other map name placements or with topographical elements, (2) the detection of placements of the same map name too close together, and (3) placement style improvement, for example the space between different text lines or the base supporting text lines.

The customised generalisation tools developed at the ICC have enabled us to derive collections of map names while optimising production costs and ensuring coherence between the original and the derived collection.

11.2.4 On-Going Work

The huge demand for updated information and the high pressure to obtain derived products for visualisation via the internet and mobile devices has introduced new requirements, especially related to topographic data, that are costly to achieve with the current data models. At this moment the ICC is assessing a solution based on the implementation of a MRDB for integrating topographic data at different scales that allows to optimise generation of derived products using generalisation, including on-the-fly processes for intermediate resolutions used mainly for visualisation. The implementation should take into account the current state of the ICC databases, the existing limitations in commercial software, the requirements to be implemented in the ICC production environment and the challenge of achieving reasonable productivity ratios.

Although the MT-10M and the BT-25M are derived by generalisation from the BT-5M, there are no explicit mechanisms linking the original and the generalised datasets. The main reason for this is that when the workflows were implemented, there was a lack of tools for managing the links during the compiling photogrammetric process. In the last few years, GIS based photogrammetric tools ready for production environments and delivering reasonable productivities have started to become common. The availability of these tools has encouraged the ICC to design a new version of the data model that would implement the main aspects of a multirepresentation database (MRDB), linked objects at different levels of detail, where the links between the objects will be established through unique identifiers (Baella et al. 2012). In the new data model, each feature instance will be characterised by an identifier that must be unique and persistent throughout the whole life-cycle of the feature instance and never reused.

The ICC MRDB will be based upon one single schema with linked features, where one feature belongs to one single resolution and has a link with one, or more, features at the other scales. The links between corresponding feature instances at different levels of detail will be determined by the various cardinalities that exist in the relations between them. Due to the complete coverage of the current version of the BT-5M and the BT-25M, the links between the feature instances at different resolutions will be established using matching techniques that will be applied in the migration of the existing data to the MRDB data model. For later updating processes on the higher accuracy data, the lower resolution data and the links will be managed during the generalisation process.

The implementation of the new ICC topographic data model will involve two main tasks, the migration of the photogrammetric data collection from a CAD system to a GIS system based on a DBMS, and the management of the data and the MRDB relationships. The migration to a GIS environment will require customisation of the commercial system and the training of the production teams in the GIS system and in the customised tools. The customisation includes the development of a set of tools that preserves the 2.5D nature of the data model and achieves a similar level of productivity to that of the existing CAD environment. The management of the MRDB links will be based on tables that store the relationships between features instances through unique identifiers. The first implementation of the MRDB will be done from the two existing databases, BT-5M and BT-25M, which do not have any explicit link between the feature instances representing the same geographical object. The matching processes to find the related feature instances will fill the tables of relationships. In further updating processes of the database, the generalisation operations to derive the smaller scale must automatically establish and manage these links.

11.3 The New Base Map Project: A Semi-Automated Production Line for Topographic Maps at IGN-France

François Lecordix and Emmanuel Maugeais

11.3.1 Important Investment in R&D at IGN-France

At the beginning of the 1990s, IGN France decided to launch a large and long term investment into research and development in generalisation. This investment was justified by the need to solve the problems of producing maps at different scales directly from databases that IGN had already produced and updated for a number of years. The main goal for the French national mapping agency was to reduce costs and, in particular, only collect the data once and use them to produce all other databases and maps. This aim stipulated that maps have to be derived from databases and the cost of deriving them should be affordable which means the process should be as automatic as possible.

In 1991, IGN placed the COGIT laboratory in charge of research on generalisation with the aim of deriving DCMs from DLMs. A team of four researchers was initially nominated and there have been between four and eight researchers working until 2013 on this project. During this period, 13 PhDs have defended their theses and many results were obtained on different aspects of generalisation:

platforms and algorithms (e.g. Lecordix et al. 1997), data enrichment (e.g. Boffet and Coquerel 2000), strategies (e.g. Ruas 1999), evaluation (e.g. Bard 2004). A network of strong collaboration was established with different universities and NMAs around the world. A part of this research was capitalised during the European project AGENT [1997–2000, see (Barrault et al. 2001)] in a generalisation prototype built on the commercial software LAMPS2 (from Laser-Scan company, now called 1Spatial, who took part in the project).

During the AGENT project, in June 1999, IGN launched the "Carto2001" development project with 4 developers, with the aim of setting up a new production flowline to derive a 1:100,000 DCM from a 10 m resolution DLM, BDCarto®. In 2004, the Carto 2001 project provided automatic solutions for networks generalisation, using agent technology for road bends coalescence and the "Beams" algorithm (Bader 2001) for overlapping in symbolised networks. The flowline was developed on LAMPS2 editor, which provides the Gothic DBMS, the AGENT prototype and many other developments, either by 1Spatial or homemade, for generalisation and updating (Jahard et al. 2003; Lecordix and Lemarié 2007). Only automatic label placement has been fully developed in-house at IGN in the form of a software called WinPAT.

This flowline was used to derive the first edition of DCM Top100 (2006–2008) with eight cartographers and then to update this DCM twice (Plessis 2011). The production, still in use, has provided for the first time experience of generalisation in production environments and has enabled to study the introduction of new solutions for generalisation problems at other scales.

11.3.2 New Base Map Project

11.3.2.1 Different Purposes

Since the end of the 1980s, IGN has been producing the BDTopo®, a DLM with a precision of 1 m. This DLM contained all information (networks, buildings, landscape, contour lines, point, etc.) for starting from 1993, the semi-automatic production of maps at 1:25,000, in a new version called Type 93. Some choices of symbolisation were made to minimise problems with generalisation. For example the widths of symbols are very thin, the buildings are only exaggerated not displaced, and the roads layer is drawn on top of the buildings layer so that the overlaps between roads and buildings are less visible. A problem with this flowline is that maps at 1:50,000 scale (used by the military) was no longer produced because of generalisation problems. By 2000, IGN had only produced 25 % of the French territory in BDTopo®. To speed up the construction of BDTopo®, new specifications and processes were selected to finish the DLM collection for the whole territory in 2007. This decision meant that the production of Type 93 would not be possible any more: the new version of the DLM would not contain all needed information. So, in 2004, IGN launched the New Base Map Project. The main challenges were:

- defining processes to collect additional information necessary for cartographic production,
- deriving 1:25,000 and 1:50,000 seamless DCMs, incorporating human operators in an user friendly environment,
- proposing an automatic process to propagate the updates collected in DLM to DCM,
- and, last but not least, the requirement that this should be slightly less expensive than previous flowlines even if there is less information in the new version of BDTopo®.

The staff of the New Base Map Project varied between three and five developers during the period 2004–2011. The first result was obtained in 2009 and it allowed the production of the collection of additional information to start. Thus, cartographic production was ready in 2010, hence the new version of the map is called Type 2010 (Maugeais et al. 2011). During that period, some modifications were introduced in IGN vector databases. Different DLMs that IGN managed were merged into one DLM called BDUni. Some modifications introduced in BDUni had a strong impact on the cartographic process under development. For example, in the new specifications for roads, lanes of dual carriageways and motorways were recorded separately (and not merged as before in BDTopo®). For buildings, the outline of a set of buildings were gradually replaced by cadastral buildings with more details and more partitions. BDUni is managed in DBMS PostgreSQL in a seamless database and the French GIS software GeoConcept (from Geoconcept company) is used to modify the data. The communication between the DBMS and GIS is done via the software GCVS that IGN had developed.

11.3.2.2 Technical Architecture and Flowline

The New Base Map Project developers decided to retain an architecture with different software, similar to BDUni's architecture, but it has an extra software specific to cartography. PostgreSQL provides the DBMS, the interactive editor is GeoConcept, with a specific layer designed for cartography, Publisher (Guislain and Lecordix 2011), and GCVS provides the link with DBMS. The software Clarity (from 1Spatial) was selected for automatic cartographic processes and generalisation by the New Base Map Project developers who developed a specific bridge between GeoConcept/Publisher and Clarity, called CLEO, which enables to export data from GeoConcept to Clarity, to launch automatically processes in Clarity and then to send data back from Clarity to GeoConcept. The IGN solution WinPAT was used for automatic label placement.

The New Base Map Project defined the flowline for the new topographic map. A new database called BDComplementaire was defined to store the data missing in BDUni but necessary to produce maps, i.e. touristic itinerary, contour lines etc. Different processes were defined to collect this information: for example, field work, information extraction from old maps or orthoimagery.

After collection, the cartographic process can start with the first step of merging BDUni and BDComplementaire and generating a new database BDRef. This BDRef is not dependent on one scale or a particular symbolisation, but some operations are applied to compute information and to introduce consistency: building and dual carriageways are merged (see below), itineraries are conflated on topographic networks, urban structures such as settlement areas or city blocks are computed between buildings and networks. The BDRef is an intermediate DLM where model generalisation has been applied from BDUni, so that DCMs at scales 1:25,000 to 1:50,000 can be derived by cartographic generalisation.

The next step consists in deriving a DCM at a defined scale, France25 at 1:25,000 or France50 at 1:50,000. The New Base Map Project worked mainly on France25, but some experiments were made on France50. In these DCMs, the symbolisation is assigned at the beginning of the process and then many automatic generalisation and label placement processes are launched. Between these automatic processes, manual cartographic editing is applied. This full process is applied on tiles of 20 by 20 km. The objects which cross many tiles are handled in each tile and a semi-automatic process is launched to manage these objects when the cartographic data are stored in the seamless database.

11.3.2.3 Main Generalisation Operations

Four main steps are necessary to ensure efficient building and network generalisation:

1. Building merging: a building is stored in the BDUni database as a plurality of objects that resulted from dividing it by cadastral plots. Buildings that are adjacent and have the same symbolisation are merged before running the generalisation process (see below). It prevents the system from assessing them too small individually, enlarging them independently from each others and as a result generating lots of overlaps.
2. Dual carriageway merging: dual carriageways, for which two lanes are captured in BDUni, are merged using a five stage process adapted from Thom (2005). Road segments that form part of a lane in dual carriageways are detected, grouped into lanes and matched by pairs, the central axis is computed by a skeleton algorithm of the central zone between each pair of lanes, and finally connections and intersections with the rest of the network are managed. To preserve bridges and tunnels and avoid inconsistent merging, this sequence is performed one by one on five categories of roads according to their nature and their position relative to the ground.

3. Network generalisation: the solution developed during the Carto2001 project was adapted for topographic objects (e.g. embankments). Firstly the process detects groups of connected objects concerned with overlapping cartographic conflicts, called "flexibility graphs" (Lecordix and Lemarié 2007). The "beams" algorithm (Bader 2001) is used to move networks to solve these conflicts. Finally a process based on the GAEL model (Gaffuri 2007) moves buildings to maintain topological consistency with the roads.
4. Building generalisation: the AGENT model proposed by Ruas (1999) and further developed during the AGENT project (Barrault et al. 2001) has proved its worth for building generalisation. This model works at two levels: a "micro" level for individual objects (in this case, buildings) and a "meso" level for groups (here "building groups", consisting of clusters of close buildings previously created with a method based on Boffet and Coquerel (2000)). Different constraints and possible actions (generalisation algorithms) to solve them are defined at those two object levels. The agent engine ensures the orchestration of the actions to solve the constraints which are defined as follows:

   - A density constraint ensures that the black to white ratio within a building group does not increase too much during scale reduction. It determines the elimination of one or more buildings based on several criteria: the building area, its distances to the neighbouring buildings and its proximity to roads surrounding the group.
   - Constraints of minimum size, granularity and shape preservation are defined at the buildings level: they are solved by exaggeration and simplification of the buildings.
   - A proximity constraint allows the movement of buildings to correct the overlap between them (due to their exaggeration at the "micro" level) or with the symbolised roads.

11.3.3 Production Launch and Results

In 2010, experimentation of the cartographic process was launched in production for the first time (an example of output is shown in Fig. 11.4). Some improvements were required and the main part of the process was validated in 2011 to start the production on rural areas. Subsequently many improvements were made such as generalising large urban areas, correcting bugs in specific cases, managing tile edge 'reconciliation', and testing at 1:50,000 scale (e.g. Fig. 11.5).

At the end of 2012, 317 tiles were collected (for additional information) over 280 h for each and the cartographic process was run at 1:25,000 on 136 tiles in 175 h. To produce the complete DCM at 1:25,000 for the French territories (overseas included), 1637 tiles will have to be produced in ten years.

Other tools are being developed in the course of the New Base Map Project but they have not been used in production yet. However they are necessary to produce

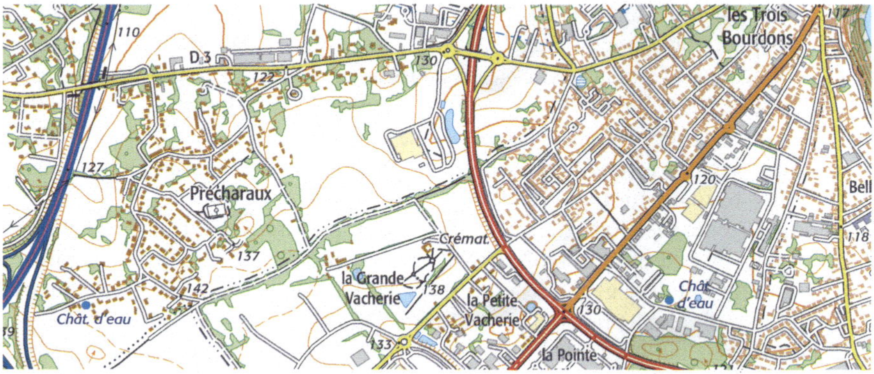

Fig. 11.4 Cut-out of base map Type 2010 at 1:25,000 scale (IGN-France)

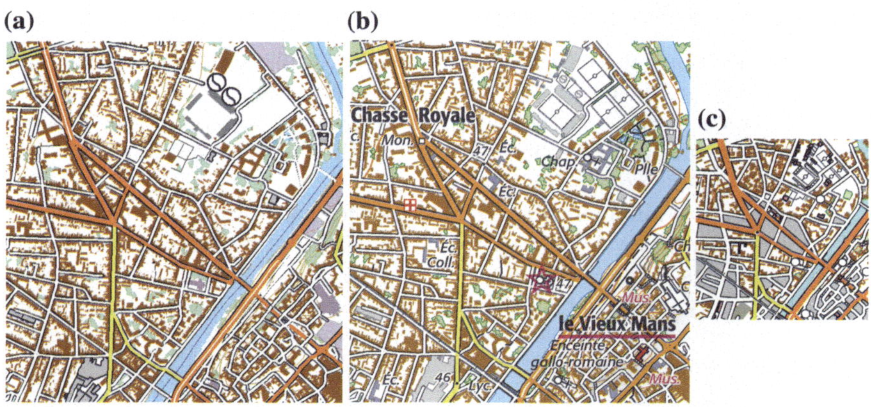

Fig. 11.5 Generalisation results of initial data (**a**) at 1:25,000 scale (cut-out of base map Type 2010) (**b**) and first trials at 1:50,000 scale (**c**)

the DCM for the whole country. These include for instance: the collapse of hedges, modelled as areas, into lines by skeletonisation (Maugeais et al. 2011), contour lines extraction from DTMs with generalisation solutions (Jaara and Lecordix 2011), and mountain representation (Gondol et al. 2008) (e.g. Fig. 11.6).

11.3.4 Conclusion and Future Works

To continue the IGN research, development and production strategy, in 2010 IGN launched the project "Map on Demand" using COGIT's research on this subject (e.g. Christophe 2011) and exploiting the results of the New Base Map project. After adaptating Type 2010 flowline, the "Map on Demand" project has been providing a

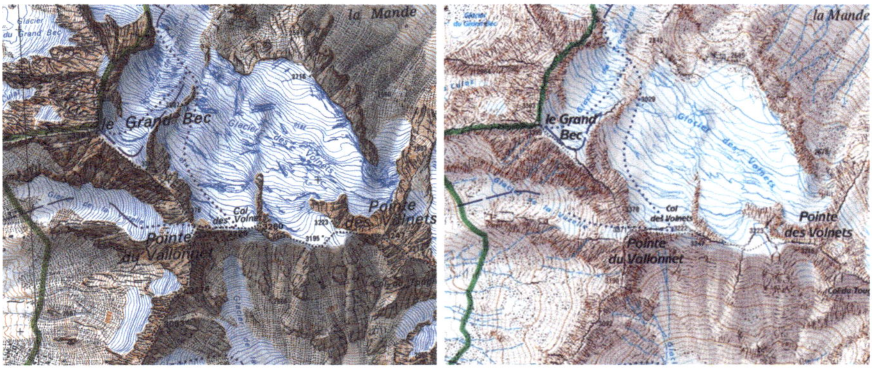

Fig. 11.6 Mountain representation: (*left*) current map obtained manually in the past; (*right*) test with numeric representation

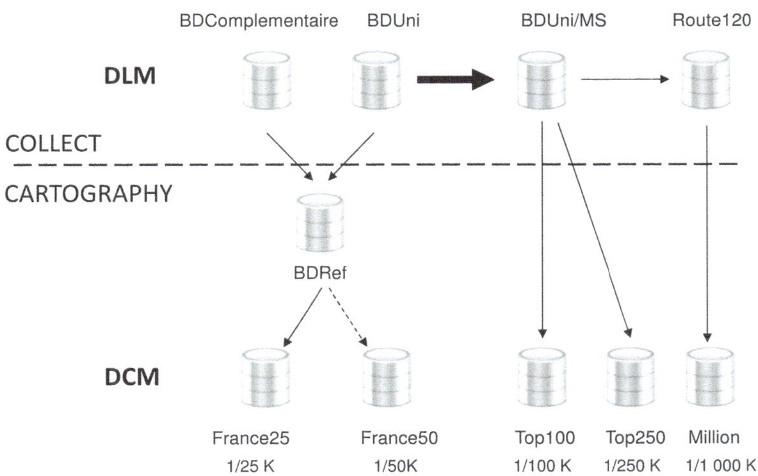

Fig. 11.7 Derivation processes at IGN in 2013 (*thin arrows*) and future work (*thick arrow*)

fully automatic process to map the BDUni for the whole of France within one week for an express product. Such an express mapping product is of a lower quality than Type 2010, particularly in terms of generalisation, but it is more up to date and offers the possibility of selecting different options of colour, and so a better customisation. After new investment into automating further the generalisation processes, we can assume that this product will be gradually improved and will obtain the same quality as Type 2010 currently obtained with semi-automatic processes.

In the future, IGN will continue to work on derivation processes. After considerable effort in cartographic derivation (partly reported in this section and summarised in Fig. 11.7), the next challenge will be to introduce the derivation between two DLMs at different level of detail. In 2010, IGN migrated the BDCarto® into the same DBMS as the BDUni: PostgreSQL. The product

BDCarto® was renamed in BDUni/MS (MS for Medium Scale) to distinguish from BDUni that corresponds to large scale. But, IGN still has to manage two different update cycles, with two separate collect processes, for these two DLMs. To further reduce production costs, it would be necessary to derive BDUni/MS from BDUni and then to be able to propagate the updates. This would be the next step in the evolution of managing derivation problems in production.

11.4 Producing Digital Cartographic Models at Swisstopo

Dominik Käuferle

«That day all the church bells of Switzerland rang in my heart» stated Professor Eduard Imhof on 21 June 1935 after the federal council had approved the law required for the creation of new National Maps of Switzerland. Now, in 2013—the year of swisstopo's 175 years anniversary—a new era begins as swisstopo completely modernises the National Map series. The new National Maps are based on Digital Cartographic Models (DCM), in seamless vector databases. The new map data are fully GIS based and are fit as well for mobile and online services, while offering the aesthetics and readability of high-quality printed maps. Automatic generalisation plays a key role in this new production environment.

11.4.1 A New Data and Database Infrastructure

Thirteen years ago, swisstopo decided to investigate new ways of producing its base data. Two projects were initiated. One to build a central topographic 3D database, the Topographic Landscape Model (TLM), and another to derive cartographic databases from the TLM, the Digital Cartographic Models (DCM). In both projects automation was key to producing a huge amount of data in a short time. The concepts were completed in 2004. The realisation began in 2005. The key components were:

- TLM (Topographic Landscape Model)
  - seamless topographic database for the whole territory of Switzerland
  - photogrammetric restitution of aerial photos
  - vector data model with more than 100 basic topographic object types, combined with a high resolution digital terrain model
  - 3D geometries with accuracies of ± 1 m
  - no cartographic generalisation

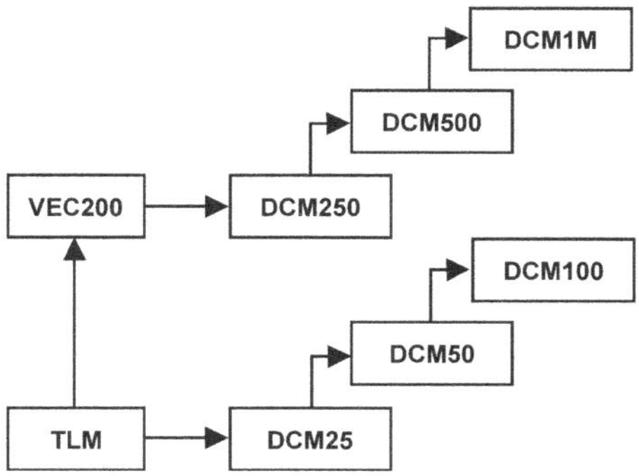

Fig. 11.8 Overview over the new TLM/DCM data infrastructure

- DCM (Digital Cartographic Models)
  - one seamless database per map scale (DCM25, DCM50, ...)
  - derivation from TLM
  - a requirement to preserve or exceed the quality as compared to the existing National Map series, which is to be achieved with the same or less human production capacity
  - 2D vector data model combined with raster layers (relief, rock and scree representation)
  - combination of automatic and manual generalisation
  - unique identifiers and feature links between DCMs and TLM to maintain database consistency and facilitate incremental updates
  - cartographic symbols adapted to TLM classifications
  - new fonts

The TLM and the large-scale maps 1:25,000, 1:50,000 and 1:100,000 are updated in a 6 years cycle. The smaller scales 1:200,000, 1:300,000, 1:500,000 and 1:1,000,000 are derived from an intermediate landscape model VECTOR200, which has a yearly update cycle.

Figure 11.8 gives an overview of the new data infrastructure and the conversions (black arrows) from one database to another. In short it can be described as a combined star and ladder approach, with two stars (TLM and VECTOR200) and two consecutive ladders (large-scale DCMs and small-scale DCMs). An attentive observer of the project gave us the feedback that the ladder approach might lead to inconsistencies in smaller scales due to error propagation. To avoid such problems, swisstopo will compare the child data (e.g. DCM50) during production not only with its parent data (e.g. DCM25), but also with the TLM. Based on the feature links, such comparisons can be automated to some degree.

11.4.2 The Role of Automatic Generalisation

In the workflow of the current National Map series, only recognised changes from reality initiate map updates. In contrast, producing DCMs means that all objects, whether they have to be updated or created for the first time, rely on the TLM. Generalisation is necessary in order to receive the required cartographic result. Findings at the concept phase of the project showed that creating the DCM would imply about a fivefold workload compared with the one needed for the manual work of a complete map update. This means, that at least 80 % of the work needs to be done automatically via available cartographic resources. Since a considerable amount of work is cartographic generalisation, automatic generalisation plays a key role. Without automatic generalisation, it will not be possible to produce the DCMs in a reasonable timeframe. The DCM production is done in map sheet perimeters and consists of four consecutive steps:

1. completion of TLM including semantic quality control;
2. model generalisation of TLM to a raw DCM, also referred to as cartographic reference model;
3. automatic generalisation of DCM vectors and automatic scree representation;
4. manual completion of the DCM including cartographic quality control (vector editing, raster editing, labelling, integration into the seamless database).

The automatic generalisation of DCM vectors in Step 3 is performed by «SysDab». This automatic generalisation system was built by *Axes Systems AG* on behalf of our project and is based on their software *axpand ng*. SysDab performs the main portion of the automatic generalisation work. The system has a client server architecture, in which data are stored in an Oracle database. It offers a variety of generalisation functions that perform specific vector generalisation tasks while maintaining topology, such as:

- building generalisation and alignment;
- parallel adjustment of line and polygon segments;
- line, area and point displacement;
- snapping polygon segments to lines and vice versa;
- point typification;
- merge functions.

The functions can be configured to reach specific generalisation goals. The core of the system is a workflow manager that allows all the tasks to be placed together in a customised workflow. The principles of this workflow management were described by Petzold et al. (2006).

As the devil is in the detail, several approaches and iterations were necessary to satisfy swisstopo's high-quality cartographic requirements. Finally, generalisation rates of 78 to 84 % were successfully generated in DCM25 samples. This means, that only 16–22 % of the features in the test map needed manual editing on their geometries. Even though these are good results, there is potential for improvements.

The processing was performed on a server containing 4 CPUs with 24 cores. Processing time ranged between 4 and 5 h per map sheet. The workflows have been redesigned and are constantly fine-tuned in order to reach even better generalisation results.

SysDab and the swisstopo workflow are built for incremental updates and generalisation and allows us to generalise the data several times. The system keeps track of the generalisation for each feature and maintains their history and relationships to its predecessors. A customised relevance check determines if changes in the input or output are relevant enough to be processed or exported.

If one defines *generalisation* as being synonymous with *abstraction*, a wider range of automatic generalisation tasks can be identified in the DCM production chain. The TLM production (Step 1) includes an automatic vegetation extraction. Digital surface models and infrared orthophotos are used to extract trees automatically, which are then generalised (abstracted) to forest polygons or 3D points for single trees in the TLM. Step 2 is performed fully automatically using *FME* by *Safe Software*. One interesting example is the automatic scree representation in Step 3. Scree is represented by small irregular black polygons (scree points) within a scree area. The size and distribution of the scree points depends on the relief shading—the darker the relief is, the larger and denser the scree points are. The software *screepainter*, engineered by Jenny on behalf of the project (Jenny et al. 2010), performs this rather complicated task in a fully automated manner. In Step 4, the *ArcGIS* based cartographic system «Genius-DB» developed by *Esri Switzerland* on behalf of our project, includes a large variety of automated tools that efficiently assist the cartographer in implementing sophisticated cartographic rules.

Even if all these automation tasks could be done without a database infrastructure, it helps a great deal to integrate all the different tasks into a robust and manageable production chain. Key elements are central data dictionaries, which are repositories for the rules, configurations and process steps performed on the different data models.

11.4.3 The Benefits of DCM's

DCMs are part of the national geodata infrastructure (NGDI) of Switzerland according to the Federal Act on Geoinformation (The Federal Authorities of the Swiss Confederation 2007). The DCMs will be included in the federal geoportal http://map.geo.admin.ch/ and in swisstopo services, such as the new «journey through time» service (Swisstopo 2013). DCMs are static datasets containing well-structured vector data. This offers the following advantageous possibilities:

- apply cartographic templates (e.g. colours);
- link cartographic data with third-party data;
- process and provide data independent of the map sheet lines;
- deploy the detailed structure by layer or object types;
- (long-term) monitoring of landscape processes at different scales.

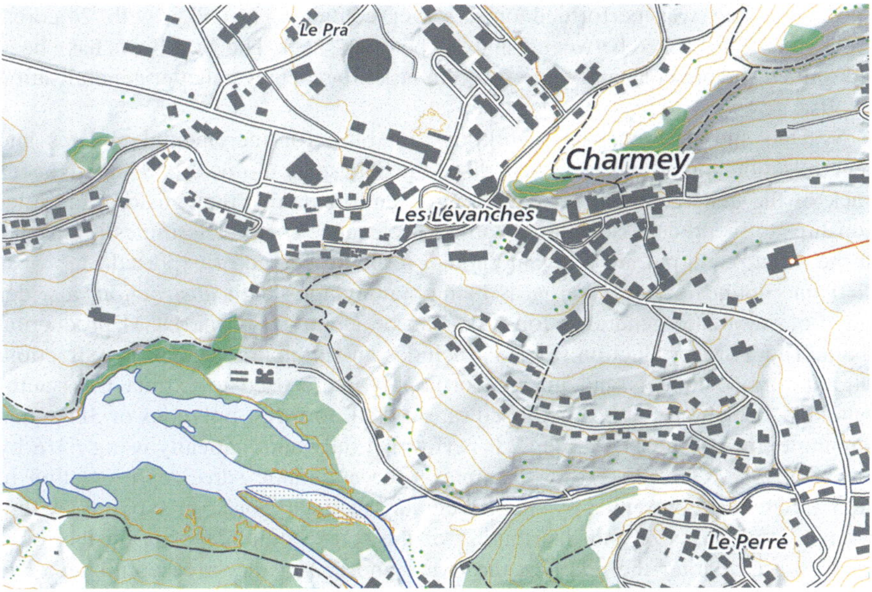

Fig. 11.9 Cut-out of a TLM extract, 1:10,000 (no generalisation applied, automatically symbolised)

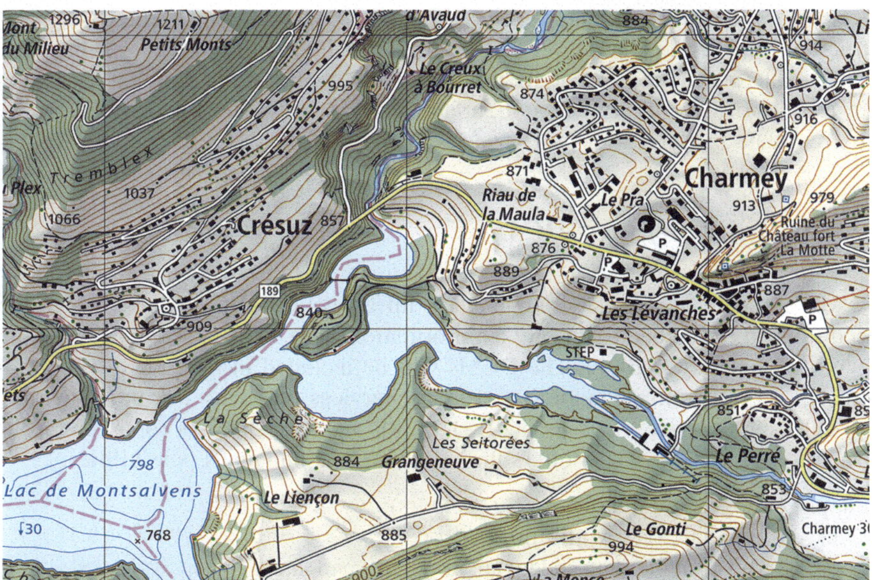

Fig. 11.10 Cut-out of a finished DCM25, 1:25,000 (generalisation applied, test sample)

These opportunities will open several doors for the use of swisstopo map data. There will be new and 'in-house' products, and tailored maps produced by external individuals and companies. Production of the DCM25 (Figs. 11.9 and 11.10) commenced in March 2013, but there remains a lot of work to do. In the sense of Eduard Imhof: our hearts are beating fast at the moment. The completion of all DCM's is planned to be accomplished within the next 8 years.

11.5 Automatic Map Derivation at Ordnance Survey GB

Nicolas Regnauld

11.5.1 Introduction

Ordnance Survey has recently started a large programme of work to change the way products are created. This was triggered by a clear change in what Ordnance Survey's customers want, as well as a need to reduce production costs. The programme is called the Multi-resolution Data Programme (MRDP), and aims at deriving products from Ordnance Survey's large scale topographic database, in a flexible and efficient manner. It also seeks to bring greater consistency across the range of products on offer. Section 11.5.2 describes the strategy developed by Ordnance Survey to derive products in the future. This includes a high level description of the database architecture used. Section 11.5.3 briefly describes the technologies used in the system. Section 11.5.4 presents OS VectorMap® District v1.0, the first product derived automatically by MRDP from Ordnance Survey large scale data. Finally, Sect. 11.5.5 discusses future plans, related to incremental updating.

11.5.2 Product Derivation Strategy

The architecture designed for this system is based on a multi-resolution database. We call it multi-resolution instead of multi-representation, as the term representation may be confused with cartographic representation, while we want this database to stay as independent as possible to the representation used in specific products. However, the concept is the same, and the multi-representation database has been widely studied (Balley et al. 2004). This architecture follows the DLM/DCM (Digital Landscape/Cartographic Models) principles, first presented in (Grünreich 1985) and now widely adopted (Trévisan 2004; Bobzien et al. 2007). The schema in Fig. 11.11 shows a simplified view of the architecture, focusing on

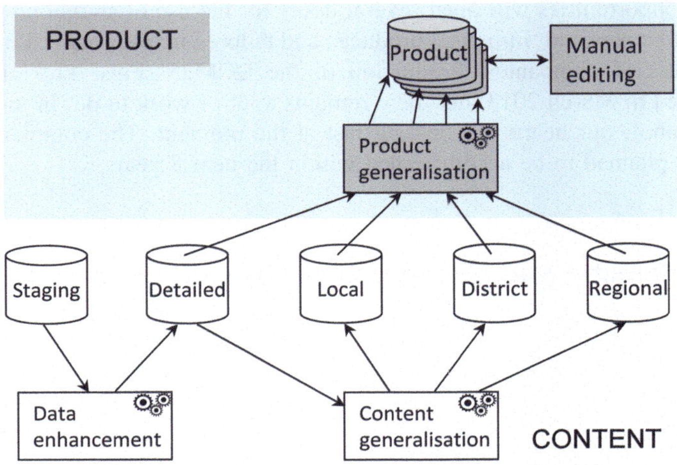

Fig. 11.11 Simplified architecture used by MRDP

how the database is organised. The staging database is a mirror of the maintenance database that contains all the large scale topographic data. It is kept separate to prevent the processes involved with deriving products from interfering with those involved with updating the main database. This database is then used as a source for the data enrichment processes, to make implicit information explicit (such as creating urban extents, deriving networks from topographic features, etc.). The results of enrichments are stored in the detailed content database. This is then generalised to populate the lower resolution databases. Each content database is then used to populate specific product databases. Content databases at each resolution (detailed, local, district, regional) are used as a single source for different products at similar resolution. The intention is to reuse the same content data for efficiency and consistency purposes.

All databases will be updated incrementally. The change data will arrive in the staging content database, and be generalised to update all the lower resolution content databases. The regime of update for the products is product dependent. The change will be processed in chunks—something we call clusters. A cluster is a set of features affected by change that need to be processed together. These are dynamically constructed, sometimes as an aggregation of smaller static clusters (such as partitions formed by road lines). This aggregation ensures that the change will be encapsulated inside a cluster and not across its boundaries. There will be several types of cluster used to propagate different types of change. This change only update mechanism is only theoretical at the time or writing, and has not been implemented beyond a quick proof of concept.

This architecture results in a lot of data redundancy. For example if the only difference between two resolutions for a particular theme is that small features have been eliminated at the lower resolution, then all the other features are present in the same form in both databases. This is not a big issue for data that does not need to be manually edited. However, if several products at the same resolution need manual finishing, we try to minimise the data redundancy to minimise the manual effort required. For example, manual editing on a common product database is done before styling them differently in separate 'child' product databases. This is only possible if the different styles have certain similarities, such as the size or width of symbols.

11.5.3 Technology Used

A number of technologies are used by the MRDP programme to build the production systems. Some have been chosen because they are already available and widely used in Ordnance Survey. This is the case with Oracle being chosen as the database provider, and ArcGIS as the editing platform. However, we needed to integrate into the system a way of developing bespoke generalisation algorithms, as no existing platform provided the generalisation tools required to meet all the requirements. Ordnance Survey therefore started a competitive tender process at the end of 2009 to choose a platform for developing its own automatic derivation tools. The contract was awarded to 1Spatial in January 2011, allowing Ordnance Survey developers, in partnership with 1Spatial, to develop processes based on Radius Studio and Radius Clarity. These two platforms have been used to perform the data enrichment, content generalisation and product generalisation tasks (Fig. 11.11). More details on these platforms and how they have been integrated for the needs of MRDP can be found in Regnauld et al. (2012).

11.5.4 Results: OS VectorMap District

MRDP has recently completed the production of OS VectorMap District v1.0. Products from the OS VectorMap family have been designed to be easily customisable to create bespoke contextual maps for websites and applications. In addition to the usual raster format, these products are also available in vector format, which allows them to be quickly loaded into a GIS, where each layer can be turned on or off, and styled as required. OS VectorMap is available as part of OS OpenData and current users include local authorities, emergency services, and the insurance industry. It is also an ideal entry-level product for market demographic displays and can be used by anyone for sharing statistics or for neocartography.

Prior to this release, OS VectorMap District had already been made available to the public in the form of alpha and beta versions. While these early versions were produced using research prototypes, the latest one was the first one produced by an MRDP system, in which all the processes from the data extraction, to the publication process have been automated. More information on this system can be

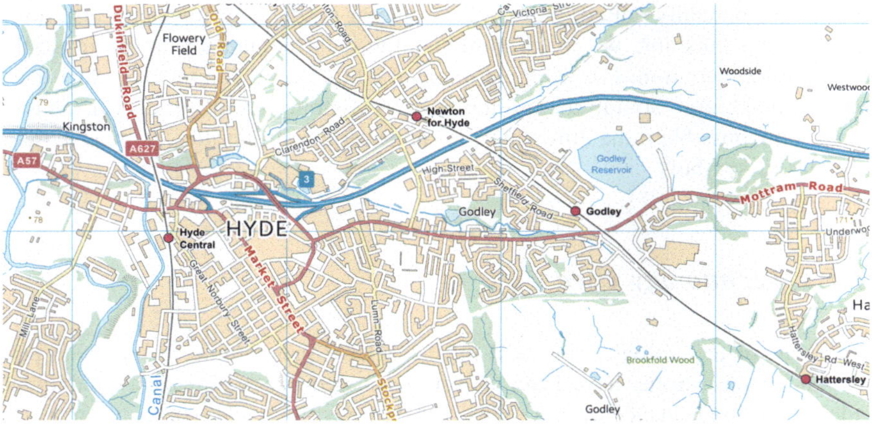

Fig. 11.12 Extract from Os VectorMap District v1.0, west of Manchester. Ordnance Survey © Crown Copyright. All rights reserved

found in Regnauld et al. (2013). Most of the generalisation processes used come from the initial research prototypes, but have been enhanced to ensure they can be maintained and reused by the programme. The detail of the generalisation processes that have been developed for deriving OS VectorMap District can be found in Revell et al. (2011). They include:

- Building generalisation (aggregation, simplification);
- Coastline: creation of a continuous coastline, followed by simplification (Zhou and Jones 2004);
- Vegetation generalisation (aggregation, simplification);
- Railways conflation and generalisation;
- Name extents conflation and filtering.

Version 1.0 of the OS VectorMap District also contains a better depiction of roads, thanks to the use of a process to automatically collapse dual carriageways (Thom 2005). Figure 11.12 shows an extract of OS VectorMap district v1.0.

Figure 11.13 shows an example of use of OS VectorMap District, where only selected layers have been used and combined with height data to produce a map illustrating flood risks in Carlisle.

11.5.5 Future Plans

Now that MRDP has delivered the first system to automatically derive a product from the base data, the next steps are to extend the capabilities of the system to tackle other products which are still maintained using old systems relying heavily on manual editing. For more complex cartographic products, manual editing will

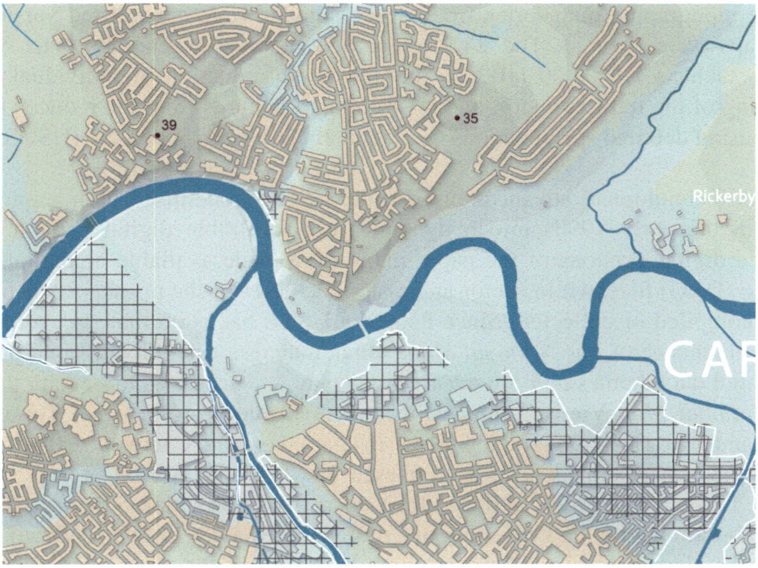

Fig. 11.13 OS VectorMap District combined with a height product to show flood risk in Carlisle. Ordnance Survey © Crown Copyright. All rights reserved

still be required, and one of the challenges will be to add the manual editing process into the workflow. The system will also be extended to perform updates incrementally. While the system has been designed with incremental updating in mind, it has not been implemented in the current system used to produce OS VectorMap District. This was left out in order to stage the development of the system, to avoid tackling too much complexity at once. OS VectorMap District is derived automatically, so incremental updating is not a critical requirement. It will become critical when the system is developed further to include manual updating, as we will want to minimise the loss of previous manual modifications.

11.6 Generalisation Methods Used for the USGS National Map and National Atlas

Lawrence V. Stanislawski, Cynthia A. Brewer and Barbara P. Buttenfield

Original 7.5 minute, 1:24,000 scale (24 k) topographic maps for the United States were handcrafted and field verified paper maps produced and subsequently revised by the United States Geological Survey (USGS) between 1945 and 1992 (USGS 1955; Moore 2011). During this time the 7.5 minute maps formed the foundation for the topographic map series from which smaller-scale generalised versions of the maps were derived. The 24 k maps were manually generalised to smaller scales, such as 1:100,000 (100 k), through photo-reduction of individual maps, panelling of the reduced maps, and re-scribing of features from the reduced colour plates using detailed specifications to adequately reduce map content (USGS 1985, 1993).

With technological advances in GIS, the USGS modernised its mapping program in the early 1990s into a dynamic web-accessible digital version. First, scanned digital versions of the paper maps were made available as digital raster graphic (DRG) files, while vector and raster databases of the primary themes were being compiled or collected. Since then, the USGS has coordinated development and maintenance of *The National Map*, which manages and delivers eight primary geospatial data themes and provides web display and access to these data through *The National Map* Viewer, along with downloadable up-to-date US Topo (24 k only) and historical topographic maps (Sugarbaker and Carswell 2011). Traditional USGS paper maps have been phased out of production.

The USGS Center of Excellence for Geospatial Information Science (CEGIS), in collaboration with the University of Colorado at Boulder and Pennsylvania State University, has coordinated research and development for automated generalisation to support multi-scale display and delivery of *The National Map* and other USGS data. A product of this research is a set of tools that automate methods to generalise the high-resolution (HR) layer of the National Hydrography Dataset (NHD) to 24 k and smaller scales. The NHD is one of the eight themes of *The National Map* and furnishes the surface water component for US Topo maps. Section 11.6.1 describes use of the tools to automatically generalise local resolution NHD to 24 k for display on the 24 k USGS Topo maps. Section 11.6.2 presents automated generalisation of high-resolution National Map data to smaller scales.

11.6.1 Generalisation of Local Resolution (Large Scale) Hydrography for 24 k US Topo Maps

US Topo maps are freely downloadable digital maps that are modelled from the traditional 24 k paper topographic map. The maps are published in Portable Document Format (PDF) with a geospatial extension that is called Georeferenced PDF (GeoPDF®) as patented by TerraGo Technologies. They are mass produced through automated procedures that access and display national GIS databases and through semi-automated annotation and editing methods (Moore 2011; Sugarbaker and Carswell 2011). Currently (2013), the USGS produces about 18,000 24 k US Topo maps per year, with a planned refresh cycle every 3 years (Moore 2013). Although US Topo maps presently have less feature content and aesthetic quality than the traditional paper map series, the maps are superior for several reasons:

they include a high-resolution orthophoto image and the best available, up-to-date data (three year refresh cycle); they are produced in a single modern coordinate system; and they are rapidly produced and freely downloadable through the web which reduces time and expense for users (Moore 2011).

Because of densification efforts implemented by state and local partner organisations, the HR NHD is a multi-scale layer that must be generalised in places from local resolution (1:12,000 or larger) to 24 k for consistent representation on US Topo maps. CEGIS NHD generalisation tools call existing ArcGIS® functions (version 9.3 and later) within customised geoprocessing scripts to generalise subsets of NHD data, which are stored in Esri file geodatabase format. As described in Chap. 6 and elsewhere (Stanislawski and Buttenfield 2011; Buttenfield et al. 2011), the tools automate enrichment, pruning, feature simplification and other generalisation operations, along with validation for subsets of NHD data, while maintaining spatial topology and the NHD model schema for use with other applications.

The NHD generalisation tools execute stratified network thinning operations by separately thinning features in assigned density partitions to target densities (Stanislawski and Savino 2011). More details about stratified network thinning for roads are presented in Chap. 6. Subsequent to network thinning, less prominent polygon features are eliminated by feature type and standards-based minimum area constraints (US EPA and US DOI 1999). Channel hierarchy is assigned to retained flowline features based on network topology and estimated upstream drainage area (Anderson-Tarver et al. 2012). Lastly, retained features are differentially generalised to reduce granularity (e.g. vertex spacing) but preserve geometric characteristics, local density, and texture. The sequence and type of generalisation operations applied in this phase of processing are tailored to local landscape and feature conditions (Stanislawski and Buttenfield 2011; Buttenfield et al. 2011).

Proper network topology and spatial relations among overlapping feature types are maintained according to NHD schema rules. However, supplementary processes are still needed to ensure coincidence among simplified features, such as among simplified coastlines and associated lake or ocean polygon boundaries. Typical subbasins, covering about 25 24 k US Topo maps, require about an hour or less for generalisation processing. Pruning parameters are automatically estimated from archived 24 k benchmark data. Selection of thinning parameters is guided by the Radical Law (Töpfer and Pillewizer 1966) in the absence of benchmarks. Default simplification parameters are available for typical density classes, but require some manual review and adjustment.

Figure 11.14 shows a section of the 24 k US Topo map for Snake Mountain, Vermont with local resolution hydrography and with local resolution hydrography generalised to 24 k using the NHD generalisation tools. Local resolution for Vermont is compiled at 1:5,000 (5 k). Four partitions having local resolution densities of 0.86, 1.69, 2.63, and 3.78 km per square kilometer (km/km^2) were defined for the flowline features within the NHD subbasin associated with the Snake Mountain map. Flowlines within these partitions were separately pruned to densities of 0.54, 1.07, 1.66, and 2.38 km/km^2, respectively, based on the 24 k

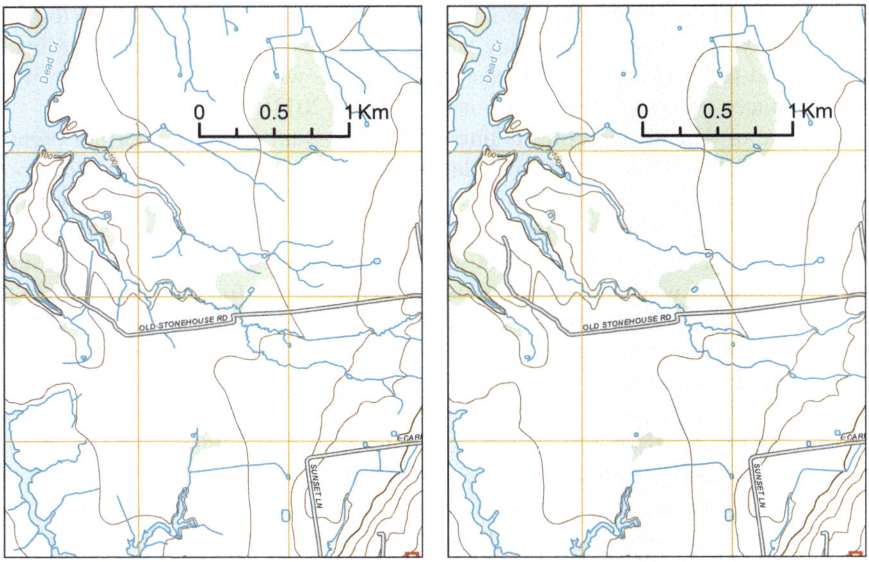

Fig. 11.14 Portion of 1:24,000-scale (24 k) US Topo map of Snake Mountain, Vermont without orthoimage displayed at 1:50,000 with local resolution hydrography (*left*) generalised to 24 k (*right*)

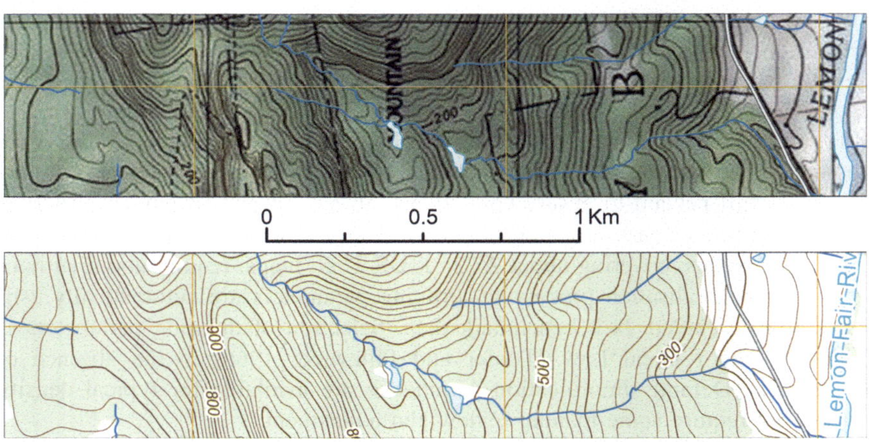

Fig. 11.15 Portion of 1:24,000-scale (24 k) US Topo map of Snake Mountain, Vermont displayed at 24 k without orthoimage. Top panel includes UTM grid and local resolution (1:5,000-scale) hydrography *over* the 1980 digital raster graphic, and the *bottom* panel shows local resolution hydrography generalised to 24 k, along with the map features typical to this product. Stream symbols have been widened to aid visualisation

NHD benchmark. Retained linear and polygonal features were differentially generalised to reduce vertices using the Bend Simplify algorithm (Wang and Muller 1998) available in ArcGIS®.

A closer view of another portion of the Snake Mountain map reveals the effect of line simplification and the quality of integration between the generalised hydrography and contour lines (Fig. 11.15). For each 24 k map, contours are vertically integrated with generalised hydrography through a fully automated hydro-enforced contour generation process (Arundel et al. 2010). This process ensures that stream lines follow valleys and that contours properly pass through double-line streams, without intersecting other waterbodies.

11.6.2 Generalisation for National Atlas (Smaller Scale) Hydrography and Multi-Scale Display

This section briefly reviews generalisation methods applied by the USGS to compile the National Atlas hydrography and presents some generalisation methods and enhanced topographic map designs being tested for multi-scale display of *The National Map* and the National Atlas data.

The 1 M hydrography of the National Atlas was recently produced for the United States by generalising the 100 k NHD through automated and semi-automated feature selection, simplification, and refinement operations. Requirements for this multi-year task were defined for the stream network, waterbodies, coastlines, and associated stream gauges. Feature selection and generalisation criteria, as well as criteria for integration between hydrography features and between hydrography and other associated National Atlas features, such as roads and boundaries, were included. Criteria for selecting which 100 k NHD stream and waterbody features to retain in the 1 M National Atlas were largely based on whether the features exist in one of several benchmark datasets. Integration criteria imposed proper snapping of connected features (e.g. stream ends to coastlines) and displacement between features (e.g. between streams and roads), which was implemented to ensure cartographic clarity. Retained streams and waterbodies were simplified using a 500 m tolerance with the Bend Simplify (Wang and Muller 1998) algorithm in ArcGIS®. Additional procedures adjusted stream densities and feature geometries. For further details see Gary et al. (2010).

The USGS has been working with Pennsylvania State University to redesign the US Topo maps for enhanced clarity over images and for multi-scale display of geospatial data. To do this, generalisation is applied to hydrography, roads, land cover, Geographic Names Information System (GNIS) point features, and elevation themes of *The National Map*. Automated and manual procedures are largely implemented through existing or customised ArcGIS® geoprocessing tools and editing environment. The following describes some of these efforts.

As demonstrated here and by others (Brewer and Buttenfield 2010; Buttenfield et al. 2011), the NHD Generalisation tools can be used to generalise the HR NHD to smaller scale Level-of-Detail (LoD) datasets that can support graphics over a limited range of scales (Cecconi et al. 2002). Figure 11.16 shows three *hydrography* LoDs.

At 50 k (Fig. 11.16a), the 24 k NHD data are shown with minor modifications. Double-line stream areas are replaced by selected primary paths available in the enriched network data, collapsing the area to a line, and small ponds are removed using a minimum area threshold. The 100 k LoD (Fig. 11.16b) has more extensive line pruning, waterbodies removed according to an increased threshold, and differentially simplified lines and polygon boundaries. At 1 M (Fig. 11.16c), National Atlas hydrography displays simplified major rivers and large waterbodies. At all scales, lines are tapered using a small number of width classes based on upstream drainage area for NHD and stream order for National Atlas data, symbolising these attributes in a simple hierarchy.

Local roads are systematically thinned using the ArcGIS® Thin Road Network tool (Sect. 6.5). The tool is applied in a ladder fashion, thinning on minimum lengths incrementing from 500 m to 30 km. Resulting visibility settings are combined into a 34-level attribute. The visibility field controls removal of the least important roads and labels, guides priorities by which the Esri Maplex Label Engine places text labels, and enables emphasis of labels for roads that are major thoroughfares within a mesh of urban roads (Brewer et al. 2013). Overlaid on this hierarchy of text labels and thinned local roads is a network of interstate and state routes, with their hierarchy established by separate road-class attributes. Comparison of Fig. 11.16a–c shows fewer local roads (light with a brown casing) that make connections across the highway network as scale decreases.

A basic landscape characterisation was produced using the National *Land Cover* Dataset (NLCD). These data were processed by raster operators to achieve a generalised, smoothed, and interpolated representation at 24 k, resampling this layer repeatedly for representations at smaller scales. Interpolation was necessary because 30 m-resolution is too coarse for 24 k display. For classified land cover data, random pixels at raster region boundaries were combined with proximal pixel values to create an "airbrushed" effect for 2 m-pixel upsampled land cover. For canopy and impervious surface data, ladder-style bilinear upsampling from 30 to 2 m pixels interpolated graded representations. Hues with similar lightness across all land cover categories produce visually vague boundaries, and these colours grade to white for high percent impervious surface and to dark greens for high percent canopy cover (Fig. 11.16a, b).

Populated places from GNIS are categorised into text label size categories based on related decennial census data from the U.S. Census Bureau. Low-population places are removed as display scale decreases. Topographic maps provide an important resource for emergency response in the United States. A selection of *structure types* from GNIS point features is shown in Fig. 11.16. Points for fire station, police, hospital, airport, and school locations are selected using feature codes, with additional queries on name content to remove point subsets. Text labels and symbols for less important structure types are removed at smaller scales; for example, schools are removed from Fig. 11.16b, c.

Terrain form is also generalised for smaller scales (Fig. 11.16). At largest scales, the 10 m-pixel (1/3 arc-second) National Elevation Dataset (NED) is smoothed once using a low-pass filter with a 3×3-pixel kernel, and then similarly smoothed

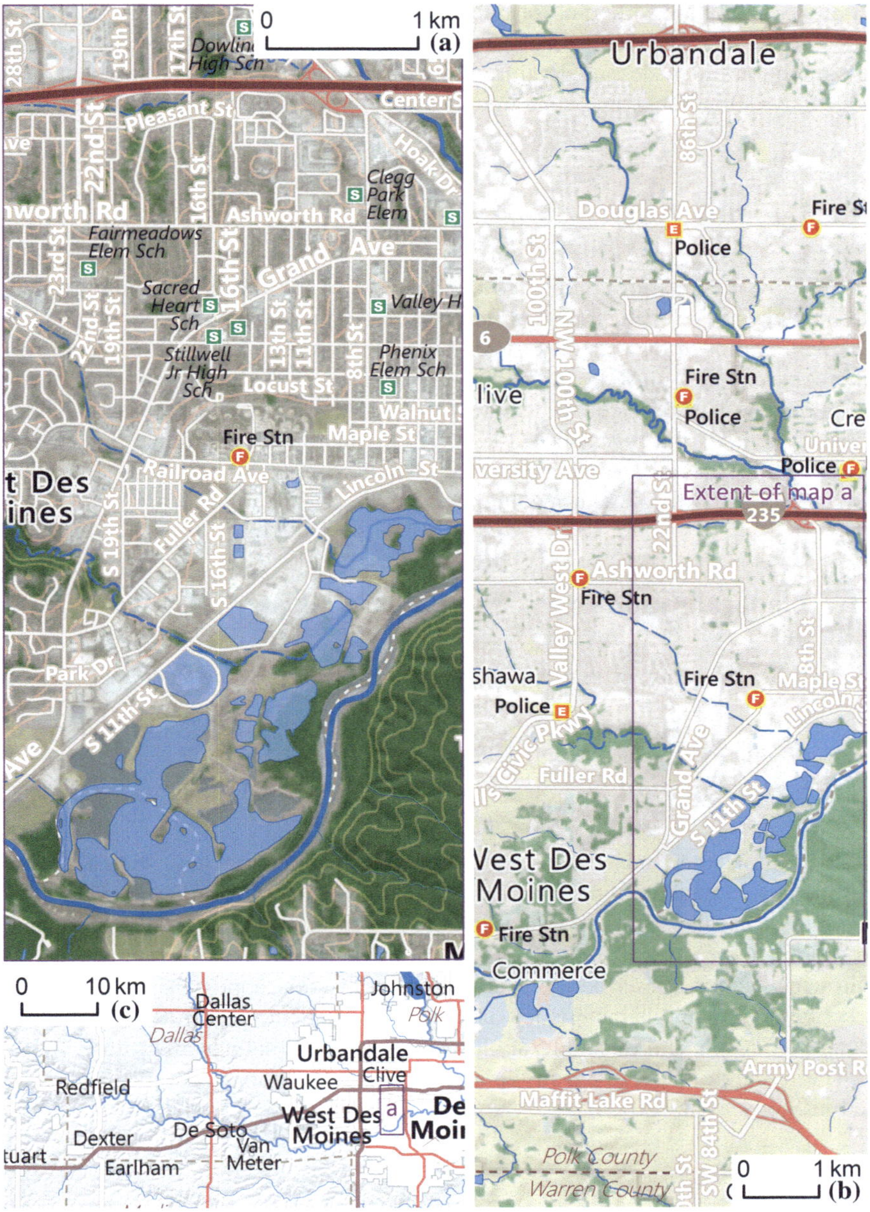

Fig. 11.16 Three example scales that anchor the continuous-scale topographic map design in development, showing a portion of West Des Moines, Iowa. Figures show generalised roads and hydrography: **a** at 1:50,000 with land cover, shaded relief, contours, and an orthoimage background; **b** at 1:100,000 with landcover and shaded relief; **c** at 1:1,000,000 with *shaded* relief only. These landscape layers can each be set off or on, and the design is intended to function for all 16 possible combinations of the four

ten times with a 5 × 5 kernel for middle scales. The 30 m (1 arc-second) NED is smoothed for a base layer at small scales. Smoothed NED data are used to produce five-direction hillshades, a curvature layer (to selectively highlight and generalise ridge and valley lines), and contours for use at large, middle, and smaller scales.

11.6.3 Future Plans

Much work remains for complete modernisation of the USGS mapping program. Research has developed strategies and tools to automate generalisation of some of *The National Map* data themes to LoD datasets required for multi-scale display. These procedures must be refined with additional automation and integration techniques to ensure adequate feature snapping between and within data themes. Further methods may enhance 24 k US Topo maps with additional and revised feature content. Plans are being designed to implement automated generalisation to expand US Topo maps with medium and small scale products.

11.7 Generalisation in Production at Kadaster NL

Jantien Stoter, Vincent van Altena, Ron Nijhuis and Marc Post

11.7.1 Introduction

In 2010 the Netherlands' Kadaster, the national mapping agency, started a feasibility study to examine how automated generalisation techniques could be introduced into its map production environment. The study led to a fully automated workflow to produce 1:50 k maps, which will come on line from 2013. The following sections describe the Kadaster generalisation approach. Section 11.7.2 describes the main characteristics of the approach, Sect. 11.7.3 describes implementation details and Sect. 11.7.4 closes with results, findings and future plans.

The research that resulted in the automated generalisation workflow is described in Stoter et al. (2013).

11.7.2 Main Characteristics of the Kadaster Approach

The feasibility study on automated generalisation first focused on the workflow from 1:10 k data to a 1:50 k map, to be extended to other scales once successful results had been achieved. From the beginning it was clear, that the automated generalisation workflow should not aim to replicate the existing map, even though the new automatically generated 1:50 k map will replace the existing map. There were several reasons for this. Firstly, starting from legacy topographic products—more than 60 years ago—might overemphasis past requirements and may ignore new opportunities of multiscale topographic information. For example, topographic information is used in more applications by a wider public than ever before and the user may prefer up-to-date maps over maps that meet all traditional cartographic principles (though the results should still be of an acceptable quality). Furthermore, automating a previously interactive process, which was designed in a previous technical and organisational context, can be very complicated (Foerster et al. 2010; Stoter et al. 2009b). Another reason to reconsider existing map requirements is that the time-saving aspect of automation makes the process well suited to the production of multiple products that meet different demands. The consequence of this is that the requirements of automatically generating multi-products for a certain scale differ from the requirements of the existing single product. A few other aspects further refined the scope of the Kadaster approach:

- The focus is on producing a map. We acknowledge that users may require mid- and small-scale data for specific themes to be applied in spatial computations. Examples are water and road networks. For those themes separate data products will be considered;
- The topographic data of Kadaster forms a partition of space at every scale. This brings specific challenges, since every object that is removed needs to be replaced by something else. In addition, the displacement process always causes other objects to be displaced as well;
- Generalisation without any interaction is the best guarantee for efficiency and consistency and the only way to produce multiple on-demand products. Therefore interactive improvements of the generalised results is not allowed;
- The most straightforward way for updates in automated generalisation is to completely replace the old version map (Regnauld 2011). At the current time, we do not maintain links between the objects at the different scale levels. If these links are required, this will be part of a subsequent study.

One of the main challenges was how to redefine specifications for automated generalisation by modifying existing guidelines while ensuring that users requirements are met. Firstly, we generated an initial 1:50 k map in a semi-automatic manner by extending the work of Stoter et al. (2009a, 2012). The aim of this first step was to see how much automation we can achieve with currently available tools and some in house developed algorithms. This work implemented existing generalisation guidelines for interactive generalisation in an automated process and improved the implementation by evaluating intermediate results.

The initial map was sent to a selection of key customers of the existing 1:50 k map to test the main principles and assumptions. Based on these insights the process was improved and refined and implemented as one integrated workflow. In the next stage the evaluation and improvement process was repeated, by asking more costumers as well as a panel of typical individual users to assess the resulting map in more detail and for different areas. Based on those evaluations and iterative testing, the optimal sequence of steps was determined as well as the most appropriate algorithms and parameter values for each step. Internal validation via cartographic experts was also undertaken. After two years, the feasibility study resulted in a fully automated generalisation process implemented as an integrated workflow in 2013.

11.7.3 Implementation Details

The following four subsections describe the software, the pre-processing of the data, the implemented automated generalisation workflow and finally the approach that was applied to automatically generalise the whole country.

11.7.3.1 Software and Technology

For the implementation we used a mixture of standard ArcGIS tools, self-developed tools within Python and a series of FME tools. ArcGIS contains some specialised generalisation tools for collapsing two lanes of a road into a single road line, displacing symbolised geometries, simplifying symbolised buildings and thinning networks (Punt and Watkins 2010).

The complete generalisation workflow is implemented within the Model builder tool of ArcGIS. The workflow consists of three main models, consisting of about 200 sub models that are responsible for each specific generalisation problem that we need to solve in the process.

11.7.3.2 Pre-Processing the Data

Since the goal is 100 % automation, the process should cover as many generalisation aspects as possible. This is accomplished by either improving the process step-by-step, or—if that did not work—by improving and enriching the source data. In addition to correcting errors, enrichment of the source data is undertaken in two ways. Either external data sources are used or the required knowledge is made explicit via cartometric analysis. Examples of enrichment of the input data are determining urban extents by defining areas with high density of buildings (i.e. higher than 10 %) and attributing TOP10NL road segments with information on highway exits so that these can be treated in a different manner in the generalisation process from the road network (derived from the TOP10NL roads).

11.7.3.3 Implemented Workflow

The implemented generalisation workflow consists of the following steps:

1. Model generalisation to reduce the data that has to be visualised;
2. Symbolisation of the data;
3. Graphic generalisation to solve cartographic conflicts.

11.7.3.4 Model Generalisation

The model generalisation process is the largest part of the process. Model generalisation is not only conversion of geometric objects to the lower density and structure of the TOP50 model, but also translation and reclassification of attributes to the TOP50 model. Examples of the operations performed in this part of the process are:

- Convert TOP10NL road centrelines (representing single lanes) into TOP50 centrelines (a complete road). The TOP50 centrelines are generated with an ArcGIS algorithm that merges two lanes of a single road into one centreline.
- Convert areas with many buildings (i.e. coverage >10 % which has been experimentally determined as an optimal threshold) to built-up areas and remove the original building objects in those areas. This conversion is only done for areas located within urban area (computed) and if they are not part of an industrial estate (external data). Important buildings such as schools, hospitals, and churches are kept as separate point symbols. This functional information is added in TOP10NL.
- Prune the road network where there is insufficient space to visualise all roads. This is not straightforward as can be seen in Chaudhry and Mackaness (2005), Thom (2007) and Thompson and Richardson (1999). Specifically because TOP10NL data does not contain many attributes for pruning. The pruning of the road network consists of several steps:

    a. Cycle paths parallel and touching roads are deleted; the others (i.e. free cycle paths) are maintained.
    b. Access roads to buildings in rural areas are detected and selected. The automated detection finds access roads "if buildings are located within 200 m of the end of the road AND no other roads are located in the neighbourhood (within 200 m) of the building".
    c. The remaining road network is pruned by the 'thin road network algorithm' available in ArcGIS. This algorithm retains connectivity and the general character while using a hierarchy of the relative importance and the minimum length of a road.

- Prune hydrographic network. First by removing small water bodies parallel and touching roads and then by applying the thin road network algorithm on the linear waterways. Lake shorelines and other water area boundaries are added in this process to maintain connectivity. Thinning is done with the road network algorithm because the Dutch water network is almost completely anthropogenic and its structure resembles a road network rather than a natural water network. Therefore algorithms available for pruning natural hydrographic networks, as studied for example in Stanislawski and Savino (2011) are less appropriate.

The last step in the model generalisation process is the reduction of the data by filtering the vertices of all linear geometries (linear objects and polygon boundaries) by applying the Douglas-Peucker algorithm (parameter value 1 m).

After the model generalisation, the symbolisation process assigns symbols to all geometries, as they should appear on the map. The symbolisation sometimes results in objects that appear larger on the map than they are in reality. The symbolisation therefore governs the next graphic generalisation process that solves the cartographic conflicts. In this process basic symbols are used which precisely correspond to the shape and outline of features, but which lack cartographic refinement. Sophisticated symbolisation and cartographic enhancement to enhance legibility (such as aligning symbols with features) are postponed to a later stage in the process.

11.7.3.5 Graphic Generalisation

The graphic generalisation process consists of the following steps:

1. Generalise (i.e. select, simplify and displace) the buildings that remain after the data generalisation process to avoid overlap and to meet a minimum building size.
2. Displace the linear objects (roads, water) and boundaries of symbolised water and terrain objects as well as all other point and linear objects (i.e. administrative boundaries, height contours, engineering constructs) with an algorithm that displaces symbolised objects and reshapes them in order to avoid overlap. The displacement algorithm uses a hierarchy of object types.
3. Rebuild terrain and water polygon-objects from the displaced boundaries and assign the former codes to the new areas by using left/right information of the boundaries.

11.7.3.6 Country Wide Coverage

To be able to generalise a map for the whole of The Netherlands, we apply the workflow on about 400 generated partitions, obtained from linear objects that must never be displaced, which are the main roads such as highways. Besides some global operations that are applied for the whole country (such as creating and simplifying the power line network), the workflow is applied per partition and partitions are connected afterwards. Because vertices of objects at and near partition boundaries are prohibited from moving in the displacement process, the objects at neighbouring partitions still fit together after generalisation. To process the generalisation of the 1:50 k map from 1:10 k source data for the whole country in a reasonable time we make use of the multiprocessing capabilities within Python. This allows us to process six partitions in parallel on each of the six available systems.

11.7.4 Results, Findings and Future Plans

Figure 11.17 shows the 1:50 k map that is generalised from the TOP10NL data using the fully automated workflow described above. Although our users could not identify significant differences with the traditional version, differences do exist because of the changed map specifications. For example the interactive generalisation guidelines prescribe that detached houses may never be converted into built-up areas. It was impossible to always meet this rule, because detached houses are often enlarged to meet the minimum size requirement (15 × 15 m). Most building blocks containing detached houses are covered with detached houses on both sides, so the widths of the building blocks should be at least 40 m in order to accommodate enlarged detached houses at a minimum distance of 10 m (required for readability). At the same time streets in TOP50 were symbolised with a line width of 20 m, which is wider than streets widths in reality. To solve this, we decided to always convert detached houses in built-up areas, if the building density threshold was exceeded.

Another example of a specification that was modified related to ditches which were not treated as separate objects but as ordinary terrain boundaries, because of their limited importance and because their visualisation is identical to terrain boundaries. All water with linear geometries in TOP10NL (i.e. smaller than 2 m) parallel to roads is eliminated in the new map to save space as we regard them as unimportant. We allow dikes to be displaced. This is because dikes are not available as single objects in TOP10NL, but only visible by hatches on the map. User consultation showed the unimportance of this rule in combination. Given the difficulty of identifying dikes, it was decided to ignore this guideline. Dikes are still present on the 1:50 k map (with hatches), but they may have been moved to make space for other objects.

Based on the results and user evaluations the Kadaster decided that a fully automated generalisation workflow was the most sustainable workflow for the future as well as the only way to produce products on-demand. The model will be implemented from early 2013. From that moment the new maps will replace the existing 1:50 k map product. With thirty-six parallel processors run via Python we are able to perform the core generalisation process for the whole of the Netherlands in approximately 50 h. The aim is to achieve a three week turnaround including pre-processing, generalisation and visualisation.

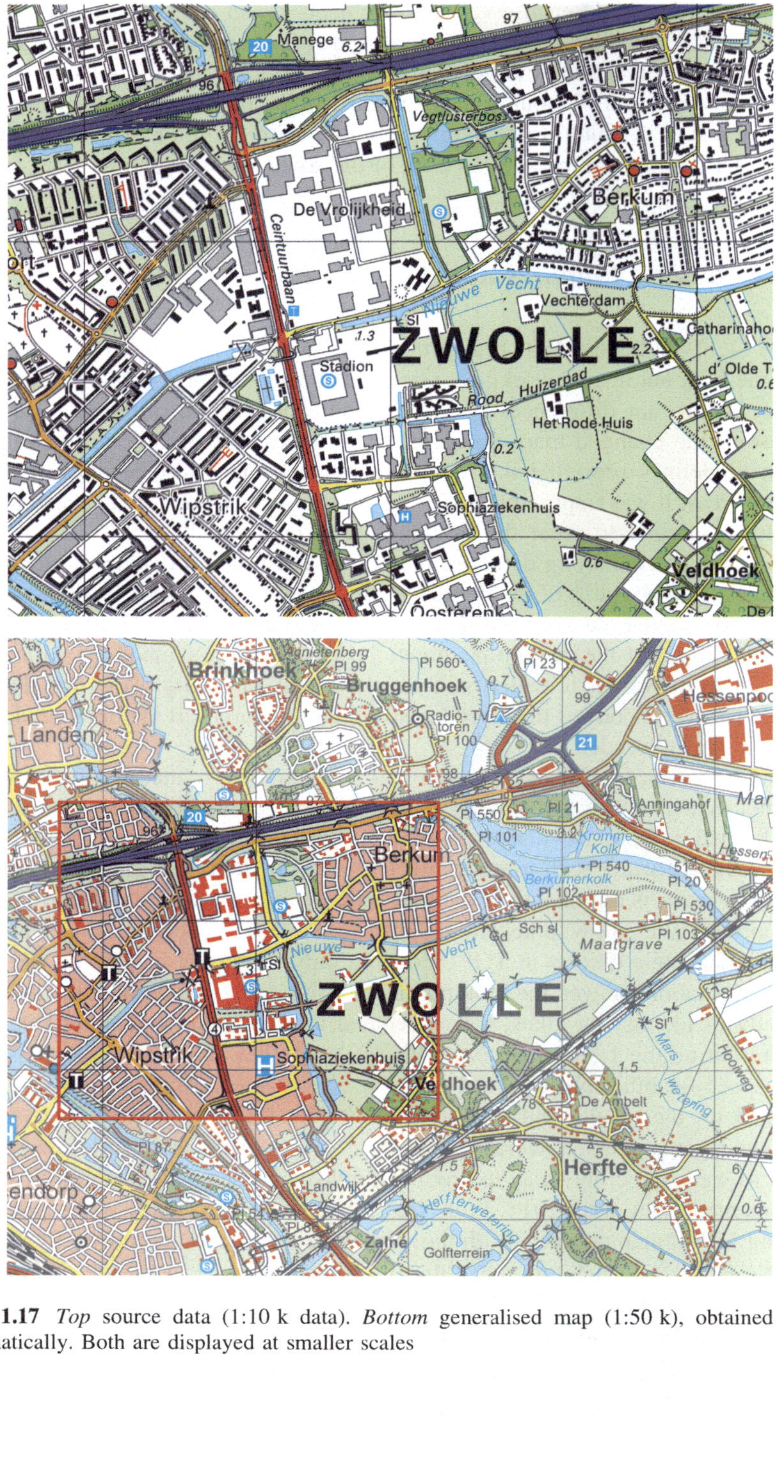

Fig. 11.17 *Top* source data (1:10 k data). *Bottom* generalised map (1:50 k), obtained fully automatically. Both are displayed at smaller scales

Based on the experiences with the new 1:50 k product, the automated generalisation approach will be extended to the 1:100 k map and to on-demand products, such as the backdrop map at multiple scales for the national geo-portal. These products will all be derived directly from TOP10NL, and therefore TOP10NL will become even more important as the base data. Future work will also study the derivation of TOP10NL from the countrywide large-scale data set (approximately 1:500), which is collected by organisations other than Kadaster (mainly municipalities) according to an information model that was established in 2012, called IMGeo (Information Model Geography). Beside scale differences this derivation process needs to cover different perspectives on topographic data, which will be an interesting challenge.

11.8 AdV-Project "ATKIS: Generalisation"—Map Production of DTK50 and DTK100 at LGL in Baden-Württemberg

Sabine Urbanke and Antje Wiedemann

The "Working Committee of the Surveying Authorities of the States of the Federal Republic of Germany" (AdV) established the "Official Topographic-Cartographic Information System" (ATKIS) which was established to map the landscape of all "German Länder" in 1998. Subsequently the automatic derivation of Digital Topographic Maps (DTK) from this database the Digital Base Landscape Model (Basis-DLM) has become increasingly important. For that reason AdV has developed the "ATKIS-Generalisation" project in 2002. The overall objective of the project was to develop software tools to enable the production of Topographic Maps (TK) within a highly automated process, based on a two-stage approach: Model Generalisation (MG) and Automatic Cartographic Generalisation (AKG). Currently twelve "German Länder" are collaborating together with commercial partners to address this task.

11.8.1 Map Production Workflow

In the Basis-DLM the real world objects are modeled as point, line and area vector objects with additional information, such as names and regions. The landscape is structured according to the ATKIS Object Catalogue (OK) in objects and differentiated by attributes. The Basis-DLM is a database with following features:

- 2D georeference,
- high positional accuracy (±3 m), digitised from Orthophotos with a reference scale of 10 k,

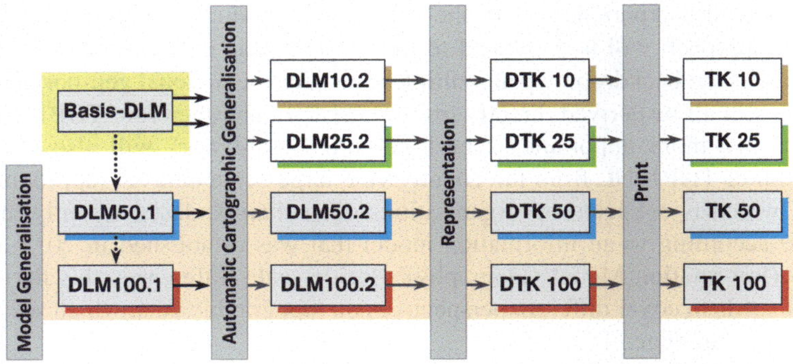

Fig. 11.18 Graphical representation of workflow

- covering the complete national area.

The Basis-DLM is the only database which is continually updated for the entire country. All other DLM are derived by automated processes from this Basis-DLM (Fig. 11.18). The map data are generated within a multi-stage process. The process starts with Generalisation of Basis-DLM data using Model Generalisation producing DLM50.1 (DLM100.1), followed by Automatic Cartographic Generalisation producing DLM50.2 (DLM100.2). The result is subsequently represented, according to the requirements of the style sheet catalogues SK50 (SK100) for DTK50 (DTK100). All the data are stored in a database and can be provided to customers in a specially defined format (NAS-format) and are printed as TK50 (TK100). The generalisation processes are entirely rule-based and can be run independently and separately from each other.

11.8.2 Generalisation Tool: Model Generalisation

The Model Generalisation (MG) is a fully automated process for converting a detailed model (e.g. Basis-DLM) into another, lower density, positionally accurate model (e.g. DLM50.1). The tool has been developed by the company 1Spatial (UK) based on the product Lamps2. The individual generalisation processes are independent of graphic design requirements. They are divided into the steps of semantic and geometric generalisation. The biggest advantage of the Model Generalisation process is to reduce the data volume, which is currently around 33 %.

Semantic Generalisation Step: The transition from the Basis-DLM to the DLM50.1 is defined in "transfer rules". This means the object types and their attributes which are defined in the object catalogue of DLM50, are transferred directly or are changed into a different object type (e.g. through elimination or aggregation). Selection criteria define when an object is to be transferred and what its minimum size must be in order to transfer it to the new database. The tool

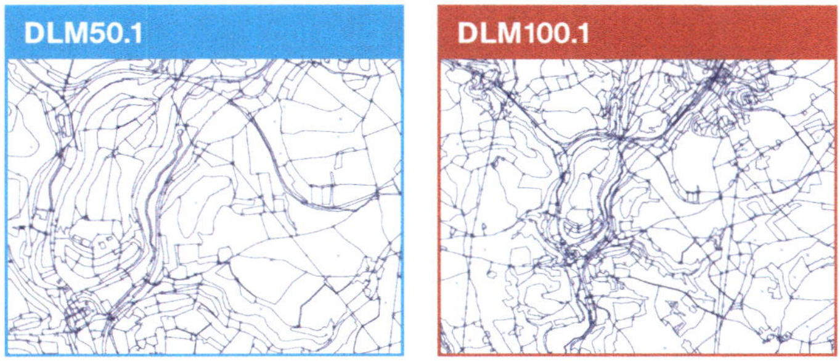

Fig. 11.19 Results of model generalisation

checks these conditions and eliminates or merges objects or attributes automatically so they are semantically correct. A similarity matrix regulates and evaluates the semantic combination of the object types.

Geometric Generalisation Step: In a geometry type change (e.g. area to line or area to point) the topological relationships between neighbouring objects must be restored, so that no topological gaps arise in the data. This is done by resolving these objects and creating new geometries according to the modelling rules. Using Douglas-Peucker algorithm, the vertices of linear objects are generalised geometrically, resulting in a thinning of the number of points. This contributes significantly to a reduction in the volume of data. Relations between origin and target objects are kept during the transfer from Basis-DLM to DLM50.1 (DLM100.1).

The production of the DLM50.1 dataset for the whole country takes 95 h. To produce the DLM100.1 dataset the Model Generalisation process is started again with a modified parameter set, running for 29 h (Fig. 11.19).

Number of objects for whole Baden-Württemberg:

- Basis-DLM (without buildings) = 5.659.421 objects
- DLM50.1 = 3.872.878 objects
- DLM100.1 = 1.385.956 objects

11.8.3 Generalisation Tool: Automatic Cartographic Generalisation

Derivation of the Digital Topographic Maps DTK50 and DTK100 is performed by a fully automated batch process, using the Cartographic Generalisation tools of RadiusClarity and Radius ClearText from 1Spatial.

The Automatic Cartographic Generalisation tool differs from the Model Generalisation tool such that the DLM50.1 (DLM100.1) dataset is presented according the style sheet catalogue SK50 (SK100). Display of the model generalised geometries causes graphical conflicts (such as overlaps), which are detected by the

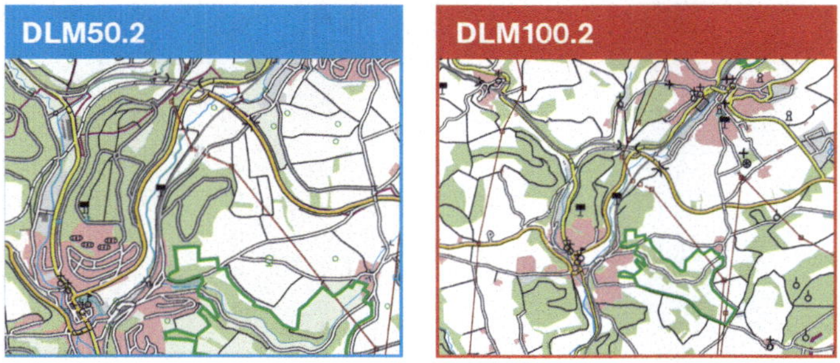

Fig. 11.20 Results of automatic cartographic generalisation

generalisation tool and solved by multiple processes. The solution to the graphical conflicts is realised by geometric and semantic generalisation rules within an adjustable workflow defining which cartographic generalisation method (displacement, simplification, typification, removing, merging, enlargement, amalgamation or enlargement) has to be used to solve a graphical conflict.

The tool uses a variety of solution methods. On the one hand there are fixed algorithms, such as the Visvalingham algorithm for simplification of areas or the Plaster algorithm to simplify lines. The AGENT-technology (Acronym for: Automatic Generalisation New Technology) has also been used. This technology allows us to model a map object as an agent which may conflict with other map objects, controlled by constraints to reach a given target. By considering these constraints, minimum size, minimum distance, minimum edge length, priority of the object, displacement and movability), the agent can search for the most optimal solution for the map object. The agent has a fixed life cycle to achieve the most suitable result to solve a graphical conflict by analysing various possible solutions (plans). Each plan is evaluated using a "happiness factor". After a specified number of trials all results are compared to a predefined required minimum value. If the agent does not reach the desired "happiness factor", the conflict is not resolved automatically and has to be edited manually.

Due to the variability of the generalisation tool it is possible to define each condition, parameter and workflow scale independently. So each map scale can be processed with these generalisation tools. The complete generalisation process consists of 50 individual sub-processes which can be built up one after the other. The sequence of the sub-process steps is variable and can be managed in many ways. Important generalisation steps can be repeated several times. The result of each sub-process is stored and can be used as a fall back to resume the process (Checkpoints) which allows a thorough review at the end. The processing run time of Automatic Cartographic Generalisation for the whole Baden-Württemberg is about 5 weeks for the DLM50.2 and about 3 weeks for the DLM100.2 (Fig. 11.20).

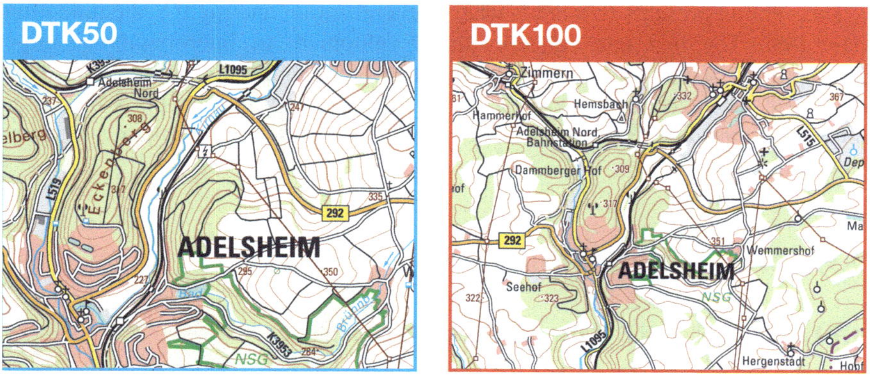

Fig. 11.21 DTK50 and DTK100 Standard Version

11.8.4 Summary

The generalisation processes for the creation of DTK100 and DTK50 (Fig. 11.21) has been used in production since summer 2011 and summer 2012 in Baden-Württemberg. Since at present not all generalisation conflicts can be solved satisfactorily in the Automatic Cartographic Generalisation process, a corresponding manual post-processing effort using the representation component is necessary. Therefore, further development and improvement of the generalisation tools are extremely important to the project. Overall, the solution already shows today that the processing and production times due to the automatic generalisation processes can be significantly reduced. So the long term aim of Baden-Württemberg is an improvement in the currency of topographic maps for all scales from 5 years to 1–2 years.

11.9 Multi-Scale Data in Spatial Data Infrastructures: Developments in INSPIRE at the JRC

Julien Gaffuri and Katalin Tóth

INSPIRE is a European directive establishing a European spatial data infrastructure (SDI). The main purpose of this infrastructure is to improve the availability and interoperability of European spatial data for environmental applications, in particular "(…) to ensure consistency between items of information which refer to the same location or between items of information which refer to the same object represented at different scales." (INSPIRE 2007, article 8.3). Different developments have been undertaken over the last few years to achieve this goal—multiple representation and generalisation have been identified as key elements for INSPIRE developments (Tóth 2007).

This section presents two fields where generalisation and multi-scale modelling were used. The first one concerns the development of interoperability target specifications, where various modelling patterns for multiple representation have been used. The second concerns a new web framework that supports the publication of vector data. For the latter we show the crucial role of generalisation and multi-scale modelling.

11.9.1 Multiple Representation in INSPIRE

11.9.1.1 The INSPIRE Data Specification Challenge

The significant achievement of INSPIRE is the release of harmonised specifications for European spatial data. These data specifications concern 34 themes such as transport networks, hydrography, geology, weather, land use, sea regions, species, etc. (INSPIRE 2007, annexes). An important challenge has been to deal with this multi-disciplinary data. For each theme, various communities with different needs and cultures have been involved. Depending on their own use cases and their specific scales of interest they often have different ways to represent the same real world objects. Overall each theme specification covers a wide thematic area and range of scales. Figure 11.22 proposes a schematic view of the INSPIRE scope and the applied scales. Cross-thematic use cases exist where two INSPIRE themes overlap—for example, the intersection between the themes transport and hydrography corresponds to water transport infrastructures.

11.9.1.2 Multiple Representation Modelling Patterns in INSPIRE

The following multiple representation modelling patterns have been used in the INSPIRE data models:

1. *Scale dependent annotations*. A commonly used modelling pattern in INSPIRE is the explicit annotation of spatial representations with their relevance to scales. For example, geographical names are characterised by a range of scales for which a given instance of a name has to be displayed. Statistical units may be represented by different geometries depending on scale extents. Some reusable scale extents are represented as lists of codes (e.g. the NUTS levels and the CityGML levels of detail). Other indirect indicators of scale are defined at the level of the datasets using specific metadata elements defined by the ISO 19000 standards series (e.g. resolution and spatial accuracy).
2. *Profiles*: When significantly diverse representations of the real world are required for various sets of use cases, different 'profiles' may be defined. When possible, the representations of the same entity in different profiles should be linked. For example, in the hydrography theme three different profiles are defined for mapping, network analysis and reporting. For the theme buildings

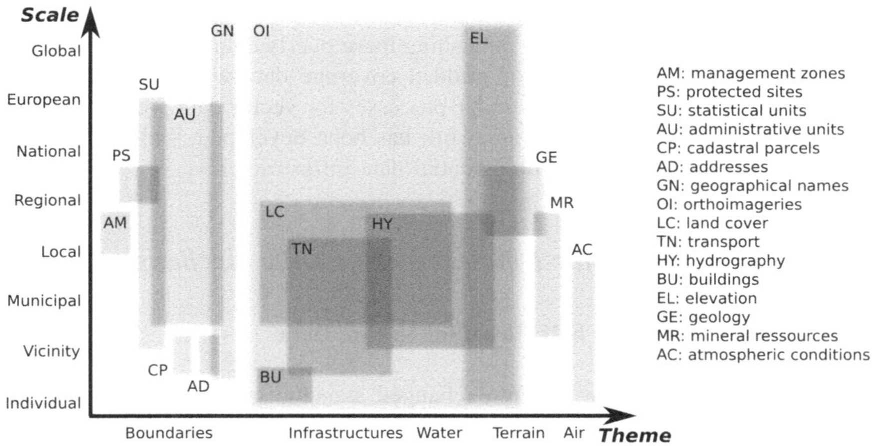

Fig. 11.22 Proposal for an INSPIRE scope representation in theme/scale space

2D and 3D data have separate profiles. Almost all data specifications are split into a simplified core profile (which contains the legally binding part of the data specification) and one or more profiles that contain additional information elements for more specific, and usually more ambitious use cases.

3. *Object and coverage views*: For many themes, depending on the scale of interest, the real world phenomena are represented as sets of objects or as coverages. For example, the themes atmospheric conditions and human health cover measurement points of physical or chemical parameters as well as interpolated coverages based on these measurements.
4. *Multi-resolution grid*: Multi-resolution geographic grids have been defined to support gridded data (orthoimagery, elevation, statistical units, etc.).
5. *Semantic level of details*: To support semantic consistency between different levels of details, hierarchical classifications are used for several themes. This includes the use of class hierarchies (specialisation/aggregation relationships) and code hierarchies (e.g. the Eurostat NACE code list on economic activities). An interesting benefit of such hierarchies is to ease the mapping of existing models with INSPIRE models: when no mapping can be found with concepts for detailed hierarchical levels, mapping are often easier to find for more generalised hierarchical levels.

11.9.1.3 Scale Gap in Spatial Data Infrastructures

As illustrated in Fig. 11.22, some parts of the scale/theme space are not yet covered by INSPIRE specifications, especially for the European scale: Some themes cover the whole scale axis, while others are limited to a very detailed range of scales. The discrepancy of scale between the data demand (pan-European applications) and offer (what is available from the national data providers) may manifest in performance issues in publishing these overly detailed datasets online. While the visualisation process of gridded coverage data widely benefits from using pyramidal data structures, similar processes for vector data are still in their infancy. For this reason a new framework has been developed, which allows an efficient usage of vector data within spatial data infrastructures.

11.9.2 A Framework for Efficient Vector Data Exchange

11.9.2.1 Next Generation SDI Based on Vector Data

Nowadays, spatial data are mainly exchanged as static images. Vector data are used only on the server side to prepare static images of the objects. Spatial data infrastructures, such as INSPIRE, are expected to support the online publication and on-the-fly utilisation of raw vector data. The reason why vector data are not used directly by the users can be explained by the performance: Vector datasets are too cumbersome to be exchanged over the network and rendered on-the-fly by the clients. In order to support the shift from static images toward dynamic vector data in SDIs, new techniques are required to ensure that only the relevant information is exchanged between the servers and the clients. For this purpose, a framework based on the following components has been developed (Gaffuri 2012):

- *Scale reference system* (*SRS*). Besides the well-known coordinate reference system, there is a need for a well-defined "scale reference system" to properly index vector data to their scale range of relevance. Taking into account the Y-axis of Fig. 11.22, twenty-one static scale levels have been defined from a local to a global scale, using a factor two between consecutive scales.
- *Generalisation methods*. In order to reduce the size of the vector data to be exchanged, model generalisation transformations are used on the server to reduce the semantic and geometric level of detail of the raw vector data. Multi-scale spatial databases are built with a suitable generalisation process for each scale level defined in the SRS.
- *Vector data tiling and spatial indexes*. In order to improve the performance of the exchange and deliver vector data only for the client's area of interest, vector tiling is used to decompose large vector objects into smaller pieces. For small and compact objects, a spatial index service is used. The tiling grid and the spatial index structure are both based on the same quad-tree structure comparable to image pyramids used in traditional raster web-mapping servers. These techniques together with generalisation allow a space/scale indexation of the data.
- *Vector and style formats*. In order to improve the vector data transfer, compact vector and style formats such as GeoJSON are used.

- *Client/server for vector data.* Specific servers and clients able to deal with these spatially indexed multi-scale vector data are used. The main capability of these components is to be able to exchange the relevant data according to the client's spatial extent and zoom level. "Scale-aware" network protocols are used. Of course, the client has the ability to render vector data on-the-fly.

11.9.2.2 Prototype and Tests

A prototype for this spatial data infrastructure based on vector data has been developed and tested. The Java4INSPIRE library was used to transform existing datasets according to the corresponding INSPIRE specifications. For model generalisation transformations the GIS Java libraries JTS and Geotools were enriched with vector tiling and spatial indexing algorithms, together with some model generalisation algorithms (object simplification, aggregation and clustering). For the GeoJSON encoding, the MapFish library was used. Two different clients were developed: one was a Java applet, and the second was based on HTML5 canvas element using the GWT library.

This prototype was successfully tested for several use cases based on Eurostat's statistical units datasets, the transport datasets of the European Environment Agency and the species distribution dataset of GBIF (http://www.gbif.org/) shown in Fig. 11.23. The raw dataset of the latter consists of 80,000 species observations represented as points. A simple clustering method based on a distance threshold is used to aggregate observations that are too close for each zoom level. The details on these observations are available only when zooming-in. In order to better visualise the distribution of these observations, a heat map visualisation method is made on-the-fly on the client side. Thanks to the generalisation and the spatial indexing, the data exchanged between the server and the client are thin; their transfer and rendering remain fast.

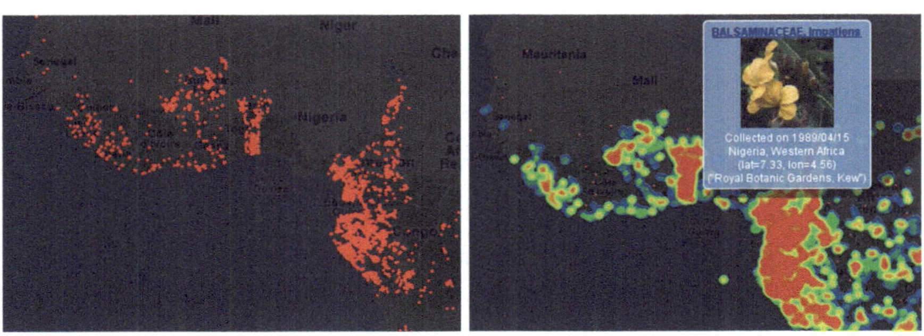

Fig. 11.23 Species distribution: raw data (*left*) and generalised data (*right*)

11.9.3 Next Steps

INSPIRE's data specification development process shows that further actions may be undertaken to progress towards a better formalisation of multiple representation modelling patterns. Standards on geographic information should better address the way to structure datasets with multiple representations. The way the level of detail of different spatial object representations is described may also be improved.

11.10 Synthesis: Recent Achievements and Future Challenges Regarding Generalisation in NMAs

This section provides a synthesis of the recent achievements, current trends, ongoing works and future challenges regarding generalisation in NMAs. This analysis is based on two sources of information. The first source of information is the set of contributions from NMAs included in this chapter—Sects. 11.2–11.7. Additional information was gathered from a larger set of NMAs during an NMA symposium held in Barcelona in March 2013, under the double cap of the ICA Commission on Generalisation and Multiple representation and the EuroSDR Commission on Data Specification, under the theme: "Designing MRDB and multi-scale DCMs: sharing experience between government mapping agencies" (ICA and EuroSDR 2013). This workshop was attended by representatives of the seven NMAs contributing to this chapter, as well as IGN-Belgium, GST-Denmark (former KMS), NLS-Finland, OSI-Ireland, IGN-Spain. Consequently, feedback from twelve NMAs is considered in total. This section is organised in two main parts followed by a conclusion. The first part summarises recent achievements and current trends regarding generalisation in NMAs, i.e. work that is already in production or has reached a certain maturity in several NMAs (11.10.1). The second part lists some challenges identified by NMAs, most of which have already begun to be tackled by work that is at a preliminary stage (11.10.2).

11.10.1 Recent Achievements and Current Trends Among NMAs

11.10.1.1 Steps in Introducing Automated Generalisation in Production Lines

Stoter (2005) and Foerster et al. (2010) identified the following steps in the introduction of automated generalisation in NMAs:

- renewing data models (from CAD-like "Map databases" to structured geographic databases, with a consistency between different levels of details),

- designing the conceptual architecture (decide what databases are derived from what databases),
- implementing generalisation processes (that actually perform automated derivation between databases), and
- managing relationships between scales.

Historically, not all NMAs have tackled these steps in the same order. The first attempts at introducing automated generalisation in production (about 10 to 15 years ago) concentrated more on the actual implementation of generalisation processes for some specific scales, widely relying on in-house developments—for example at ICC, Catalonia for 1:25,000 (Baella and Pla 2003) or at IGN, France for 1:100,000 (Jahard et al. 2003). Feedback from those first experiments (showing that automated generalisation in production was indeed possible), the availability of more and more advanced research results, and the increasing maturity of GIS software, have resulted in more NMAs taking up the challenge of automating their production lines. They have been able to do so with a much more mature and rational approach: by considering from the beginning the renewal of their data models (including consistency between levels of detail), the design of their conceptual architecture for generalisation (hereafter called "derivation scheme"), and the management of relationships between scales. The steps proposed by Stoter (2005) reflect such a rational approach to automation. Using such an approach, renewing data models was the first step. Therefore the survey by Foerster et al. (2010) logically revealed that this first step had already been completed (or mostly completed) by many of the NMAs in the survey. West-Nielsen and Meyer (2007) give a comprehensive review of how this process has been accomplished by GST (formerly KMS), Denmark. All NMAs contributing to this chapter (Sects. 11.2–11.7), or present at the Barcelona NMAs symposium in March 2013 (ICA and EuroSDR 2013), have now completed this step. However, it is interesting to note that ICC, IGN-France or OSGB have completed this step quite recently (Sects. 11.2, 11.3 and 11.5, respectively): this is because their first effort focused on the implementation of actual cartographic generalisation processes for some specific scales using in house developments, which necessitated well structured source databases, but not necessarily consistency between data models at different levels of detail.

11.10.1.2 Architectures for Generalisation and "Derivation Schemes"

The survey by Stoter (2005) was one of the first to analyse the conceptual architectures for generalisation in NMAs, which also resulted in promoting some already existing but not so widely used terms within the NMA community, which have now become widely accepted: DLM vs. DCM, model vs. cartographic generalisation, or star vs. ladder approaches.

DLM stands for "Digital Landscape Model", a term initially proposed by Grünreich (1985), which describes a geographic database where feature geometries and attributes are stored without consideration of any cartographic symbols.

This is different from DCM which stands for "Digital Cartographic Model": a database where the geometry of the features has been distorted or displaced to take into account the scale and cartographic symbols with which they will be displayed (Weibel and Dutton 1999, and Sect. 5.6 of this book). Model generalisation occurs between one DLM and another DLM. It is driven by the data model and resolution of the source and target data. Cartographic generalisation seeks to produce a DCM from a DLM of the same resolution or from a DCM of lower resolution, while taking into account symbol widths, legibility and separation thresholds. Finally, the terms "star" and "ladder" derivation have been introduced by Eurogeographics (2005) to refer to the derivation schemes according to which the databases of different resolutions are derived from each other within a NMA. In every case a "base DLM" exists, that corresponds to the finest level of detail. In the "star" approach every DLM or DCM is derived directly from the base DLM through independent generalisation processes. In the "ladder" approach, the base DLM is generalised into a coarser DLM (resp. DCM), which is in turn generalised into a coarser DLM (resp. DCM), etc. Again, in comparison with the survey of Foerster et al. (2010), progress has been made within NMAs regarding the choice of derivation schemes, since all NMAs contributing to this chapter or present at the Barcelona workshop have now completed this step.

OSGB is the only NMA among the contributors to this chapter that chose a star derivation scheme (Sect. 11.5): all coarser DLMs are derived from the so-called "detailed" DLM, which indeed includes not only the features that are captured from the real world, but also computed enrichments that will support the generalisation processes.

All other NMAs have chosen an approach that mixes star and ladder approaches, which confirms the trend already noticed by Foerster et al. (2010). Some NMAs still maintain several DLMs of different levels of detail independently (each of them acting as a base DLM in an independent derivation scheme), and identify the merging of their derivation scheme as a necessary further step. This for example, is the case of ICC, and IGN France (Sects. 11.2 and 11.3 respectively).

11.10.1.3 Linking Databases and Managing Incremental Updates

In addition to choosing a derivation scheme, NMAs also have to decide whether they want to maintain links between their different DLMs and DCMs, thereby resulting in a multi-representation database (MRDB). Creating and maintaining links presents several advantages, summarised by Mustière and van Smaalen (2007) as being: (1) ease of database maintenance and propagation of updates, (2) assessment of quality (when existing databases are linked, knowing which is of better quality), (3) potential for more efficient applications using these databases (including more powerful analysis and visualisation tools).

The proportion of NMAs that have decided or begun to implement the MRDB approach at least for a part of their DLMs and DCMs has increased between the survey by Foerster et al. (2010) and the 2013 Barcelona NMAs workshop: eight

NMAs out of twelve in 2013, vs. five NMAs out of eleven in 2010. The links between datasets can be built in two ways. The first way is to keep derivation links during the generalisation process, when one database is derived from another. This approach has been chosen by swisstopo, OSGB and AdV for all their DLMs and DCMs (Sects. 11.4, 11.5 and 11.8 respectively), and by IGN-France for the links between its DLMs and the DCMs that are semi-automatically derived from them (Sect. 11.3). The second way to build links is to use data matching techniques between two existing datasets. This approach has for instance been chosen by ICC (Sect. 11.2). Both techniques can be combined: NLS Finland designed an automated process to derive its 1:100 k DLM from its 1:10 k DLM, but guided the automated hydrographic network selection via its existing, interactively derived 1:100 k DLM. Only segments that match a feature of the existing 1:100 k DLM are retained (Pätynen and Ristioja 2009).

Although more and more NMAs have chosen the MRDB approach, feedback from MRDB management show mixed results to date. A few NMAs have successfully implemented automated update detection between DLMs at different levels of detail (IGN Belgium, GST Denmark). Updates are introduced in the more detailed DLM and the relevance of propagating them to coarser DLMs is automatically checked according to derivation rules, before the propagation is interactively performed. Automated update propagation between a DLM and a DCM of the same level of detail has been implemented (for example at IGN-France for the 1:100 k (Sect. 11.3). But the feedback from these update propagation experiments are not without reservations due to the lack of tools that can deal with links and objects life-cycle management in existing commercial GIS software. NMAs would like to define their own unique identifiers, while most commercial GIS work with internal identifiers. Human operators capturing updates in the base DLM often prefer to delete and re-create an object that had its geometry modified instead of modifying it, which results in missing update information. This could be partially overcome by more ergonomic tools to assist updates capture, although the definition of an update as opposed to a deletion and creation is in itself quite fuzzy. For IGN Belgium, who had set up an automated updating process from DLM 1:10,000 to DCM 1:50,000, these challenges meant that they went back to an interactive propagation approach supported by a semi-automated detection (Féchir 2013). Transformations performed on the data model of the base DLM are also difficult to propagate, which also led GST Denmark to reapply a complete derivation of their base DLM to coarser DCMs in 2013, after several successful updating cycles between 2010 and 2013 (Danish Geodata Agency 2013). We also note that fully automated incremental update propagation requires the NMA to decide what specific computed enrichments should be kept in target DLMs and DCMs (e.g. partitioning information), and what information should be stored regarding the applied generalisation process, so that this process can be locally re-applied while minimising the modifications to surrounding features. To date, only partial solutions have been experimented with.

As discussed in the 2013 Barcelona NMAs workshop, the decision to apply incremental updates rather than regularly perform a complete re-derivation is dependent on several factors, among which: (1) the cost of the derivation (the more manual edits, the more costly re-derivation becomes), and (2) the willingness to provide change only updates to users so that they can in turn propagate the updates on thematic data that they have captured based on the NMA dataset. As an example, Kadaster NL chose the complete re-derivation approach for their 1:50 k DCM, since the derivation process is fully automated and they do not consider the 1:50 k DCM as a suitable backdrop dataset to which users can add their own data—although it can be used as a backdrop map (Sect. 11.7).

11.10.1.4 Implementing the Generalisation Process

Of all steps needed to introduce automated generalisation in NMAs, implementing the generalisation process for the production of topographic maps has been the most comprehensively tackled topic of NMAs and researchers. While the survey by Foerster et al. (2010) stated that "full automated generalisation processes do not exist" and that only five out of the eleven considered NMAs had made major steps towards automation, in 2013 major progress has been achieved. Eleven of the twelve NMAs attending the 2013 NMAs workshop in Barcelona have implemented automated or semi-automated generalisation processes. Fully automated generalisation processes now exist in production. This is the case in OSGB Great Britain in the derivation of a "light" 1:25 k DCM from a mixed 1:1.25–1:10 k DLM (Sect. 11.5), at IGN France for the derivation of a "light" 1:25 k DCM from a 1:10 k DLM (Sect. 11.3), and at Kadaster Netherlands (Sect. 11.7) for the derivation of the 1:50 k DCM from a 1:10 k DLM, a scale change that is traditionally considered difficult to achieve.

This full automation was achieved while accepting compromises in terms of cartographic quality and differences compared to existing manually derived products. Another currently existing approach, sometimes in the same NMAs, is to include limited manual editing in order to reach different standards of cartographic quality. One of the factors that affect the feasibility of complete automation is the kind of features present in the target DCM and their representation. To date, manual edits are still needed to reach a good cartographic quality when individual buildings are kept in dense urban areas, which mainly explains the inclusion of manual edits in the production lines of the standard 1:25 k of ICC, IGN France or swisstopo (Sects. 11.2, 11.3 and 11.4 respectively). It is also worth to notice that preserving the topological relations in the target DCM enables its further generalisation in a ladder scheme or its automated incremental updating, but potentially increases the amount of manual editing required.

Automated generalisation processes currently in use in NMAs are based on commercial GIS software distributed by 1Spatial, Axes-Systems, ESRI and the University of Hanover, that have been customised to take into account the specifics of each NMA, either by the vendors or by the NMAs themselves, relying partly on recent research results. These efforts in development, particularly intensive over recent years, have helped close the gap identified by Stoter (2005) and Foerster et al. (2010) between research and production.

11.10.1.5 Producing Maps for Delivery Over the Web

Until recently, producing paper maps was the main focus of NMAs. But in parallel most of them have developed solutions that produce raster versions of their paper maps, either sold to customers or delivered over the web through geoportals. The derivation of vector DCMs has created new opportunities. The main focus is still on cartographic products similar to traditional paper maps (e.g. traditional scales are still considered), but all NMAs understand the importance of delivering both paper and digital cartographic products. Vector web versions based on the produced DCMs are now available or under development in most NMAs, who also work on homogenising the symbolisations across scales or on delivering services that enable users to customise the display, at least by interactively selecting or unselecting layers. For a few NMAs, the main delivery channel is via the web, with printing left to the users or only purpose made, possibly outsourced to third party vendors (USGS, GST Denmark).

11.10.2 On-going Work and Future Challenges Around Generalisation in NMAs

Section 11.10.1 highlighted the fact that the introduction of automated generalisation has made significant progress over the last years. The continuously increasing availability of geographic data over the web and on mobile devices, the increasing variety of associated uses and the evolution of related economical models (Chap. 1 of this book) raise new challenges for NMAs regarding generalisation. NMAs are no longer the only producers of topographic DLMs. People can use data produced by commercial producers such as Google or from collaborative projects such as OpenStreetMap. Open Street Map data suffer from strong heterogeneities (e.g. Girres and Touya 2010), and cartographic services based on both OpenStreetMap and Google data offer limited cartographic quality (only DLMs are available, which are displayed with cartographic symbols without any cartographic generalisation). Despite this, they are sometimes more up-to-date than the data produced by NMAs and can be obtained for free. This, combined with the open data movement, is raising the expectations of users in terms of currency and low cost: they expect data to be up-to-date, and free. Producing DCMs of high cartographic quality is no longer sufficient—although cartographic quality also contributes to better decision making. Therefore NMAs have begun to adapt their economic models while minimising manual post-processing and decreasing update cycles, and to study ways of delivering new services that build upon their know-how and cartographic expertise. At the same time, the economic context is forcing administrations to decrease their number of employees. As a result, NMAs are facing the challenge of producing more, with fewer resources. From this we can distil four major challenges that have been identified and partially tackled by NMAs.

11.10.2.1 Increasing Effectiveness of DLM and DCM Production

The first challenge currently identified by NMAs is to produce faster with fewer personnel and less financial resource; one ambition being to reduce the update cycles of DLMs and DCMs. This challenge already led to greater automation in production environments, and explains the willingness expressed in all NMAs contributions to this chapter to go further with automation, whilst reducing manual intervention. Producing faster with fewer resources can be achieved at least in four ways: by improving the effectiveness of the automated generalisation processes, by adopting an incremental updating approach and therefore limiting the amount of generalisation required, by improving evaluation so that manual edits can be well targeted, and by reducing the intended level of cartographic quality (traditionally very high in most NMAs) in order to privilege fitness for use.

All four approaches can be variously combined. As stated in Sect. 11.10.1.4, one of the use cases that still requires intensive manual edits is the generalisation of 1:50 k DCMs where individual buildings are kept even in dense zones. However, recent research results show significant progress in the automated generalisation of complete topographic datasets at this scale (for example see Touya and Duchêne 2011, and Sect. 7.2 of this book). In the same way, recent research has provided good results in the automated evaluation of cases that typically require manual edits (for example Zhang 2012, and Sect. 9.7 Case study I). It is likely that further performance and quality enhancement can be expected in production in the near future.

An approach using incremental updates appears promising and has now been adopted by several NMAs which has led to a real need for new tools and methods. Tools for change detection have been successfully tested in production, and the needs for further tools and methods are well identified: further work is needed to develop tools for data management in commercial software and to set up methods for completely automated generalisation of updates with minimal modifications on surrounding objects. These improvements will take time.

It is also important to consider adjustment in the quality of outputs, especially regarding the cartographic details within DCMs, while recognising that it may be worthwhile reasonably decreasing the intended cartographic quality of a DCM compared to traditional manually produced maps, in order to increase its currency. The challenge is in deciding what is deemed to be a "reasonable decrease". Although users are not necessarily aware of it, cartographic quality enables better decision making. Seeking for maps that are cartographically perfect is counter-productive. Decreasing the intended cartographic correctness of produced maps too much would decrease their usability, and result in a vicious circle where the cartographic expertise of NMAs would be lost. To find a satisfactory compromise, better knowledge of how the users actually use the produced DCMs is necessary (as was evident from the NMAs present at the Barcelona workshop). Surveys involving user panels, such as the one reported by Kadaster NL in Sect. 11.7, are a good example of user requirements acquisition.

11.10.2.2 Building SDIs that Support the Integration of Heterogeneous Data

In addition to improving the automation of their derivation processes, NMAs are working on the rationalisation of their update collection processes. Indeed, they become more and more integrators of data collected by other providers, especially administrations. NMAs are also responding to user feedback and the activities of VGI contributors. For example the USGS has recently announced experiments to use crowd sourced POIs captured by volunteers through dedicated platforms. From the generalisation point of view, this raises the question of how to describe the level of detail of the data and how to ensure relevant interactions between integration and generalisation processes. Indeed, both processes are interrelated. Generalisation can be used in support of integration to homogenise the level of detail among multi sourced data. Integrating data is a pre-requisite to better generalisation. The same issue arises in the context of the creation of the European SDI based on data provided by European states, as required by the INSPIRE Directive. Creating this framework is in itself a real challenge. An analysis of these questions was presented in Sect. 11.9, together with some proposals and preliminary results.

11.10.2.3 New Services: Automated On-Demand Derivation of Customised Cartographic Products

With the increasing availability of geographic data, more and more users (professionals as well as the public) want to combine their own or third party data with NMA data—for visualisation and automated analysis. NMAs have begun to investigate solutions to this need. One approach is to deliver specific "light" DCMs intended to act as backdrop data on which users can capture their own data or overlay existing data, as proposed by OSGB with VectorMap District (Sect. 11.5). As a complement to this, online services can be offered to customise vector backdrop maps (conventional DCMs produced by the NMA or specific light DCMs) by selecting themes to display and adjusting symbol colours, and to overlay them with third party data (via a geoportal).

An alternate, more sophisticated solution, is to build services that enable the creation of so-called "analytical maps" integrating topographic and user data in a consistent manner. Such an approach automatically assists the user in the choice of backdrop data, integrates the user's data with the NMA data, and generalises the map whilst taking into account the user's data and their relations with topographic data, and the user's requirements. Such processes are still at the research stage (e.g. Chaps. 2, 5 and 7 of this book).

Even when no additional data are considered, the production of vector DCMs instead of just paper or raster maps opens up a lot of possibilities for NMAs in terms of on-demand customisation of their printed maps. They can enable the user to choose or build an original portrayal instead of the using the standard NMA symbols—for example portrayals with customised colours, or similar to an old map. Work in this area is on-going (e.g. Christophe 2012), and some results have already been achieved (Sect. 2.7 of this book). Their incorporation in production environments is under study at IGN-France. Such forms of customisation could also include generalisation, ideally on-the-fly, either to adapt the scale or to display using the chosen symbols.

11.10.2.4 Maintaining Composite Datasets Including User Data

In addition to integrating their own data with NMA data, users may wish to maintain such composite data over time. This use case is well known by NMAs, who all have customers who already use their DLMs for such purposes, usually by capturing their own data manually with the NMAs data acting as a reference. NMAs could support such use by offering automated integration services to enable the creation of consistent composite data from existing user data. And they could support the maintenance of such composite data by delivering updates in an incremental manner, and by offering services to semi-automatically propagate the delivered updates to the user data. This again demands better models and tools to manage incremental updates in GIS and map production software.

11.10.3 Conclusions

The utilisation of automated generalisation in NMAs has recently made significant progress. Further developments are ongoing. Longer term challenges for NMAs are currently being tackled via several research studies, although they are not solely dedicated towards NMAs. The last and probably most important challenge regarding generalisation in NMAs concerns communication. It is important to carry on encouraging communication between researchers, NMAs and GIS vendors, and increase communication with users of cartographic products and services, so that future research and development efforts keep pace with user needs and expectations. Only then can we ensure that results are converted as quickly as possible into operational tools that can eventually benefit the public at large.

Acknowledgments The authors would like to thank all participants of the NMAs symposium "Designing MRDB and multi-scale DCMs: sharing experience between government mapping agencies" held in Barcelona on 21–22 March 2013 for the rich exchanges they enabled regarding current practices of generalisation and future challenges for NMAs. They represented the NMAs contributing to Sects. 11.2–11.8 of this chapter, as well as: IGN-Belgium, GST-Denmark (former KMS), NLS-Finland, OSI-Ireland, IGN-Spain.

References

Anderson-Tarver C, Gleason M, Buttenfield BP, Stanislawski LV (2012) Automated centerline delineation to enrich the National Hydrography Dataset. In: Xiao N, Kwan M-P, Goodchild M, Shekhar S (eds) GIScience 2012. 7th international conference on geographic information science, Columbus, September 2012. Lecture Notes in Computer Science, vol 7478. Springer, Berlin, p 15

Arundel ST, Whitaker S, Phillips L, Thiem P (2010) An overview of automatic digital contour production for the USGS US Topo graphics product. Paper presented at AAG annual meeting, Washington DC, 14–18 Apr 2010

Bader M (2001) Energy minimization methods for feature displacement in map generalization. Dissertation (PhD Thesis), University of Zürich

Baella B, Pla M (2001) Map names generalization at the Institut Cartogràfic de Catalunya. In: Proceedings of the 5th ICA workshop on progress in automated map generalization, Beijing, 2001. http://generalisation.icaci.org/images/files/workshop/workshop2001/baella-pla_v0.pdf. Accessed 25 June 2013

Baella B, Pla M (2003) An example of database generalization workflow: the Topographic Database of Catalonia at 1:25,000 In: Proceedings of the 7th ICA workshop on progress in automated map generalization, Paris, 2003. http://generalisation.icaci.org/images/files/workshop/workshop2003/baella_et_al_v1.pdf. Accessed 29 May 2013

Baella B, Palomar Vázquez JM, Pardo Pascual JE, Pla M (2007) Spot heights generalization: deriving the relief of the Topographic Database of Catalonia at 1:25,000 from the master database. In: Proceedings of the 11th ICA workshop on generalisation and multiple representation, Moscow, 2007. http://generalisation.icaci.org/images/files/workshop/workshop2007/Blanca_ICAWorkshop2.pdf. Accessed 29 May 2013

Baella B, Lleopart A, Pla M (2012) ICC topographic databases: design of a MRDB for data management optimization. In: Proceedings of the 15th ICA workshop on generalisation and multiple representation, jointly organised with EuroSDR commission on data specifications, Istanbul, 2012. http://generalisation.icaci.org/images/files/workshop/submissions2012/genemr2012_session4_sub4.pdf. Accessed 29 May 2013

Balley S, Parent C, Spaccapietra S (2004) Modelling geographic data with multiple representations. Int J Geograph Inf Sci 18(4):329–354

Bard S (2004) Quality assessment of cartographic generalisation. Trans GIS 8(1):63–81

Barrault M, Regnauld N, Duchêne C, Haire K, Baeijs C, Demazeau Y, Hardy P, Mackaness W, Ruas A, Weibel R (2001) Integrating multi-agent, object-oriented, and algorithmic techniques for improved automated map generalization. In: Chinese Society of Geodesy Photogrammetry and Cartography (eds) Proceedings of the 20th international cartographic conference, Beijing, 2001

Bobzien M, Petzold I, Burghardt D (2007) Automated derivation of a 1:300,000 topographic Map from Swiss DLM VECTOR 200. In: Proceedings of 23rd international cartographic conference, Moscow, 2007

Boffet A, Coquerel C (2000) Urban classification for generalization orchestration. In: ISPRS 2000. Proceedings of the 19th congress of the international society for photogrammetry and remote sensing, Amsterdam, 2000

Brewer CA, Buttenfield BP (2010) Mastering map scale: balancing workloads using display and geometry change in multi-scale mapping. Geoinformatica 14:221–239

Brewer CA, Stanislawski LV, Buttenfield BP, Raposo P, Sparks KA, Howard MA (2013) Automated thinning of road networks and road labels for multiscale design of The National Map of the United States. Cartography and Geographic Information Systems (in press)

Buttenfield BP, Stanislawski LV, Brewer CA (2011) Adapting generalization tools to physiographic diversity for the United States National Hydrography Dataset. Cartography Geogr Inf Syst 38(3):289–301

Cecconi A, Weibel R, Barrault M (2002) Improving automated generalization for on-demand web mapping by multiscale databases. In: Proceedings international symposium on geospatial theory, processing, and applications, Ottawa
Chaudhry OZ, Mackaness WA (2005) Rural and Urban road network generalization deriving 1:250,000 from 1:1250. In: Proceedings of the 22nd international cartographic conference, a Coruña, 10–16 July 2005
Christophe S (2011) Creative colours specification based on knowledge (ColorLegend System). Cartographic Journal—ICC Paris 48(2):138–145
Christophe S (2012) Cartographic styles between traditional and original (towards a cartographic style model). In: AutoCarto'12. Proceedings of the 17th AutoCarto international symposium on automated cartography, Columbus, 16–18 Sep 2012
Danish Geodata Agency (2013) ICA Barcelona—Generalisation and MRDB. Presentation at the ICA/EuroSDR NMA Symposium, Barcelona, 2013. http://generalisation.icaci.org/images/files/workshop/symposium2013/presentations/2013_03_21_BarcelonaNMAMeeting_GST.pdf. Accessed 25 June 2013
Eurogeographics (2005) Generalisation processes: a benchmark study of the Expert group on Quality. Eurogeographics internal report, 2005
Féchir A (2013) The topo-geographic inventory. Presentation at the ICA/EuroSDR NMA Symposium, Barcelona, 2013 http://generalisation.icaci.org/images/files/workshop/symposium2013/presentations/2013_03_21_BarcelonaNMAMeeting_IGNBelgium.pdf. Accessed 25 June 2013
Foerster T, Stoter JE, Kraak M (2010) Challenges for automated generalisation at European mapping agencies: a qualitative and quantitative analysis. Cartographic J 47(1):41–54
Gaffuri J (2007) Field deformation in an agent-based generalisation model: the GAEL model. In: GIDAYS'07. Proceedings of the 5th geographic information days, Münster, 10–12 Sept 2007
Gaffuri J (2012) Toward web mapping with vector data. In: Xiao N, Kwan M-P, Goodchild M, Shekhar S (eds) GIScience 2012. 7th international conference on geographic information science, Columbus, September 2012. Lecture Notes in Computer Science, vol 7478. Springer, Berlin, p 87
Gary RH, Wilson ZD, Archuleta C-AM, Thompson FE, Vrabel J (2010) Production of a national 1:1,000,000-scale hydrography dataset for the United States—feature selection, simplification, and refinement. Scientific Investigation Report 2009-5202, U.S. Department of Interior, U.S. Geological Survey
Girres JF, Touya G (2010) Quality assessment of the french OpenStreetMap dataset. Trans GIS 14(4):435–459
Gondol L, Le Bris A, Lecordix F (2008) Cartography of high mountain areas—testing of a new digital cliff drawing method. In: Proceedings of the 6th ICA mountain cartography workshop, Lenk, 11–15 Feb 2008
Grünreich D (1985) Computer assisted generalisation: Papers CERCO-Cartography Course. Frankfurt am Main, Institut für Angewandte Geodäsie
Guislain P, Lecordix F (2011) The ever improving GIS map: evolution of cartographic representations and map production processes—Lessons learnt from the IGN topographic map production project. In: Proceedings of the 25th international cartographic conference, Paris, 3–8 July 2011
ICA and EuroSDR (2013) ICA/EuroSDR NMA Symposium 2013. http://generalisation.icaci.org/index.php/activities/12-next-events-details/48-nma-symposium-2013-general. Accessed 17 June 2013
INSPIRE (2007) Directive 2007/2/EC of the European Parliament and of the Council of 14 March 2007 establishing an Infrastructure for Spatial Information in the European Community (INSPIRE). http://inspire.jrc.ec.europa.eu/. Accessed 29 May 2013
Jaara K, Lecordix F (2011) Extraction of cartographic contour lines using digital terrain model (DTM). Cartographic J 48(2):131–137

Jahard Y, Lemarié C, Lecordix F (2003). The implementation of new technology to automate map generalisation and incremental updating processes. In: International Cartographic Association (eds) Proceedings of the 21st international cartographic conference, Durban, 2003
Jakobsson A, Ostensen O, Lovell D, Hopfstock A, Mellum R, Kruse D, Portele C, Urbanas S, Hartnor J, Bray A, Aslesen L (2013) European location framework—One reference geo-information service for Europe. In: Proceedings of the 26th international cartographic conference, Dresden, 25–30 Aug 2013
Jenny B, Hutzler E, Hurni L (2010) Scree representation on topographic maps. The Cartographic Journal 47(2):141–149. http://www.screepainter.com. Accessed 29 May 2013
Kereneur R, Thomas J, Grémeaux N (2011) Generalisation rules for ExM medium to small scale. ESDIN Project, Deliverable D10.6. http://www.esdin.eu/project/summary-esdin-project-public-deliverables. Accessed 7 June 2013
Lecordix F, Lemarié C (2007) Managing generalisation updates in IGN map production. In: Mackaness WA, Ruas A, Sarjakoski LT (eds) The generalisation of geographic information: models and applications, Chapter 15. Elsevier, pp 285–300
Lecordix F, Plazanet C, Lagrange JP (1997) A platform for research in generalization: application to caricature. GeoInformatica 1(2):161–182
Maugeais E, Lecordix F, Halbecq X, Braun A (2011) Dérivation cartographique multi échelles de la BDTopo de l'IGN France: mise en œuvre du processus de production de la Nouvelle Carte de Base. In: Proceedings of the 25th international cartographic conference, Paris, 3–8 July 2011
Moore L (2011) US Topo—A new national map series. Direction Magazine. 16 May 2011. http://www.directionsmag.com/articles/us-topo-a-new-national-map-series/178707. Accessed 29 May 2013
Moore L (2013) US Topo—A new national map series, 2012 update. Directions Magazine, 14 Jan 2013. http://www.directionsmag.com/articles/us-topoa-new-national-map-series-2012-update/300690. Accessed 29 May 2013
Palomar-Vázquez JM, Pardo Pascual JE, Baella B, Pla M (2011): Generalizing the altimetric information of the Topographic Database of Catalonia at 1:5,000: classification and selection of break lines. In: Proceedings of the 14th ICA Workshop on Generalisation and Multiple Representation, jointly organised with ISPRS Commission II/2 Working group on Multiscale Representation of Spatial Data, Paris, France, 2011 http://generalisation.icaci.org/images/files/workshop/workshop2011/genemr2011_Palomar-Vazquez.pdf. Accessed 29 May 2013
Pätynen V, Ristioja J (2009) Vector based generalization application in the production of small scale databases of the NLS of Finland. In: Proceedings of the 24th international cartographic conference, Santiago, 15–21 Nov 2009
Petzold I, Burghardt D, Bobzien M (2006) Workflow management and generalisation services. In: Proceedings of the 10th ICA workshop on generalisation and multiple representation, Portland, 2006. http://generalisation.icaci.org/images/files/workshop/workshop2006/ICA2006-Petzold.pdf. Accessed 29 May 2013
Plessis O (2011) New cartography of France at 1:100,000 scale: an IGN mapped vector database. In: Proceedings of the 25th international cartographic conference, Paris, 3–8 July 2011
Punt E, Watkins D (2010) User-directed generalization of roads and buildings for multi-scale topography. In: Proceedings of the 13th ICA workshop on generalisation and multiple representation, Zürich. http://generalisation.icaci.org/images/files/workshop/workshop2010/genemr2010_submission_22.pdf. Accessed 29 May 2013
Regnauld N (2011) OS VectorMap District: automated generalisation, text placement and conflation in support of making pubic data public. In: Proceedings of the 25th international cartographic conference, Paris, 3–8 July 2011
Regnauld N, Plews M, Martin P (2012) An enterprise system for generalisation. In: Proceedings of the 15th ICA workshop on generalisation and multiple representation, jointly organised with EuroSDR commission on data specifications, Istanbul. http://generalisation.icaci.org/images/files/workshop/submissions2012/genemr2012_session4_sub2.pdf. Accessed 29 May 2013

Regnauld N, Lessware S, Wesson C, Martin P (2013) Deriving products from a Multi Reso-lution database using automated generalisation at ordnance survey. In: Proceedings of the 26th international cartographic conference, Dresden, 25–30 Aug 2013

Revell P, Regnauld N, Bulbrooke G (2011) OS VectorMapTM District: automated generalisation, text placement and conflation in support of making Pubic Data Public. In: Proceedings of the 25th international cartographic conference, Paris, 3–8 July 2011

Ruas A (1999) Modèle de généralisation de données géographiques à base de contraintes et d'autonomie. Dissertation (PhD), Université de Marne-la-Vallée

Ruas A (2001) Automatic generalisation project: Learning process from interactive generalisation. OEEPE Official Publication n 39

Sarjakoski T, Sester M, Sarjakoski LT, Harrie L, Hampe M, Lehto L, Koivula T (2005) Web generalisation services in GiMoDig—towards a standardised service for real-time generalisation. In: Toppen F, Painho M (eds) AGILE 2005. Proceedings of the 8th AGILE conference on geographic information science, Estoril, 26–28 May 2005

Stanislawski LV, Buttenfield BP (2011) Hydrographic generalization tailored to dry mountainous regions. Cartography Geograph Inf Sci 38(2):117–125

Stanislawski LV, Savino S (2011) Pruning of hydrographic networks: a comparison of two approaches. In: Proceedings of the 14th ICA workshop on generalisation and multiple representation, jointly organised with ISPRS Commission II/2 Working group on Multiscale Representation of Spatial Data, Paris, 2011. http://generalisation.icaci.org/images/files/workshop/workshop2011/genemr2011_Stanislawski.pdf. Accessed 29 May 2013

Stoter J (2005) Generalisation within NMA's in the 21st century. In: Proceedings of the 22nd international cartographic conference, A Coruña, 10–16 July 2005

Stoter JE, van Smaalen J, Bakker N, Hardy P (2009a) Specifying map requirements for automated generalisation of topographic data. Cartographic J 46(3):214–227

Stoter JE, Burghardt D, Duchêne C, Baella B, Bakker N, Blok C, Pla M, Regnauld N, Touya G, Schmid S (2009b) Methodology for evaluating automated map generalization in commercial software. Comput Environ Urban Syst 33(5):311–324

Stoter J, Baella B, Blok C, Burghardt D, Duchêne C, Pla M, Regnauld N, Touya G (2010) State of the art of automated generalisation in commercial software. EuroSDR official publication no. 58 http://www.eurosdr.net/projects/generalisation/eurosdr_gen_final_report_mar2010.pdf

Stoter JE, van Smaalen J, Nijhuis R, Dortland A, Bulder J, Bruns B (2012) Fully automated generalisation of topographic data in current geo-information environments. In: Zlatanova S, Ledoux H, Fendel EM, Rumor M (eds) Urban and regional data management—UDMS annual 2011. CRC Press, Leiden, pp 109–120

Stoter JE, Post M, van Altena V, Nijhuis R, Bruns B (2013) Fully automated generalisation of a 1:50 k map from 1:10 k data. Cartography and Geographic Information Science, In press. (expected to be published in January 2014)

Sugarbaker LJ, Carswell WJ (2011) The National Map: U.S. Geological Survey Fact Sheet 2011-3042. http://pubs.usgs.gov/fs/2011/3042. Accessed 29 May 2013

Swisstopo (2013) Swisstopo journey through time http://www.swisstopo.admin.ch/internet/swisstopo/en/home/current/newproducts/20130117.html. Accessed 29 May 2013

The Federal Authorities of the Swiss Confederation (2007) Geoinformation Act. http://www.admin.ch/ch/e/rs/5/510.62.en.pdf. Accessed 29 May 2013

Thom S (2005) A strategy for collapsing OS integrated transport network dual carriageways. In: Proceedings of the 9th ICA workshop on generalisation and multiple representation, La Coruña. http://generalisation.icaci.org/images/files/workshop/workshop2005/Thom.pdf. Accessed 25 June 2013

Thom S (2007) Automatic resolution of road network conflicts using displacement algorithms orchestrated by software agents. In: Proceedings of the 11th ICA workshop on generalisation and multiple representation, Moscow. http://generalisation.icaci.org/images/files/workshop/workshop2007/Thom_ICA_Workshop.pdf. Accessed 29 May 2013

Thompson RC, Richardson DE (1999) The good continuation principle of perceptual organization applied to the generalization of road networks. In: Proceedings of the 19th international cartographic conference, Ottawa, 14–21 Aug 1999
Töpfer F, Pillewizer W (1966) The principles of selection: a means of cartographic generalization. Cartographic J 3(1):10–16
Tóth K (2007) Data consistency and multiple-representation in the ESDI. In: Proceedings of the 11th ICA workshop on generalisation and multiple representation, Moscow. http://generalisation.icaci.org/images/files/workshop/workshop2007/Toth-ICAWorkshop.pdf. Accessed 29 May 2013
Trevisan J (2004) From DLM to multi representation DCM—Modelling an application on buildings. In: Proceedings of the 8th ICA workshop on generalisation and multiple representation, jointly organized with EuroSDR Commission on Data Specifications, Leicester. http://generalisation.icaci.org/images/files/workshop/workshop2004/Trevisan-v2-ICAWorkshop.pdf. Accessed 29 May 2013
US EPA and US DOI (1999) Standard for National Hydrography Dataset, United States Environmental Protection Agency, United State Department of the Interior, USGS, National Mapping Program Technical Instructions. July 1999. http://nationalmap.gov/standards/pdf/NHDH0799.PDF. Accessed 29 May 2013
USGS (1955) Map publication scales, book 1. Part B, Geological Survey topographic instructions, March 1955
USGS (1985) Part 2: Base preparation. Standards for 1:100,000-scale quadrangle maps. Jan 1985
USGS (1993) Part 3: feature specification and compilation. Standards for 1:100,000-scale quadrangle maps. Feb 1993
Wang Z, Muller JC (1998) Line generalization based on analysis of shape characteristics. Cartography Geograph Inf Syst 25(1):3–15
Weibel R, Dutton G (1999) Generalising spatial data and dealing with multiple representations. In: Longley PA, Goodchild MF, Maguire DJ, Rhind DW (eds) Geographical information systems—Principles, techniques, application and management, 2nd edn, vol 1. pp 125–155
Zhou S, Jones CB (2004) Shape-aware line generalisation with weighted effective area. In: Fisher PF (eds) SDH 2004: Developments in spatial data handling, proceedings of the 11th international symposium on spatial data handling, Leicester, Springer, pp 369–380

Chapter 12
Conclusion: Major Achievements and Research Challenges in Generalisation

Dirk Burghardt, Cécile Duchêne and William Mackaness

Abstract The core ambitions of map generalisation remain the same-as do a set of inter connected research activities concerned with algorithm developments, user requirements modelling, evaluation methodologies, and the handling and integration of multi scale, global datasets. Technology continues to afford new paradigms of data capture and use, in turn requiring generalisation techniques that are capable of working with a wider variety of data sources (including user generated content), and web friendly mapping services that conceal the complexities of the map design process from the lay user. This summarative chapter highlights yet again, the truly inter disciplinary nature of map generalisation research.

12.1 Major Achievements

Within the last generalisation book six topics of generalisation research were listed in the research agenda (Mackaness et al. 2007):

1. Modelling user requirements
2. Evaluation methodologies
3. Generalisation and global datasets
4. Generalisation in support of maintenance of multiple representation databases

---

D. Burghardt (✉)
Dresden University of Technology, Dresden, Germany
e-mail: dirk.burghardt@tu-dresden.de

C. Duchêne
Laboratoire COGIT, IGN, Paris, France
e-mail: cecile.duchene@ign.fr

W. Mackaness
School of GeoSciences, The University of Edinburgh, Edinburgh, UK
e-mail: william.mackaness@ed.ac.uk

D. Burghardt et al. (eds.), *Abstracting Geographic Information in a Data Rich World*, Lecture Notes in Geoinformation and Cartography, DOI: 10.1007/978-3-319-00203-3_12,

5. Generalisation and interoperability
6. Map generalisation and geoinformatics

Several chapters within this book have presented successful research related to these topics. For example Chap. 2 refers directly to "Modelling user requirements" (1) and Chap. 9 is focused on the topic of "Evaluation methodologies" (2) Some of the research topics have became even more important in the light of recent developments within cartography. This is especially true of "Generalisation and interoperability" (5) given the current focus of research on the integration of user generated content and multi sourced data—now available through all kinds of geosensors (a topic covered in Chap. 5). Developments in MRDB research has led to successful implementations within NMAs (Chap. 11), however maintenance issues (4) have not been thoroughly researched though they remain an important topic. The next subsections will describe progress made against this research agenda, summarising the major achievements presented within the chapters of this book.

12.1.1 Automated Generalisation for Topographic Map Production Within NMAs

The development of automated generalisation solutions for topographic map production has been a long-standing research field in cartography. One measure of success has been the relatively recent implementation of fully automated generalisation solutions within the production environments of national mapping agencies (Chap. 11). One can speak of something of a breakthrough in the automated derivation of topographic maps that only a few years ago was not considered possible because of its complexity (Foerster et al. 2010). Adoption has been fuelled by (1) progress in research and development in automated generalisation, and (2) the need to make cost savings, whilst acknowledging that it is perhaps unrealistic to suppose that automated solutions can be as good as the human hand. Given the cost of manual intervention, the involvement of the cartographer has either been reduced or removed.

12.1.2 Capturing of User Requirements for On-Demand Mapping

In addition to the advances made in the automated derivation of predefined topographic map series, on-going research has especially focused on a user oriented perspective, that is to say, systems that have the flexibility to support on-demand mapping solutions. In this context, the user and the application governs cartographic design and the degree of generalisation. Prerequisites are: (1) the collection and interpretation of user requirements, and (2) their automated transformation into a map specification. Influencing factors are modelled through a variety of user variables such as context of use, user preferences and visual disabilities. These all have an influence on the map specification. One approach presented in Chap. 2 illustrates how the derivation of low level cartographic constraints can be used to guide the generalisation process (based on high level map specifications defined by the user). Currently map specification models designed for on-demand mapping use constraint based methodologies for data integration and generalisation. Another case study illustrates the collection of user requirements through map samples—utilising the COLorLEGend system for making creative colour choices.

12.1.3 Generalisation Process Modelling

The impact of on-demand mapping for generalisation process modelling has been detailed in Chap. 7. These approaches illustrate how the generalisation process can be decomposed into components that have the potential to be developed independently of each other and shared over the web. A lot of generalisation tools have been designed and implemented; the challenge of creating systems capable of using them automatically is still a very real one. Web generalisation services have been proposed as a way to encapsulate and publish generalisation tools for further re-use. Prerequisite is the formalisation of procedural knowledge (for example through rules, constraints and ontologies) with the aim of chaining generalisation operators together in order to offer web generalisation services. Successful implementations of the automated chaining of generalisation components were presented with the collaborative generalisation (CollaGen) approach documented in a case study in Chap. 7. In CollaGen, the generalisation functionalities of the components are described together with pre-conditions for data input and post-conditions for the expected data modifications caused by the generalisation processes.

12.1.4 Contextual Generalisation

Another important generalisation domain in which substantial progress has been made in recent years is in contextual generalisation—which requires us to make explicit the spatial relationships that exist among geographic entities. Chapter 3 provides an overview of existing classifications and modelling frameworks for spatial relations. On-going research continues to explore the utilisation of a 'spatial relations ontology' to support automated generalisation processes. Case studies illustrate the spatial relation modelling for the extraction of groups of objects, the migration of topographic data with thematic user generated data, as well as the extension of CityGML for modelling spatial relations in 3D environments. In Chap. 8 relations between field and object type data are modelled for the purposes of terrain generalisation. Here, the focus has been on maintaining topological relationships between rivers, contours and spot heights. The GAEL model presented in one case study of this chapter offers a solution that handles object-field relations within the generalisation process and thus allows us to find a balance between the deformation and preservation of features. Generalisation of relief and hydrographic networks based on generic deformation algorithms are carried out in parallel in order to have their common outflow relation preserved. The GAEL model has been further developed from a research prototype to a software component that forms part of the production line for the derivation of a 1:25,000 scale base map of France.

12.1.5 Generalisation of Object Groups and Networks

In the past research and development of generalisation algorithms has focussed on isolated objects. In recent years the generalisation of groups of objects and networks has attracted more and more attention. Ideas related to the generalisation of individual objects have been extended to the generalisation of groups of objects. Several approaches were proposed for the generalisation of road and stream networks, island and building groups (case studies in Chaps. 3, 6 and 9). The strategies illustrate the importance of data enrichment to support generalisation that is tailored to specific landscape and settlement contexts. As networks vary with local geographic conditions, network thinning operations now take into account local line density variations. For river networks, 'prominence estimates' enable a ranking of 'strokes' which are based on stream order, stream name, longest path, drainage area, straightness and the number of upstream branches. In the case of road networks combined stroke and mesh-density strategies have been successfully applied to the task of 'thinning'. Research on building alignments has concentrated on the development of strategies for preserving alignment properties. This requires corresponding relations between source and target groups to be made explicit, which requires implementation of both pattern recognition and matching strategies. Evaluation of this work shows that group characteristics of building alignments are not just preserved but also becoming more homogenous.

12.1.6 Evaluation in Automated Generalisation

Chapter 9 presents developments in evaluation methodologies. The two main strategies are visual and automated evaluation. While visual evaluation is qualitative, time consuming and focused on map readability, automated evaluation is quantitative and is particularly appropriate for assessing specific feature representations. Research has also been undertaken to define map readability by using map readability formulas assessed through empirical user studies. Measures of readability include: the amount of information, spatial distribution, object complexity and graphical resolution. The results revealed that some measures of content and spatial distribution corresponded well with participants' opinions, while measures of object complexity did not show the same correspondence.

12.1.7 Generalisation of User Generated Content and Multi-Source Data

Cartographic visualisations and mashups in the context of web mapping and mobile applications continue to evolve (demonstrating further application of automated generalisation techniques). The need for this type of mapping has been fuelled by the increasing availability of volunteered geographic information and geosensor data over the web. The integration and visualisation of such multi-source data presents new generalisation challenges arising from the heterogeneity of data, the differences in quality and the opportunities afforded by tag-based semantic modelling. These issues are discussed in Chap. 5. New opportunities lie in combining VGI data with topographic and other spatial reference data in order to provide meaningful contexts. Generalisation is needed in order to present information with clear semantics, consistent topology and free of graphical conflicts—the overall ambition being to avoid misinterpretation and incorrect decisions. One case study in this chapter compared generalisation of OpenStreetMap data with the generalisation of authoritative data. Generalisation in OSM mainly utilises selection and re-classification operations, with an emphasis on performance and short update intervals between data capturing and map rendering. The ambition is to avoid any manual post processing. In contrast the officially controlled generalisation of NMA data uses the complete set of generalisation operations, with a focus on cartographic quality, over longer update cycles, supported by manually controlled interactions. These activities reflect the importance of quality control in NMA data.

12.1.8 On-the-Fly and Continuous Generalisation

Dynamic use of web and mobile mapping applications need to give immediate response to user interactions and adapt to changing contexts. Thus methods of continuous zoom and on-the-fly generalisation are essential, particularly when it comes to the visualisation of thematic foreground data. A case study in Chap. 6 described the utilisation of quadtrees for on-the-fly generalisation of point data. Chapter 4 introduced the idea of the topological generalised area partitioning (tGAP) structure and the corresponding 3D space-scale cube (SSC). This is an encoding of variable-scale data structures developed especially for continuous

generalisation. Advantages are the tight integration of space and scale with guaranteed consistency between scales. Research on the constrained approach (constraint tGAP) showed significant improvement in generalisation quality (as compared to unconstrained tGAP). The tGAP is well suited to web and mobile environment as it requires no further client side processing and supports progressive refinement
of vector data. Ongoing research will explore the derivation of mixed scale representations through non-horizontal slices within the SSC and a usage of the vario-scale data structure for the modelling of 3D data.

12.1.9 Generalisation of Schematised Maps

Building on a long tradition of designing schematised maps, there is growing interest in their presentation in interactive environments—particularly the schematised presentation of navigational information on mobile devices, or as a way of presenting highly summarised thematic data. Chapter 10 illustrates recent research on the role of map generalisation techniques in the automated derivation of schematised maps in general and chorematic diagrams in particular. For automated solutions, the challenge is in finding the right balance between displacement, line simplification and smoothing. Despite the complexity in their design, in some cases, the quality of automated solutions is such that it is becoming increasingly hard to differentiate hand drawn solutions from automated ones.

12.2 Future Challenges

Various challenges have been identified throughout this book. In this section we try to frame those challenges in the context of increasing and ready access to geographical information in all its various forms and sources.

12.2.1 Generalisation of Tailored 2D Maps

Generalisation in the context of on-demand mapping is not a new topic (Chaps. 2 and 7). However, the challenge of building a system that is able to understand a user's needs, and respond with appropriate generalisation techniques to produce a tailored map remains a considerable challenge. Any solution requires more complete models, and knowledge to instantiate them, of (1) map uses (in terms of tasks, user profiles and contexts of use), and (2) how the design of maps, and by how much they are generalised, govern their usability for a specific task. Until recently, the main focus of generalisation was the production of topographic maps by NMAs. The aim, inherited from the time when topographic maps governed what was digitally stored, was to produce general purpose maps with all topographic information relevant to a given scale. The way maps were actually used and the fitness for use of the available maps were rarely studied. This was because it was not considered feasible to automatically adapt maps for various specific uses. NMAs now recognise that they should allocate more resources in order to better understand the needs of their users. Additionally, such knowledge provides useful case studies for continuing research.

A second research topic concerns the development of user-friendly interfaces that facilitate the capture of user needs, preferences, and context of use. Mackaness et al. (2007) had already identified this as a research challenge, suggesting the idea of "generalisation wizards" that could help the user figure out the consequences of their choices by illustration of generalised sample areas. Progress has been made in the modelling of user variables that influence the on-demand mapping process, and the use of data samples has recently led to good results for symbols design (Chap. 2); nevertheless there is room for further development in generalisation interface design. Interfaces of various complexities could be imagined, depending on the level of expertise of the user. They could range from complex interfaces with large numbers of adjustable parameters, through to interfaces with a simple, single slider-bar, graded from "more detailed" to "less detailed". Perhaps the ultimate solution is one that is able to infer all the required parameters depending on how the user interacts with the map.

A third research topic concerns the translation of formalised user needs, preferences, and contexts of use, into a parameterised generalisation process built from existing components. This requires a language by which we can formalise procedural knowledge. One approach is to describe any generalisation component in terms of its ability to handle various situations (Chap. 7). Another approach could be to identify typical parameterisations and 'chaining' of components that appear to fit some given need and contexts of use, and to build solutions to new needs and contexts based upon those partial solutions. Such an idea (a collaborative filtering approach) was proposed by Burghardt and Neun (2006). Tools like ScaleMaster (Brewer and Buttenfield 2010), and its extension by Touya and Girres (2013) that includes an object oriented modelling approach to the storage of sequences and parameters, are the first steps in this direction.

Another important topic is concerned with the re-use of code (Chap. 7). How do we standardise generalisation components and make them available as interoperable generalisation services that could then be combined (either by a human or a machine)? In this context, is it more efficient to send code through the network rather than data? This idea was explored by Müller and Wiemann (2012) who presented the idea of "moving code" since data transfer can be slow and expensive, but further work is needed.

All these issues lead to the requirement for shared semantic frameworks that enable us to formally describe all concepts upon which generalisation depends. Some early solutions to this problem have been presented in Chaps. 2, 3 and 7;

again further work on the topic is required. In addition, in terms of 'contexts of use' for 2D tailored maps, two particular areas are worthy of greater focus:

- Firstly the generalisation of maps intended for digital display, where we take into account the legibility constraints of the screen, and the potential for interaction. This context is quite different from the generalisation of traditional paper maps.
- Secondly, the need for greater research into on-the-fly generalisation for mobile devices—requiring us to address smooth data transfer, over limited bandwidth (Weibel 2010). Though this book has presented partial solutions (Chap. 4, and a use case in Chap. 6), more work is required in this area.

12.2.2 Dealing with Different Communities of Users

By examining the increasing popularity of web 2.0 technologies and services and the growth in VGI, we can discern two communities of users of generalisation technologies[1]—each with their own expectations. The first group are the professional mapmakers such as the traditional cartographers of NMAs who have a long history of map making. This group makes products of high quality (often at high cost), and are somewhat wary of the idea that higher levels of automation are achievable without sacrificing some elements of quality. GIS vendors and the Development Departments of NMAs, who provide automated solutions to this group based on research results, seek highly sophisticated solutions. The second community of users are non-professional mapmakers who make maps using web mapping applications based on open source data, (for example via OSM or Google maps APIs). This enables them to render data at different scales and also to collect third party data and share the resulting maps. This group is typically free of the constraints imposed on NMAs in terms of completeness, precision and accuracy of the resulting maps. This second community requires solutions that are quick, easy-to-use and available for free. Developers of web mapping applications (i.e. entrepreneurs or volunteers) are aware that map compilation and design requires abstraction at lower levels of detail, but they have limited resources to invest in the development of sophisticated generalisation technologies. Therefore, current web mapping applications provide simplistic generalisation solutions mainly based on attribute-based selection and rendering. Whilst they contribute to the collection of geographic data and its free sharing, the cartographic quality of such maps is not always sufficient.

Weibel (2010) argues that it is this second community of users who are increasingly more deserving of our focus. But the gap between these two communities is narrowing. As Chap. 11 illustrated, NMAs are increasingly considering the balance between cost, currency and cartographic quality, and are also increasingly

[1] Note that the users of generalisation are the mapmakers, not to be confused with the users of the map (even if in some cases both roles are undertaken by the same person).

developing viewers that enable online map customisation. Therefore, ideas of ease of data integration, map comprehension, simplicity in specification, and rapid data transfer are common goals. Perhaps the difference narrows to one of finding compromise between "quick and simplistic" and "elaborate and expensive"—delivering a range of solutions that can be chosen depending on the context of use.

12.2.3 Generalising Data Stemming from Multiple Sources

In addition to adapting the generalisation process to particular needs, we also need to acknowledge that data are increasingly multi sourced. Two particular cases can be distinguished. In the first case the source data includes VGI data retrieved from the web. The major characteristic of this kind of data is that they are often not well described in terms of meta-information and are heterogeneous. As discussed in Chap. 5, methods are needed to automatically acquire metadata associated with VGI data, in order that it can be integrated with "institutional" data, and both can then be generalised together.

In the second case, it is assumed that part of the source data comes from the user, themselves. In this instance, the user usually knows their data; methods are needed to help the user to record metadata, such that the system can propose tailored sequences and parameterisations of operators.

In both cases, we have "thematic" data overlaid on background, reference data that need to be integrated and then handled together, while taking into account the important relations that exist between them. A use case in Chap. 3 examined this issue, but further research is needed to develop generalisation methodologies that are aware of the relations and interdependencies between geographic phenomena.

12.2.4 Generalisation of Higher Dimensional Data

Generalisation has relevance beyond the two dimensional. There are research challenges around the egocentric perspective, and the handling of higher dimensional data (3D and temporal data). Location aware technology coupled with new forms of media (such as augmented reality glasses) tends to overload the user with information. There is therefore an ever greater need for data abstraction and for ways of varying the level of detail depending on the user's location. There is also a need for research that better connects abstracted data with other augmented information. In such cases we need to take account of the user's position, the objects in their field of view, and tailor the information in a dynamic way in response to the various tasks that the user is trying to complete. The best solution may well not be a single abstracted form but instead delivered through a range of media, in a variety of forms (map, textual, sonic, graphic, etc.) ranging from the highly abstracted to something akin to augmented reality. A few studies are beginning to emerge on this topic (e.g. Harrie et al. 2002; Dogru et al. 2008; Chap. 4), but much more work needs to be done.

Generalisation of higher dimensional data extends to the temporal domain—a topic of research that has been previously identified (Mackaness et al. 2007), yet to date there are few studies that have addressed this topic. The ambition is to detect trends and spatio-temporal patterns and summarise these changes in a visual form, giving emphasis to the major events that have affected the data over time. One of the few reports is by Andrienko et al. (2010) who proposed the anonymisation of spatio-temporal movement data by means of generalisation.

12.3 In Conclusion

A research agenda is not something set in stone—it changes in response to new opportunities and findings. The underlying ambition of generalisation remains unchanged; ultimately it is concerned with ensuring quality in our geographic decision making. The core methodologies remain the same with focus on manipulation and abstraction of map features through the varying combination and application of generalisation techniques. The research agenda compared with the one proposed in 2007 is not significantly different however it does reflect the following changes: (1) the high expectations of users that they should be able to integrate, combine and explore distributed data sources, and share them with their own data in mapped form; (2) that they should be able to do so intuitively, instantly, anywhere, in a variety of media and web enabled devices, and (3) that the map should provide data at the appropriate level of abstraction, but also support interaction to access further levels of detail. In order for generalisation research to continue to progress, these changing contexts and expectations have to be taken into account. It is also important to investigate the opportunities and limitations of new, often mobile, technologies—technologies that have the potential to make easier the creation of interactive and immersive maps.

All of the research activities in this book point to the inter-disciplinary nature of research in map generalisation—whether it be cognitive modelling of the user, the role of ontologies in semantic modelling, or the design of web interfaces to support service driven cartography. Future advances in this field will surely depend on continuing close collaboration with researchers working in such fields as visual analytics and cartography, spatial reasoning and informatics, computational geometry and geospatial semantics, databases and web services (to name but a few!). Its contribution and continued relevance to the display of geographic information will depend upon its continued close coupling between these complementary disciplines.

Acknowledgments The authors are grateful to their colleagues from the COGIT Laboratory of IGN, the Dresden University of Technology and the University of Edinburgh for the discussions around the future research challenges in generalisation that indubitably made this chapter all the richer!

References

Andrienko G, Andrienko N, Giannotti F, Monreale A, Pedreschi D, Rinzivillo S (2010) A Generalisation-based approach to anonymising movement data. In: Painho M, Santos MY, Pundt H (eds) Geospatial thinking, AGILE 2010 CD-ROM Proceedings, pp 1–10

Brewer CA, Buttenfield BP (2010) Mastering map scale: balancing workloads using display and geometry change in multi-scale mapping. Geoinformatica 14(2): 221–222

Burghardt D, Neun M (2006) Automated sequencing of generalisation services based on collaborative filtering. In: Raubal M, Miller HJ, Frank AU, Goodchild M (eds) Geographic information science, 4th International Conference on Geographical Information Science (GIScience), IfGIprints 28, 41–46

Dogru A, Duchêne C, Mustière S, Ulugtekin N (2008) User centric mapping for car navigation systems, In: Proceedings of the 12th ICA Workshop on Generalisation and Multiple Representation, jointly organised with EuroSDR Commission on Data Specifications and the Dutch program RGI, Montpellier, France, 2008. http://generalisation.icaci.org/images/files/workshop/workshop2008/02_Dogru_et_al.pdf

Foerster T, Stoter JE, Kraak M (2010) Challenges for automated generalisation at european mapping agencies: a qualitative and quantitative analysis. Cartogr J 47(1): 41–54

Harrie L, Sarjakoski T, Lehto L (2002) A variable-scale map for small-display cartography. Int Arch Photogrammetry Remote Sens Spatial Inf Sci 34(4):237–242

Mackaness W, Ruas A, Sarjakoski T (2007) Observations and research challenges in map generalisation and multiple representation. In: Mackaness W, Ruas A, Sarjakoski T (eds) Elsevier, 315–323

Müller M, Wiemann S (2012) A framework for building multi-representation layers from OpenStreetMap data. In: Proceedings of the 15th ICA workshop on generalisation and multiple representation, jointly organised with EuroSDR commission on data specifications, Istanbul, Turkey

Touya G, Girres J-F (2013) ScaleMaster 2.0: a scalemaster extension to monitor automatic multi-scales generalizations. Cartogr and Geogr Info Sci 40(3): 192–200

Weibel R (2010) From geometry, statics and oligarchy to semantics, mobility and democracy: some trends in map generalization and data integration. In: Workshop on generalization and data integration (GD1 2010) Boulder, USA, June 2010. Postprint available at: http://www.zora.uzh.ch

Index

A
Adaptive generalisation, 18
Agent based modelling, 248–251, 252–255
AGENT project, 205, 266, 269, 340, 343, 372
ATKIS, 369–373

B
Binary Line Generalisation tree (BLG-tree), 85–86
Building generalisation, 276–283

C
Cartographic generalisation, 2, 114, 144
Chorematic maps, 304, 314, 316–319
CityGML, 66–67, 395
Cloud computing, 206
Collaborative generalisation, CollaGen, 207–212
COLorLEGend/COLLEG, 43, 46–48
Commercial generalisation software, 31, 163–164, 287
Component based modelling, 199–205
Constrained tGAP, 101–107
Constraint modelling, 36–37, 39–42, 159–160, 198, 209, 247, 248, 253, 268–269, 288–291
  cartographic, 23–25, 33–34, 46
  relational, 62–63
  representation, 39–40
  symbolisation, 40–41
Content measures, *see* Radical law
Content measures, 174, 181, 260, 264, 276
Context-sensitive generalisation, 26, 158, 167, 170–175, 180–181
Continuous generalisation, 84
Customised maps, *see* On-demand mapping

D
Data enrichment, 6, 166–167, 175–180
Data integration, 74, 120, 124, 145–150, 374–378
Data matching, 124–128, 145–150, 272–273, 280
Data quality, 136–137
  visualisation, 129
Density analysis, 171–172, 182, 184, 189
Digital Cartographic Model (DCM), 144, 349, 369, 379–383
Digital Earth, 217
Digital Landscape Model (DLM), 141, 351, 369, 379–383
DTM, *see* Relief generalisation

E
EuroSDR project, 163, 277, 287–291
Evaluation methodologies, 209, 259–292, 287–288, 313, 396
  linear objects, 265
  polygonal objects, 265

E
EuroSDR project, 163, 277, 287–291
Evaluation methodologies, 209, 259–292, 287–288, 313, 396
  linear objects, 265
  polygonal objects, 265

G
GAP-tree, 84, 86–91, 97–98
Generalisation methodologies, 316, 342–343
Generalisation operators, 160–162
Generalisation workflows, 160, 333–390, 349, 362–364

D. Burghardt et al. (eds.), *Abstracting Geographic Information in a Data Rich World*, Lecture Notes in Geoinformation and Cartography, DOI: 10.1007/978-3-319-00203-3,

Geoprocessing, 217–218
Geosensor networks, 121–122
Gestalt, 3
GPS trajectory, 145–150
Graph theory, 321–324
Grünreich model, *see* DCM/DLM

H

Hierarchical structures, 84–91, 97–98, 106–107, 138, 234–235, 322
Hill climbing optimisation, 310
History of generalisation, 159–160
Hydrological network generalisation, *see* River network generalisation
Hypsometric colouring, 240–241, 242–245

I

IGN-France, 339–346
Information theory, 264
INSPIRE, 373–377
Institut Cartografic de Catalunya, 333–339
Interoperability, 201, 202, 205, 206, 207–208

K

Kadaster NL, 362–369
Knowledge formalisation, 27, 28, 203–204, 209, 211–217

L

Label placement generalisation, 311, 340
Landform modelling, 230–235

M

Map legibility/map readability, 273–276, 283–286, 291
Map matching, 145–150
Map specifications, 21–23, 27
  formalising, 33–35, 268
Mash-ups, 18
Measures of information content, 269, 275, 285, 286, 305, 317, 318
Mixed Integer optimisation, 311
Mixed-scale map representations, 96–97, 309
Model generalisation, 2, 142, 365–366
Multiple representation data structures, 85, 351–352
Multi-scale mapping, 160
Multi-scale/variable-scale data structures, *see* Hierarchical structures

N

National Mapping Agencies, 199, 330–386
Nautical charts, 238, 239, 245–250
Network generalisation, 158, 169–181, 308

O

Object generalisation, 141
On-demand mapping, 18–20, 199, 385, 394
  architecture, 199–203
On-the-fly generalisation, 181–189, 218
Ontology constraints, 62–63
Ontology generalisation domain, 23, 38, 203, 209, 213–217
Open source environments, 202
OpenStreetMap (OSM), 138, 139–145
Optimisation methods, 105–106, 309, 311
Ordnance Survey, 351–355
Orthophoto generalisation, 336–337

P

Partitioning, 168
Pattern analysis, 69–73, 170, 172, 181–182, 189, 301
  buildings, 278–280
Perceptual grouping, 70
Point data generalisation, 169, 181–189
Procedural knowledge, *see* Knowledge formalisation
Process modelling, 198, 200, 203–217, 221
Progressive data transfer, 85, 93–96, 218, 219

Q

Quadtree-based generalisation operators, 184

R

Radical law, 170, 172–174, 184, 264
Raster generalisation, 162, 163
Real-time data, 217
Real-time processing, *see* On-the-fly generalisation
Real-time processing, 217–221
Relations modelling, 3–7, 22
  binary, 56, 57
  foreground/background, 74–77, 251–255

groups, 69
multiple representation, 61, 62
quantitative, 56, 57
Relations modelling, *see* Spatial relations modelling
similarity, 71, 72
Relief generalisation, 228–256, 238, 240–241, 242–245
Relief modelling, 241, 242, 251–255
Relief representation, 227
River network generalisation, 175, 178, 180, 396
Road network generalisation, 169–170, 171–172, 174, 189
Rule-based generalisation, 198, 206

S
Schematic maps, 302–325
Semantic modelling, 119
Semantic zoom, 320
Service chaining, 205
Shape metrics, 270–271
Simulated annealing, 310
Smooth line simplification, 110
Space Scale Cube (SSC), 91, 107–112
Spatial relations classification, 55, 56
Spatial relations modelling, 54, 58–60, 74–76
Spatial relations ontologies, 58, 59, 67, 68
Spot height selection, 236
Stratified thinning, 168–170, 357
Stream-based processing, 218
Swisstopo, 346–351

T
Terrain generalisation, *see* Relief generalisation
Text placement, 337
tGAP-tree, 86, 88, 97–98
Transportation map generalisation, 319–325
Transportation maps, 311
Typology of generalisation operators, 160–162

U
User requirements, –collecting, 25, 29–32, 43
User-generated content, *see* VGI
USGS, 169–175, 355–362

V
VGI (Volunteered geographic information), 120–124
data quality, 136–137
generalisation, 130, 137–138
integration, 127–128, 137–138, 145–150
challenges, 150
data sources, 131–135, 402
visualisation, 128–129, 138
aggregation, 126–127

W
Web processing services (WPS), 216, 218
Web services, 197–203, 207, 210, 216–219, 221
WebGen, 201, 202
Workflow, 205

The manufacturer's authorised representative in the EU is Springer Nature Customer Service Centre GmbH, Europaplatz 3, 69115 Heidelberg, Germany. If you have any concerns regarding our products, please contact ProductSafety@springernature.com

Printed and bound by CPI Group (UK) Ltd, Croydon, CR0 4YY
30/09/2026
02225567-0019